Chocolate and Cocoa

Health and Nutrition

Edited by

Ian Knight

Knight International

International Cocoa
Organization London

Blackwell
Science

International Cocoa Research
and Education Foundation

Editorial Offices:
Osney Mead, Oxford OX2 0EL
25 John Street, London WC1N 2BL
23 Ainslie Place, Edinburgh EH3 6AJ
350 Main Street, Malden
MA 02148 5018, USA
54 University Street, Carlton
Victoria 3053, Australia
10, rue Casimir Delavigne
75006 Paris, France

Other Editorial Offices:

Blackwell Wissenschafts-Verlag GmbH
Kurfürstendamm 57
10707 Berlin, Germany

Blackwell Science KK
MG Kodenmacho Building
7–10 Kodenmacho Nihombashi
Chuo-ku, Tokyo 104, Japan

First published 1999

Set in 10/13 Times
by DP Photosetting, Aylesbury, Bucks
Printed and bound in Great Britain by
MPG Books Ltd, Bodmin, Cornwall

DISTRIBUTORS

Marston Book Services Ltd
PO Box 269
Abingdon
Oxon OX14 4YN
(*Orders:* Tel: 01235 465500
Fax: 01235 465555)

USA
Blackwell Science, Inc.
Commerce Place
350 Main Street
Malden, MA 02148 5018
(*Orders:* Tel: 800 759 6102
781 388 8250
Fax: 781 388 8255)

Canada
Login Brothers Book Company
324 Saulteaux Crescent
Winnipeg, Manitoba R3J 3T2
(*Orders:* Tel: 204 837 2987
Fax: 204 837 3116)

Australia
Blackwell Science Pty Ltd
54 University Street
Carlton, Victoria 3053
(*Orders:* Tel: 03 9347 0300
Fax: 03 9347 5001)

A catalogue record for this title
is available from the British Library

ISBN 0-632-05415-8

Library of Congress
Cataloging-in-Publication Data

Chocolate and Cocoa: Health and Nutrition/
edited by Ian Knight.
p. cm.
Includes bibliographical references and index.
ISBN 0-632-05415-8
1. Chocolate – Health aspects. 2. Cocoa – Health aspects.
I. Knight, Ian.
QP144.C46R48 1999
612.3'96 – dc21 99-17982
CIP

For further information on Blackwell Science, visit our website: www.blackwell-science.com

Contents

Contributors

Mary M. Amann MS, RD
TAS-Environ, Arlington
VA 22203, USA

Joan L. Apgar BS
Hershey Foods Corporation
Technical Center
Hershey, PA 17033, USA

Gary K. Beauchamp PhD
Monell Chemical Senses Center
Philadelphia, PA 19104, USA

David Benton PhD
Department of Psychology
University of Wales
Swansea SA2 8PP, Wales, UK

Ronald G. Bixler BS
Ex-M&M/Mars Inc.
Palmyra, PA 17078, USA

Caroline Bolton-Smith BSc (Hons), PhD, FRSH
Nutrition Research Group
Cardiovascular Epidemiology Unit
Ninewells Hospital and Medical School
University of Dundee, Scotland, UK

Janette C. Brand Miller PhD
Human Nutrition Unit
University of Sydney
NSW 2006, Australia

Louise M. Burke PhD
Nutrition Programme
Department of Sports Nutrition
Australian Institute of Sport
Bruce, ACT 2617, Australia

Susan L. Cheney MS
Technical and Regulatory Affairs
Chocolate Manufacturers Association
McLean, VA 22102, USA

Matty Chiva PhD
Department of Psychology
University of Paris X, Nanterre
France

Martin E.J. Curzon PhD
Department of Paediatric Dentistry
University of Leeds
W. Yorks, LS2 9LU, UK

Judith S. Douglass MS, RD
TAS-Environ, Arlington
VA 22203, USA

Terry D. Etherton PhD
Dairy and Animal Science
Penn State University
University Park, PA 16802, USA

Paul-Andre Finot PhD
Nestlé Research Centre, Lausanne
CH-1000, Switzerland

Susan L. Hefle PhD
Institute of Agriculture and Natural Resources
University of Nebraska
Lincoln, NE 68583, USA

Marion M. Hetherington BSc (Hons), DipEd, DPhil, AFBPsS
Department of Psychology
University of Dundee, Nethergate
Dundee, Scotland, UK

Nicholas J. Jardine BSc, PhD
Nestlé R&D Centre
York, YO1 1XY, UK

Ian Knight BSc (Hons)
Knight International
Tappahannock
VA 22560, USA

Penny Kris-Etherton PhD
Department of Nutrition
Pennsylvania State University
University Park, PA 16802, USA

David Kritchevsky PhD
Wistar Institute
Philadelphia, PA 19104, USA

R. Anthony Lass BSc, DTA
Agriculture and Environmental Affairs
Cadbury Limited, Bournville
Birmingham, B30 2LU, UK

John R. Lupien MSc
Food and Nutrition Division
Food and Agriculture Organization
00100 Rome, Italy

Dawn A. Marcus MD
Pain Evaluation and Treatment Institute
University of Pittsburgh Medical Center
Pittsburgh, PA 15213, USA

Jeffrey N. Morgan
M&M/Mars Inc.
Elizabethtown, PA 17022, USA

Marcia L. Pelchat PhD
Associate Member
Monell Chemical Senses Center
Philadelphia, PA 19104, USA

Lisa Scharff PhD
Pain Treatment Service
Children's Hospital of Boston
Boston, MA 02115, USA

Stanley M. Tarka Jr PhD
Technical Center
Hershey Foods Corporation
Hershey, PA 17033, USA

Steve L. Taylor PhD
Institute of Agriculture and Natural Resources
University of Nebraska
Lincoln, NE 68583, USA

Pierre Würsch PhD
Nestlé R&D Centre
Orbe, Switzerland

Foreword

Cocoa, referred to as the Food of Gods as far back as 1662 by Dr Joseph Bachot, a Paris physician, and formally classified as such (Theobroma cacao) by the Swedish botanist, Linnaeus, in 1753, has successfully conquered all countries and continents of the world in just over 500 years since its discovery in the ancient civilisation of the Mayas and Aztecs in South America.

Thanks to its many combinations in a variety of confectionery items, it is consumed by peoples from all walks of life. It suits the opulent taste of the business executive and brings a smile to the face of a child. It provides quick energy to the athlete and prolongs the pleasurable moments of a tender dinner and, not least, it expresses so delicately our gratitude or love for someone.

Many myths have grown up around cocoa and chocolate through attempts to explain or temper the long-standing enthusiasm for these products whilst numerous studies and scientific articles have been produced to acclaim their positive effects or warn against their allegedly detrimental consequences.

As part of its continuous efforts to keep the world at large informed about cocoa and chocolate, the International Cocoa Organization (ICCO) has decided to undertake the publication of an exhaustive review of the health and nutrition aspects of these products.

The International Cocoa Organization is an intergovernmental organization of some 40 cocoa exporting and importing countries created in 1972/73 under the auspices of the United Nations to promote the development and strengthening of international co-operation in all sectors of the world cocoa economy.

International co-operation is a keyword for ICCO – co-operation between producers and consumers, between the private and public sectors and between trade associations and research institutes, with the ultimate objective of enhancing the well-being and satisfaction derived from cocoa and chocolate by the general public. It is, therefore, with a sense of pride and achievement that the International Cocoa Organization presents this study on health and nutrition in cocoa which has been produced in close co-operation with the International Cocoa Research and Education Foundation (ICREF).

ICREF is a non-profit foundation formed in 1994 for educational purposes, including the transfer of scientific research and technology to the public sector. The American Cocoa Research Institute (ACRI) is an associated organization of ICREF, and is dedicated to the performance of basic research studies related to cocoa and chocolate, including areas of nutrition, health, and the agronomic aspects of cocoa growing. Additionally, both ACRI and ICREF seek to work closely with organizations such as the ICCO to bring important knowledge on cocoa and chocolate to the attention of the public sector.

This review covers existing literature published over the last 25 years on the nutritional and health aspects of cocoa and chocolate and is the first time that such a comprehensive review has been conducted. The findings explored in these pages show that not only are the nutritional and health benefits of cocoa and chocolate extensive, but they also go a long way towards explaining why cocoa and chocolate have been so successful in conquering the hearts and palates of the majority of men, women and children throughout the world.

Edouard Kouamé
Executive Director
International Cocoa Organization

Carol Knight
President
International Cocoa Research and Education Foundation

Preface

The need for an updated text outlining the latest research on the health and nutritional properties of cocoa and chocolate is glaring. Never before has such a comprehensive compendium been assembled for a product of such interest to both health professionals and consumers alike. Although much information on chocolate and its component ingredients – such as cocoa – has been published over the years, many areas of importance related to the health aspects of cocoa and chocolate have *never* been the subject of an in-depth scientific evaluation.

Because so much of the traditional thinking regarding chocolate and cocoa has changed over the last decade due to much research interest, and since so many new areas of potential health benefits are being elucidated, such as antioxidants in cocoa, it is of great interest to prepare and publish this review.

To conduct this review, the best and most knowledgeable worldwide scientific experts in their respective fields were approached. In this way, a critical evaluation of the research was possible, putting the progress in each field into the proper scientific context, rather than simply reporting research findings.

Accordingly, the book is a collection of chapters dealing with the various subjects relevant to nutrition and health properties of cocoa and chocolate. It is not presented as a continuous narrative and so it will be clear that the styles and indeed the level of science presented in the individual chapters will reflect each author's perspective. Rather than attempt to standardize this, I have decided to allow this diversity. I have also left the style of English to reflect the country of origin of the author.

The reader will determine the worth of this publication; however, it is hard to imagine this being anything less than a valuable text for the industry, health professionals, media, educators, researchers, and food scientists.

I wish to express my gratitude to the 28 authors who wrote these chapters for their hard work and dedication to this effort. I also thank the members of the Scientific Advisory Committee for their valuable assistance on this review, without whose help this would have been onerous indeed.

Ian Knight
Editor

Acknowledgements

The International Cocoa Research and Education Foundation (ICREF) gratefully acknowledges the contributions of the Scientific Advisory Committee in preparing this review.

Scientific Advisory Committee

E. Maureen S. Edmondson BSc, PhD, FIFST
International Scientific Affairs, Mars Incorporated, Slough, UK

Carol A. Knight PhD
President, International Cocoa Research and Education Foundation, Vice President, Scientific Affairs, American Cocoa Research Institute, Chocolate Manufacturers Association, McLean, VA, USA

Ian Knight BSc (Hons)
President, Knight International, Tappahannock, VA 22560, USA

Penny Kris-Etherton PhD
Professor, Department of Nutrition, Pennsylvania State University, State College, PA, USA

Kenneth Mercurio PhD
Director, Nutrition and Labeling, Nestlé USA Inc., Glendale, CA, USA

Reginald Ohlson
President, International Office of Cocoa, Chocolate and Sugar Confectionery, Brussels, Belgium

Dan Rosenfield PhD
M&M/Mars Inc., McLean, VA, USA

Bruce R. Stillings PhD
President, Food and Agriculture Consultants Inc., Haymarket, VA, USA

Stanley M. Tarka Jr PhD
Senior Director, Food Science and Technology, Hershey Foods Corporation, Technical Center, Hershey, PA, USA

Section I

Overview

Chapter 1

Overview of the Nutritional Benefits of Cocoa and Chocolate

John R. Lupien

Chocolate has a long and rich history dating back much further than its introduction to Spain and then the rest of Europe by the Spanish explorers Christopher Columbus and Hernando Cortés in the early 1500s. The Mayas of the Yucatán and the Aztecs of Mexico cultivated cocoa and the Aztec emperor Montezuma is said to have regularly consumed a preparation called *chocolatl*, a mix of roasted cocoa nibs, maize, water and spice (1). In fact, cocoa beans comprised a kind of unified monetary system in the middle Americas of the Aztecs and Mayas (2). There are ancient records chronicling price lists at that time. Although the first Latin name of the tree – *Amygdalae pecuniariae* – meant 'money almond' in recognition of its status as currency, it was the Swedish botanist Linnaeus, himself a regular consumer, who named the genus *Theobroma*, which translates as 'food of the gods'. This was prompted by natives' belief that the cocoa tree was of divine origin; this also resulted in a holy ritual being performed whenever cocoa trees were planted. The Mayas had a god of their own who was worshipped, together with the sacrifice of a dog having a cocoa brown colored patch on its skin, to ensure a rich crop. The farm workers responsible for planting the beans were kept away from their mates for several days in order to harness their sexual energy for the growth of the plant, and the actual beans were exposed to moonlight for several nights before planting.

When the Spanish became interested in cocoa, some 20 years after it had been brought back by the early explorers, its consumption was confined to the nobility. Predictably, because of its scarcity, many claims were made about this new drink, one of which was its being an aphrodisiac. Interest in it grew dramatically, and it is said that when Pope Pius V tasted a cup of chocolate, he was so disgusted with the taste that he gave up all thoughts of banning it under church rule, believing that no one would habitually consume such a product. Notwithstanding further attempts by the clergy to prevent its spreading, cocoa became popular across

Europe and the first chocolate house opened in London in the 1660s. Cocoa was mentioned in Pepys' diaries in 1664.

Historical texts (3) speak at length of the need to pay attention not only to the nutritious effects of various foods, but equally to those which affect the 'nervous energy'. Consumption of cocoa almost exclusively as a hot drink in 19th-century England is credited with providing nutrition more akin to milk than to, for example, tea. The fat content of cocoa was highly prized particularly for the poor, who were susceptible to hunger. Even with its relatively high price, cocoa was felt to produce a feeling of well-being in addition to its nutritional value.

The same text describes cocoa as 'the richest and most nutritious form of vegetable food ... although somewhat poor in gluten or nitrogenous flesh-formers is excessively rich in heat-givers, especially in fat' – a prized finding. It goes on to compare a cocoa drink with tea and milk, as having the advantages of both – the *exhilarating* properties of tea and the strengthening and *ordinary body-supporting qualities* of milk.

We can see that cocoa and chocolate have long been prized not only for their appealing sensory qualities, but for their potential nutritional benefits. In keeping with the historical traditions of chocolate, its past conjures up great and recognizable names, many of whom were philanthropists in their own right – Milton Hershey, Elizabeth Fry, George Cadbury, Joseph Rowntree – as well as other pioneers of the industry – Conrad van Houten, Henri Nestlé, Lindt and Jacob Suchard – all household names today.

It is said that chocolate contains sugar and fat in exactly the right proportions to provide a sensory experience like no other. Add to that a taste rich in complex flavors and we arrive at a truly unique product. The combination of a 'snappy' texture in its solid form with a smooth, fluid form in the mouth is characteristic of chocolate and of nothing else. It is the cocoa butter that intrinsically has the property of being solid at room temperature, yet melting at slightly less than body temperature that gives this unmistakable texture.

Another aspect of cocoa is its value in the developing countries where cocoa is produced. At present, the Food and Agriculture Organization (FAO) estimates that about 840 million people, mostly in developing countries, do not get enough to eat each day to enable them to lead fully productive lives. Food-deprived adults are unable to work over extended periods, and children cannot grow or learn to their full potential. Poverty is the root cause of lack of access to food and consequent malnutrition and increased susceptibility to disease (4, 5).

Cash crops such as cocoa beans can help overcome problems of economic access to food, and revenues from cash crops can also help support better food supplies, health care and education. Cocoa is an excellent cash crop for many farmers and is an important part of the national economy of cocoa-growing countries. It thus helps to increase income and support access to food for cocoa bean farmers and their families and promotes better national food security and economic development for developing countries. In fact, interest in growing cocoa beans is spreading beyond the traditional growing countries, and demand is also likely to grow as economic conditions continue to improve in many developing and developed countries. It should be remembered, however, that producing cocoa requires skill and good support services from government and the cocoa industry.

While many people in developing countries do not have regular access to adequate amounts of varied, good quality and safe foods, for many people in the main countries where cocoa products are consumed the problem is not the source of their next meal, but rather how to consume less calories, particularly fat calories and those saturated fats that are linked with health risk factors. Is it a surprise, then, that chocolate, a mixture of refined sugar and mostly saturated fat, has been a favorite target? It is only now that research is demonstrating yet again that there are no absolutes – not all saturated fats are created equally harmful; some fats such as stearic acid can be benign.

This review examines all major areas of interest from a nutrition and health standpoint and explores many of the popularly held beliefs about cocoa and chocolate. Chocolate has been accused of causing headaches and migraines, promoting heart disease, being addictive, causing dental decay, being a potent allergen, causing outbreaks of acne and unreasonably leading diabetics to abandon dietary common sense, amongst many other things. In most cases, this review of the scientific literature finds that these simply do not stand up to careful scrutiny. The fact that many dietitians tend to recommend their patients to avoid chocolate products seems to have its basis on emotional rather than scientific grounds (6).

The nutritional benefits of cocoa and chocolate are extensive. Although historical texts may have lacked the sophistication of today's scientific methodology, they recorded benefits known at the time. Most of their findings still hold some merit even today. One of the most serious dietary concerns today relates to cardiovascular disease and the impact of dietary fat on blood lipids. Of particular concern is serum cholesterol, which is generally raised by increased saturated fat intake. However, the recent work of Kris-Etherton *et al.* (7) has demonstrated stearic acid (the main fatty acid in cocoa butter) to elicit a neutral cholesterolemic response, which may be accompanied by an increase in high-density lipoprotein (HDL) cholesterol along with a decrease in plasma triglycerides (8).

There is no question that cocoa is a veritable storehouse of natural minerals, more so than almost any other food item. In fact, cocoa and chocolate have been shown to be a major source of dietary copper for North Americans (9).

Antioxidants are present in profusion in cocoa (10) and it has been shown that their presence is protective of many potential problems such as low-density lipoprotein (LDL) oxidation and stress effect alleviation (11). Antioxidants have been proposed as one of the components responsible for the so-called *French paradox*, in which lower heart disease is apparently prevalent despite a diet high in saturated fats.

Whether any of these effects are operative when cocoa is incorporated into chocolate remains debatable since the data on chocolate is scant. In any event, it is the flavonoid phenolics in chocolate that prevent rancidity of the cocoa butter in chocolate, thereby giving it such a long shelf-life. Of course, chocolate is calorically dense, but because of this, it represents a highly concentrated source of energy in a convenient and portable form and so is used by certain types of athletes and armed forces in supply rations (12).

Mapping the glycemic response of common foods presents a different picture than classical nutrition theory would suggest. Starchy foods such as certain common breads and many cereals emerge as eliciting a high glycemic response, whereas chocolate elicits only a moderate glycemic response (13). This has prompted Brand-Miller to advise against treating hypoglycemia in type 1 diabetic patients with chocolate. This suggests that chocolate has a place, albeit limited to occasional use, in the diet of people with diabetes.

Neither cocoa nor chocolate has any reproducible correlation with headaches, although stress – a factor endorsed by a high percentage of sufferers – is often associated with sweet craving. Fasting or skipping meals, similarly, is also a factor correlated with headache onset. It is perhaps a combination of these factors, from which snack food consumption may result, that seem to implicate chocolate when there exists no scientific evidence of any direct, causative effect (14).

Food allergy and intolerance are far less common than most people believe. Allergy to chocolate and cocoa has rarely been documented, but allergies to milk, egg, peanut and tree nut, which are often components of chocolate-based snack foods, are more frequent (15).

With regard to dental caries, research measuring plaque pH following consumption of common food items gives a measure of their acidogenicity, which is an indicator of the potential to cause caries. Another important factor for dental caries is clearance time. The Cariogenic Potential Index (CPI) can be constructed for different foods using animal studies, with sucrose as the anchor at 1.0. On this scale, milk chocolate scored 0.8. This score is lower than would normally be assumed, but is consistent with measurements of acidogenicity and clearance time. The inhibitory effect is probably due to the presence of protective chemicals naturally present in cocoa and in milk used to make milk chocolate, which limit the rate of cariogenesis (16).

As to the reasons for the almost universal appeal of chocolate, there are in chocolate a number of naturally occurring phytochemicals, some of which have the potential to stimulate the brain. However, they exist in such small concentrations as to render their properties negligible (17). The most plausible explanation is in its sensory characteristics. It simply provides a unique and wonderful sensory experience, extending far beyond just taste. Not only that, it somehow has the ability to provide an overall feeling of well-being, which in itself is beneficial to the consumer (18).

From a nutrition perspective, chocolate, like every other food, is neither good nor bad *per se*. In general, when consumed as part of a balanced and varied diet, chocolate can be both a source of nutrients as well as pleasure, and can be considered as being part of a healthful, wholesome diet. This is especially true in light of the contribution that the enjoyment of one's food makes to overall well-being.

References:

1. Cook, L.R. (1982) Historical notes. In *Chocolate Production and Use* (Rev. by Meursing, E.H.). Harcourt Brace Jovanovich, Inc., New York.
2. Nielsen, N. (1995) *Chocolate*. Trevi, Stockholm.
3. Johnston, J.F.W. (1880) The cocoas – the beverages we infuse. In *The Chemistry of Common Life* (Ed. and Rev. by Church, A.H.). Blackwood & Sons, Edinburgh.
4. FAO (1992) FAO Report of the International Conference on Nutrition. Food and Agriculture Organization, Rome.
5. FAO (1996) FAO Report of the World Food Summit. Food and Agriculture Organization, Rome.
6. Rossner, S. (1997) Review: chocolate – divine food, fattening junk or nutritious supplementation? *Eur. J. Clin. Nutr.* **51**, 341–345.
7. Kris-Etherton, P.M., Derr, J., Mitchell, D.C., *et al.* (1993) The role of fatty acid saturation on plasma lipids, lipoproteins and apolipoproteins: I. Effects of whole food diets high in cocoa butter, olive oil, soybean oil, dairy butter and milk chocolate on the plasma lipids of young men. *Metabolism* **42**, 121–129.
8. Kris-Etherton, P.M., Derr, J., Mustad, V.A., Seligson, F.H. and Pearson, T.A. (1994) Effects of milk chocolate bar per day substituted for a high-carbohydrate snack in young men on an NCEP/AHA Step 1 diet. *Am. J. Clin. Nutr.* **60** (suppl. 6), 1037S–1042S.
9. Joo, S-J. and Betts, N.M. (1996) Copper intakes and consumption patterns of chocolate foods as sources of copper for individuals in the 1987–88 nationwide food consumption survey. *Nutr. Res.* **16** (1), 41–52.
10 Waterhouse, A.L., Shirley, J.R. and Donovan, J.L. (1996) Antioxidants in chocolate. *Lancet* **87**, 311–315.
11. Jardine, N.J. (1999) Phytochemicals and phenolics. In *Chocolate and Cocoa: Health and Nutrition* (Ed. by Knight, I.). Blackwell Science, Oxford.
12. Burke, L.M. (1999) The role of chocolate in exercise performance. In *Chocolate and Cocoa: Health and Nutrition* (Ed. by Knight, I.). Blackwell Science, Oxford.
13. Brand-Miller, J.C. (1999) Chocolate consumption and glucose response in people with diabetes. In *Chocolate and Cocoa: Health and Nutrition* (Ed. by Knight, I.). Blackwell Science, Oxford.
14. Scharff, L. and Marcus, D.A. (1999) Chocolate and headache: is there a relationship? In *Chocolate and Cocoa: Health and Nutrition* (Ed. by Knight, I.). Blackwell Science, Oxford.
15. Taylor, S.L. and Hefle, S.L. (1999) Food allergy, intolerance and behavioral aspects. In *Chocolate and Cocoa: Health and Nutrition* (Ed. by Knight, I.). Blackwell Science, Oxford.

16. Curzon, M.E.J. (1999) Chocolate and dental health. In *Chocolate and Cocoa: Health and Nutrition* (Ed. by Knight, I.). Blackwell Science, Oxford.
17. Benton, D. (1999) Chocolate craving: biological or psychological phenomenon? In *Chocolate and Cocoa: Health and Nutrition* (Ed. by Knight, I.). Blackwell Science, Oxford.
18. Chiva, M. (1999) Cultural and psychological approaches to the consumption of chocolate. In *Chocolate and Cocoa: Health and Nutrition* (Ed. by Knight, I.). Blackwell Science, Oxford.

Section II

Introduction and Background

Chapter 2

Cacao Growing and Harvesting Practices

R. Anthony Lass

Cocoa was a well-established crop and article of commerce in the early 16th century in Central America. In 1520, when Cortés discovered Mexico City (then the capital of the Aztec peoples) and met their leader (Montezuma), he found that cocoa beans were used in the preparation of a luxury drink – *chocolatl* – made by roasting the whole cocoa beans, grinding them and mixing with maize meal, vanilla and chilli. They were then stirred with a special whisk, rather in the fashion still adopted today in Colombia, the Philippines and elsewhere. These cocoa beans had not been actually grown by the Aztec peoples but by Mayas who gave them as tribute to Montezuma. At that time cocoa had more significance than merely being the main ingredient of a drink; as the cocoa beans were easy to count and were relatively valuable, they were widely used as currency. In Mexico, this use for cocoa beans appears to have continued until at least 1840.

From 1520 through to the middle of the 17th century, the main cocoa areas were all in, or around, present-day Mexico, extending as far as Honduras. All the cocoa cultivated at that time was Criollo (see below), probably as this type gives a palatable drink with little fermentation. Columbus encountered cocoa from Honduras in 1502 and this represented the first contact of the Old World with cocoa beans. He transported some to Spain and introduced cocoa drinks to the Spanish court, where it was much valued. The drink soon became popular amongst the aristocracy in Spain, later in Italy, Flanders, France and England. Spain maintained a monopoly on the trade in cocoa until the Dutch took over Curaçao in 1634, enabling the trade and use of cocoa beans to then expand rapidly, though still only amongst the most wealthy as the duties and cost of transport were very high.

In the mid-16th century, cocoa cultivation of Criollo types spread in the West Indies (Jamaica, Martinique and Trinidad), having in addition been transported in about 1560 across the Pacific to the Philippines, thence a little later to Sulawesi and Java, and perhaps as well to India and Sri Lanka (1). By 1700, cocoa was being grown throughout Central America, in many of the Caribbean Islands and in areas adjacent to the Andes in South America, but it still remained a great luxury.

Early in the 1800s, duties were reduced and consumption increased, though only as a chocolate drink – high in fat, being still made from the whole cocoa bean. In 1828, Van Houten designed a press to remove some of the fat and opened up a vast range of new products – including chocolates as we know them today! This was the first of many major technical advances that have led to the wide variety of products based on the cocoa bean now available. Very modest quantities of cocoa are used in cosmetics, the rest being as chocolate and other foodstuffs.

There was considerable trade between Brazil and West Africa in the 19th century and so the introduction of cocoa into Africa could be seen as inevitable, and 1822 seems to be the date generally given for the movement of cocoa to Principe, a small volcanic island just off the West African coast, then under Portuguese control. Plants were then soon moved to all the other islands off that coast, but large-scale cultivation in West Africa only started in Nigeria in 1874 and Ghana in 1879. However, from 1857 missionary groups had been attempting introductions into Ghana although with almost no success. Some data on cocoa production over the last 150 years are presented in Table 2.1.

In 1876, Daniel Peter mixed milk solids with cocoa and sugar to make milk chocolate and this led to very rapid growth in chocolate popularity from the start of the 20th century. With this, cocoa plantings on a significant scale began and there was a shift in the balance of production from South America and the Caribbean to West Africa (see Table 2.1 and Fig. 2.1). The other great change then was the move from plantings of the Criollo types to the Forastero types, because of their higher yield potential and greater resistance to pests and diseases. In the 1850s, beans from Criollo types accounted for almost 80% of total global production of cocoa, by 1900 it had fallen to 40–45% and since then it has continued to fall steadily to perhaps only 1 or 2% in 1998.

Present pattern of global cocoa production

On a global level, cocoa is one of the smaller commodities, with only an estimated 2.7 million tonnes being produced in 1996/97, in comparison to current estimates of global wheat production at 6000 million tonnes, rice at 380 million, oilseeds at 260 million, cotton at 90 million and coffee at 6.15 million tonnes (2). Cocoa is a classic example of an agricultural commodity that, whilst being produced almost totally by developing countries in the tropics, is largely consumed in the (cooler) industrialized economies. Seven countries – Brazil, Cameroon, Côte d'Ivoire, Ghana, Nigeria, Indonesia and Malaysia – presently account for about 86% of world production, with the four West African countries amounting to over 55%

Table 2.1 Growth of cocoa production for major cocoa producers for selected years, 1850–1997/8.

Country	Production (thousand tonnes) in each year						
	1850	1900	1920	1940	1960	1980	1997/98F*
Brazil	3.5	18	36	131	124	349	162
Colombia	—	3	4	12	19	39	45
Ecuador	5.5	23	41	14	42	81	45
Mexico	—	1	2	2	27	30	33
Venezuela	5.4	9	22	17	17	14	12
Dominican Republic	—	7	26	20	35	32	50
Grenada	—	5	4	3	2	3	1
Trinidad & Tobago	1.7	12	35	8	6	3	1
Cameroon	—	1	3	23	74	120	120
Equatorial Guinea	—	1	5	13	25	8	3
Ghana	—	1	118	241	440	403	390
Côte d'Ivoire	—	—	1	43	94	258	1105
Nigeria	—	17	18	103	198	156	155
São Tomé and Principe	—	—	28	5	10	8	2
Indonesia	—	1	1	1	—	8	345
Malaysia	—	—	—	—	—	43	95
Papua New Guinea	—	—	—	—	7	28	30
Others	1.9	16	27	36	69	77	79
World total	**18**	**115**	**371**	**672**	**1189**	**1660**	**2673**
Proportion (%) from							
Central and S. America	80	47	28	26	19	31	11
West Indies	9	21	18	5	4	2	2
Africa	—	17	39	63	70	57	66
Asia & Oceania	—	1	8	1	1	5	18
Others	11	14	7	5	6	5	3

After Gill and Duffus Cocoa Statistics, November 1989; E.D. & F. Man Cocoa Report No. 360
* Forecast

of the world total (see Table 2.1 and Fig. 2.1). From 1973 to 1985, Africa's share of the world market declined, while producers in the Americas (mainly Brazil) and Asia (Indonesia and Malaysia) expanded their shares. The major losers of market share were Ghana and Nigeria, and the expansion of Ivorian production was not sufficient to offset this. After 1985, Côte d'Ivoire's production continued to increase and Ghana's production began to recover, causing a gradual increase in the African share of world cocoa production. During the same period, the Asian share of the world market also increased while American production, especially that of Brazil, declined. In summary, during the last 20 years Asian cocoa producers expanded their market share, mainly at the expense of American, but also African, cocoa producers.

It is estimated that some 80% of world cocoa production is produced by growers with less than 5 ha and 84% by growers with less than 40 ha, though there is an absence of reliable data. Cocoa therefore essentially remains a smallholder

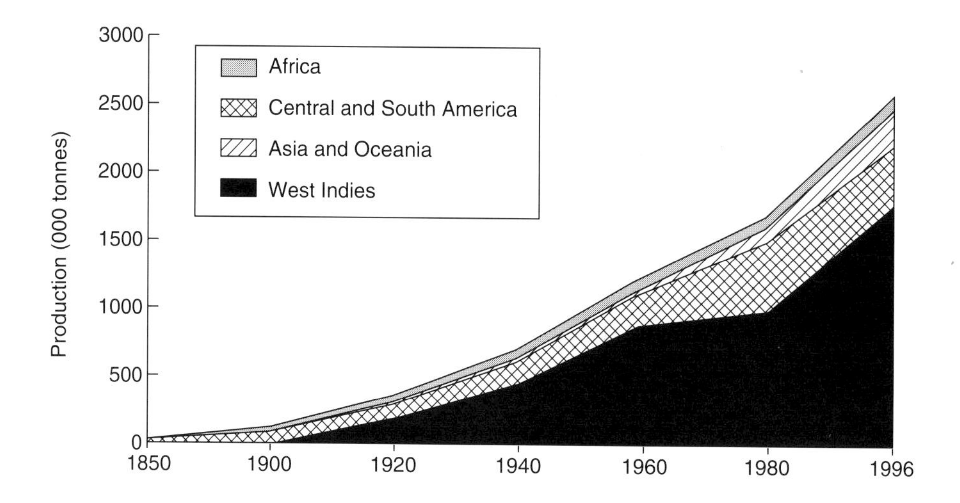

Fig. 2.1 World cocoa production by region, 1850–1996. After Gill and Duffus Cocoa Statistics November 1989, ED&F Man Cocoa Report No. 360. Copyright Cadbury Limited.

crop; the often-used 'tag' of cocoa being a 'plantation' or estate crop is actually very misleading.

In South and Central America, cocoa has been, and continues to be, cultivated on plantations and smallholdings with units of about 20 ha being the customary size. In Brazil and Ecuador, there have been some much larger holdings, often planted by individuals or families, rather than by estate companies. It now seems that these are being broken up or are having cocoa removed, largely due to the crop's unprofitability following the devastation caused by witches' broom disease in both countries.

In the 1930s, there were large plantings of cocoa in Costa Rica by the United Fruit Company after earlier problems with bananas due to Panama disease, though these areas seem to have reverted to bananas once more. Production in Brazil increased after the cocoa price boom of the late 1970s as the Brazilian Government provided incentives (mainly subsidized credit) for planting cocoa. This programme ended in the mid-1980s, and new plantings were reduced significantly. Devastation by witches' broom disease, unstable macroeconomic conditions, high cost of credit and high cost of production along with declining world cocoa prices caused significant and dramatic reductions in cocoa production in Brazil in the early 1990s.

In Ecuador, cocoa production has suffered for many decades from a number of devastating diseases (including witches' broom) which have frequently rendered cultivation barely viable and yields are still often very low, though some high yielding fields exist in areas with favourable conditions and good management.

In West Africa, cocoa is grown almost entirely on smallholdings and usually each farm is very small. While individual plantings representing 1 year's clearing are generally small – less than 1 ha – there is often little relationship between such plantings and the total size of one farmer's cocoa holdings (3) which can be at a number of different locations around a village or even around different villages. Nevertheless, very few African cocoa growers have more than 8 ha of cocoa under cultivation and almost none have more than 50 ha. However, there are again a few (mostly historic) exceptions: in Cameroon, several estates were started by German companies before World War I, but these were all soon converted to other crops; in Equatorial Guinea at one time, all the cocoa was produced on estates; cocoa was planted quite extensively on some private estates in Zaire as well as on some state-owned ones in Nigeria; in Côte d'Ivoire, there are a few private sector cocoa estates, originally created by European investors.

Production (by smallholders) in Côte d'Ivoire increased significantly during the 1980s, due to the maintenance of high producer prices and incentives to increase cocoa planting; production is forecast to be over 1.1 million tonnes again in 1997/98. Production in Ghana, which declined dramatically during the 1970s and early 1980s, started recovering from about 1986, in large measure due to economic reforms that aimed, among other objectives, to restore macroeconomic stability and incentives for farmers. In Nigeria, production declined throughout the 1970s and into the early 1980s, but recovered somewhat following the liberalization of cocoa marketing in 1986; with a subsequent brief upheaval of the cocoa subsector, production has stabilized well below the levels of the 1960s. Cameroon production has been stagnant during the last 25 years.

In South-East Asia, large-scale agriculture has been actively encouraged for many decades and has made very substantial contributions to the economic welfare of Malaysia and now increasingly of Indonesia. As world cocoa prices rose dramatically in the late 1970s and the profitability of large-scale cocoa cultivation looked assured, estate companies in Malaysia turned to cocoa and uprooted much rubber and oil palm. Huge areas of cocoa (see Fig. 2.2) were planted by estates and a wealth of smaller growers in the early 1980s – at great speed but with a very high degree of technical competence.

Cocoa production in Malaysia (and later in Indonesia) increased significantly, mainly because of low production costs (Indonesia), high producer prices as a result of market competition and the virtual absence of taxation (Indonesia and Malaysia). Vigorous market competition in both countries resulted in very low marketing costs and margins which contributed to a high share of the export price going to growers (about 85–90% in Indonesia and over 90% in Malaysia). By 1998, a high percentage of the estate cocoa in Malaysia had been removed (usually excluding all but the very highest yielding fields) and replaced again with rubber and oil palm, because of the much greater profitability of these crops combined with the difficulties of managing cocoa on an estate scale – a task presently made particularly difficult in Malaysia due to the serious shortage of agricultural labour; cocoa has a higher labour requirement – and a very large peak demand.

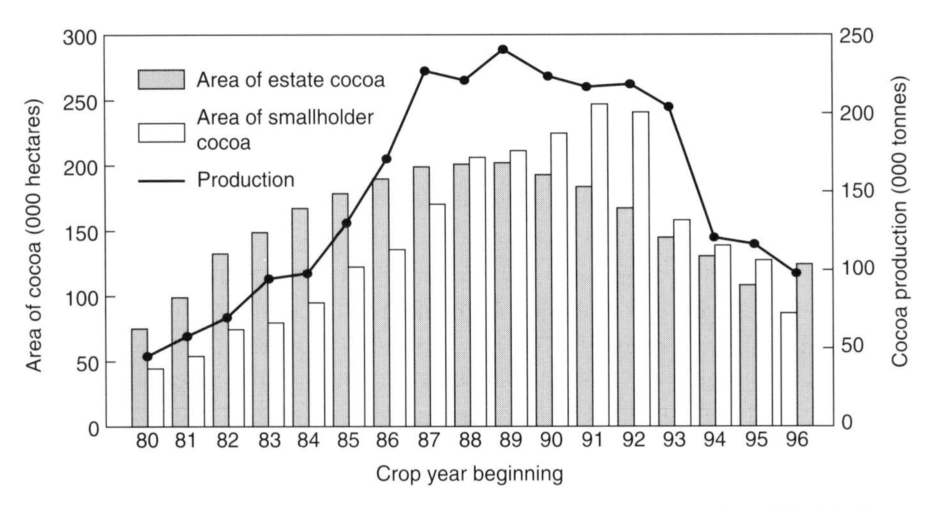

Fig. 2.2 Area of cocoa by holding type and cocoa production in Malaysia, 1980–96. After Malaysian Cocoa Monitor, June 1996, Malaysian Cocoa Board, Kota Kinabalu, Malaysia. Copyright Cadbury Limited.

Some large-scale plantings of cocoa continue in Indonesia, but the Indonesians have replaced all but their highest yielding fields with rubber and oil palm for the same reasons as in Malaysia. It is now clear that cocoa is not an easy crop to cultivate on a large-scale basis, but seems well suited to smallholder cultivation.

For some nine decades, global cocoa production has increased steadily and consistently to keep pace with the ever-increasing needs for cocoa beans. Consumption has increased on average by 3.5% per annum over recent years and is projected to increase by 1.5–3.3% per annum over the coming 5 years (4). A promising future for cocoa can be expected, as demand should show a steady rise from traditional, existing chocolate consumers with the possibility of substantial increases in usage in newer chocolate markets over coming years – in South America, Eastern Europe, Russia, India, China and other parts of South-East Asia.

Botany of the cocoa plant

The cocoa tree belongs to the genus *Theobroma*, a group of about 20 species of small trees occurring wild in the Amazon basin and other tropical areas of Central and South America. *T. cacao* is the only one extensively cultivated, although increasing interest is being shown in cultivation of *T. grandiflorum* (Cupuassu) in certain parts of the Amazon, for use as flavouring for juices, ice creams and other foods (5). The centre of diversity of cocoa, and thus perhaps the origin of the species, seems to be the headwaters of the Amazon River, where wide morphological and physiological variation is demonstrated in the wild trees collected.

Early writers such as Morris (6) identified two broad botanical groups (Criollo and Forastero) in the species *T. cacao* which are still valid today. Later another group was added (7), which is a cross between the other two and is called Trinitario. The characteristics of these groups are shown in Table 2.2 and in Fig. 2.3. The Criollo types ferment quickly and in earlier times were considered to have a highly regarded, but usually weak, chocolate flavour. These types were probably domesticated by the Mayas and Criollo pods appear in their early stone reliefs, perhaps even as early as the 6th century. They have been found in diverse locations; as far apart as Guatemala, Honduras, Nicaragua and Veracruz in Mexico. The other main group is the Forasteros of which the Amelonados are the most extensively planted, being established in a large (though decreasing) percentage of both Brazilian (State of Bahia) and West African cocoa areas. The planting material for the State of Bahia originated from the lower Amazon in about 1700, while that of West Africa came from Bahia after establishment in Principe in about 1822. There is great uniformity amongst much of the cocoa produced in West Africa, partially because the original planting material had this common origin.

Table 2.2 Some distinctive characters of Criollo, Forastero and Trinitario types.

Characteristics	Criollo 'fine or flavour'	Forastero 'bulk'	Trinitario
Pod husk			
Texture	Soft, crinkly	Hard, smooth	Mostly hard
Colour	Red occurs	Green	Variable
Beans			
Average no. per pod	20 to 30	30 or more	30 or more
Colour of cotyledons	White, ivory or very pale purple	Pale to deep purple	Variable; white beans rarely
Agronomic			
Tree vigour	Low	Vigorous	Intermediate
Pest and disease susceptibility	Susceptible	Moderate	Intermediate
Quality			
Fermentation needs	1 day maximum	Normally 5 days	4 to 5 days
Flavour	Weak chocolate; mild and nutty	Good chocolate	Good chocolate; full cocoa
Fat content	Low	High	Medium
Bean size (g/100 beans)	85	94	91[1]
Percentage of world production, 1996/97	1.5	93.5	5.0[2]

Source: after Toxopeus (9).
[1] Some typical values from Crespo (8).
[2] Mostly Cameroon, PNG.

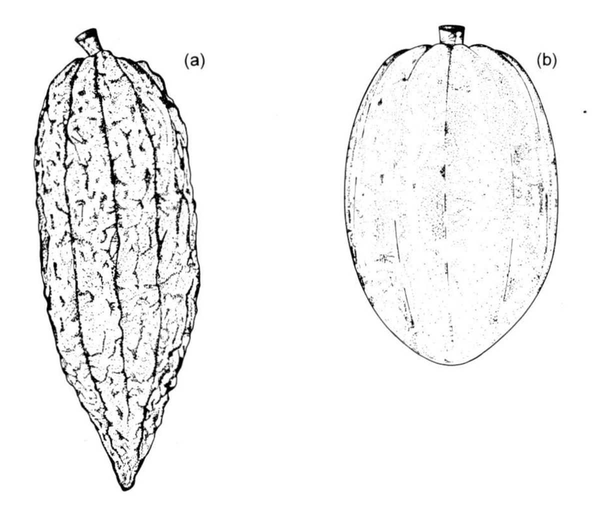

Fig. 2.3 Examples of (a) Criollo and (b) Forastero (Amelonado) cocoa pods. After Cuatrecasas (10).

The natural habitat of the genus *Theobroma* is the lower storey of the evergreen tropical rain forest, where rainfall and shade is heavy, temperature is fairly uniform and there is constant relative humidity through the year. This indicates a latitude range for successful cultivation of cocoa (*T. cacao*) from about 10° above to about 10° below the Equator, though growing has been successful in India at 14°N and attempted in Sao Paulo State, Brazil at 24°S. The need for an amount of overhead shade for cocoa (usually a dappled shade is to be recommended) – at least in its earliest years – is a major constraint, limiting both the locations where cocoa can be grown as well as the systems to be adopted for successful establishment.

A mature cocoa tree has a deep tap-root with an extensive system of lateral feeder roots and so grows best in deep soils. As the trees develop, the foliage and branches of neighbouring cocoa trees will grow together to form an integrated canopy – vital for high yields, as the canopy has to trap the maximum amount of light energy. Cocoa has a requirement for large amounts of biochemical energy for the conversion of the carbohydrates produced by photosynthesis into fat which is accumulated in the developing beans; cocoa beans contain a little over 50% cocoa butter (also called cocoa fat). The cocoa tree can easily grow to a height of over 15 m if allowed to do so unchecked, but under the more intensive cultivation systems this is prevented by careful pruning to keep the plants to heights of 2.5–3.0 m. A stylized transverse section of an ideal cocoa farm is shown in Fig. 2.4.

The fruit of the cocoa tree is a pod, which arises from flower cushions directly on the trunk/branches. Flowering is profuse: a tree may produce 50 000, though less than 5% will be set as pods. Pollination is undertaken by a myriad of minute flying insects, which breed in rotting vegetation and require cool, moist and damp conditions to survive. The cocoa pods, containing 30–40 or so seeds, take some 5

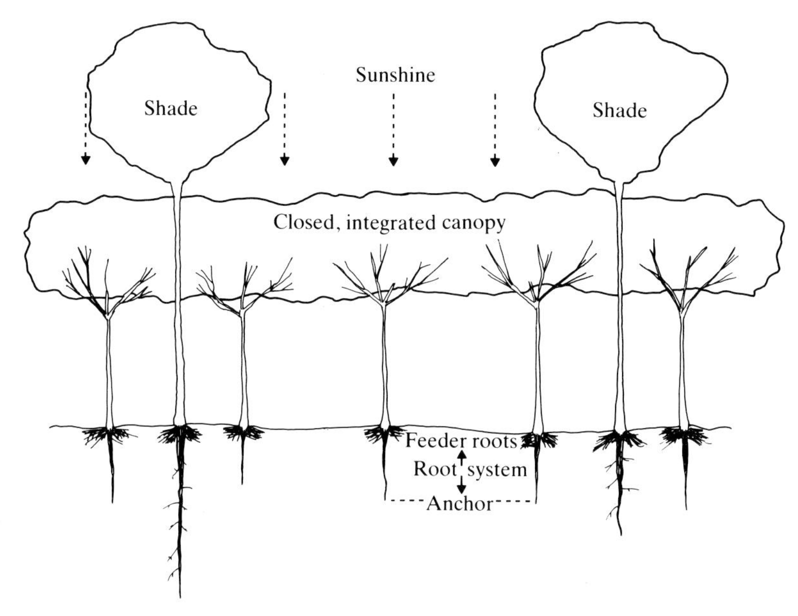

Fig. 2.4 A stylized transverse section of an ideal cocoa farm. After Toxopeus (9).

months from pollination to ripeness; when ripe, the seeds are embedded in a sweet mucilaginous pulp comprised mainly of sugars and when dried are the cocoa beans of commerce (see Fig. 2.5).

When the pods are opened, the sweet pulp rapidly breaks down and the conditions under which this happens have a major influence on the flavour of the chocolate from these beans. This process is called fermentation, though it is not a

Fig. 2.5 An open Amelonado pod from West Africa. *Source:* Toxopeus (9).

true fermentation as acids rather than alcohols are the final product. When ripe, the Amelonado pods (see Fig. 2.5) – typical of West Africa and much of Brazil – are a pleasing bright yellow colour. Pods of other types are of different colours and shapes (see Table 2.2).

Climate, growth and cropping of cocoa

Areas where rainfall is limiting are unlikely to be successful for cocoa: it requires 1250–3000 mm of rainfall per annum, with a dry season of no more than 3 months (a dry month meaning one when less than 100 mm of rain falls in that month). The mean maximum temperature should vary between 30–32°C and the mean minimum between 18–21°C with an absolute minimum of 10°C; at 5°C growth probably ceases and death of the plant may result from a short exposure to 0°C. Cocoa prefers areas with high relative humidity. A hot, moist climate will favour the growth of cocoa and in areas with no dry season, cocoa will develop more quickly than in locations where growth may be slowed, or even stopped, by low temperatures or drought conditions.

Countries offering continuous growing conditions (Malaysia and Indonesia, for example) have a considerable advantage over those in which growth is regularly checked due to dry seasons, which can sometimes be severe, even in areas favourable for cocoa cultivation. Cocoa does not thrive on poorly drained soils, very sandy soils, shallow soils or soils of very low fertility. The soil depth should be at least 1.5 m, with a good structure, being a mixture of clay and sand and about neutral in terms of acidity. Soils with a good surface layer of organic matter are to be preferred. Reasonable yields require reasonable levels of calcium, magnesium, potassium and phosphorus to be available in balanced quantities.

For young cocoa, some temporary shade will always be needed for the first few years to ensure the right growth of the young cocoa trees. Low light (high shade) levels encourage tall, thin trees; high light intensities tend to cause the opposite. Different species of temporary shade at different planting plans will be successful in different cocoa growing environments. Suitable permanent shade can be provided by forest trees left standing at the time of land preparation – the so-called thinned forest shade. Alternatively, permanent shade can be provided by selected fast growing tree species planted on a regular pattern; legumes are ideal as they provide nitrogen. Again, the spacing and the species will depend on the local situation. In some circumstances, shade can be dispensed with altogether for mature cocoa, as long as the cocoa can obtain adequate nutrients and moisture throughout the year. Higher yields (and probably a shorter economic life of the cocoa tree) can be expected under such conditions, though in cocoa areas where sap-sucking insect pests are a problem it may well be that some shade should still be provided as this seems to often reduce insect attacks.

The cropping pattern for mature cocoa is clearly related to the rainfall distribution and various studies have shown that bean size is affected by rainfall during the development of the crop. The climate also strongly influences the methods of drying which can be utilised; it has been of considerable advantage to many West African growers to be able to reliably dry their cocoa in the sun.

The presence of overhead shade and a complete cocoa canopy conserves nutrients, protects the soil from erosion, and conserves rainfall by preventing soil run-off, while the breaking down of the leaf litter recycles nutrients to the soil. In addition, the use of shade trees and the fact that so many plots of cocoa (though still often a monoculture) are very small, increases biodiversity in and around many cocoa farms. In the light of these observations, a number of authorities are suggesting that cocoa is an environmentally beneficial crop, potentially having a useful role in systems of sustainable agriculture. Further research is required to quantify these benefits and make comparisons with other annual and perennial tropical crops.

Cocoa cultivation and some of its constraints

There are a large number and a wide variety of constraints facing cocoa growers; some are location specific, but many are common to nearly all cocoa growing regions. A number of the latter are now briefly described.

Low yields

Average yields of cocoa are low due to extensive systems of cultivation, ageing tree populations, high incidence of pests and diseases, poor systems of their control, ageing farmer populations, shortage of affordable labour, lack of easily available inputs, poor extension services and, above all, the use of poor/average quality planting material. In many cocoa growing areas the average yields have not increased over recent decades; in fact, as trees age and are not replaced, the average yield is probably declining.

Planting material

This is the most important input in any cropping system; in the case of cocoa, trees need to give a good yield of dry beans under a range of growing conditions, having a number of desirable agricultural, commercial and local characters (see Table 2.3 for an outline of some desirable characters in these groups). Sadly, much of the cocoa planting material currently planted does not meet many of these desirable criteria. Some observers suggest that farmers will only be interested in replanting fields with new planting material when the yield increase is clearly visible to them and that to achieve this, new material needs to offer a

Table 2.3 Some characteristics desired in future generations of cocoa planting material.

Desirable agricultural characteristics	Desirable commercial characteristics	Desirable local characteristics
Vigorous early growth	Good chocolate flavour	Tolerance/resistance to pests
Drought tolerance	Good value for bean weight[1]	Tolerance/resistance to diseases
Early bearing	Shell content[2]	Local adaptation (if appropriate) to:
Low, open branching habit	High fat content[3]	drought tolerance
High yield	High number beans per pod	flavour tolerance
	Good weight of beans[4]	local flavour[5]
		soil types
		wind tolerance, etc.

[1] Average bean weight of 93 beans/100 g (or 107 g per 100 beans).
[2] Maximum shell content should be 10–12%.
[3] Cocoa butter/fat content in the cotyledon should be above 55%.
[4] Objective should be that a maximum of 15/16 pods are needed to produce 1 kg of dry, fermented cocoa; the successful breeding programme of Freeman in Trinidad has achieved results of 8/9 pods/kg (11).
[5] For example, in case of a 'fine or flavour' cocoa producer.

minimum yield improvement of 25% over the existing material. In the past, this level of yield improvement has only rarely been achieved in new generations of cocoa planting materials and may explain why farmers often seem to prefer to continue with the old material – Amelonado is still being planted in West Africa.

Corley (12) studied the physiological status of a number of tree crops and from this analysis estimated the potential yield of cocoa could be as high as 11 tonnes of dry bean per hectare, or a massive 23 times the current average yield of the smallholder in Ghana (probably 480 kg/ha) and 3.4 times the yield of some of the most productive plantings in South-East Asia; the best cocoa fields in Malaysia still only have 5-year mean yields of 3.2 tonnes. On this basis, there remains substantial scope for yield improvement of cocoa through breeding. In recent decades, there have been dramatic advances in the yield potential and productivity of a wide range of other tropical, temperate, annual and perennial crops; no such advances have been seen in cocoa – average cocoa yields in West Africa are little higher than those recorded 20–30 years ago. This is putting cocoa as a crop at a competitive disadvantage.

Age of cocoa trees

A high percentage of the world cocoa tree stock is of advancing age. Although there have recently been substantial new plantings in Côte d'Ivoire and Indonesia, there has been insufficient new planting elsewhere in West Africa. It seems that a significant proportion of the world's cocoa plantings are reaching, or have already reached, the end of their economic life. Economic life is hard to calculate, but a consensus opinion would probably be that in a field where most trees are around 40 years of age, there will be insufficient gross income to give a real rate of return on the capital employed. Some typical yield profiles are presented in Fig. 2.6.

As many of these smallholder farmers are the very people who initially planted the cocoa farms all those years ago, it is hardly surprising that many are now also of advanced years and are finding it increasingly difficult to maintain their farms in the way they should.

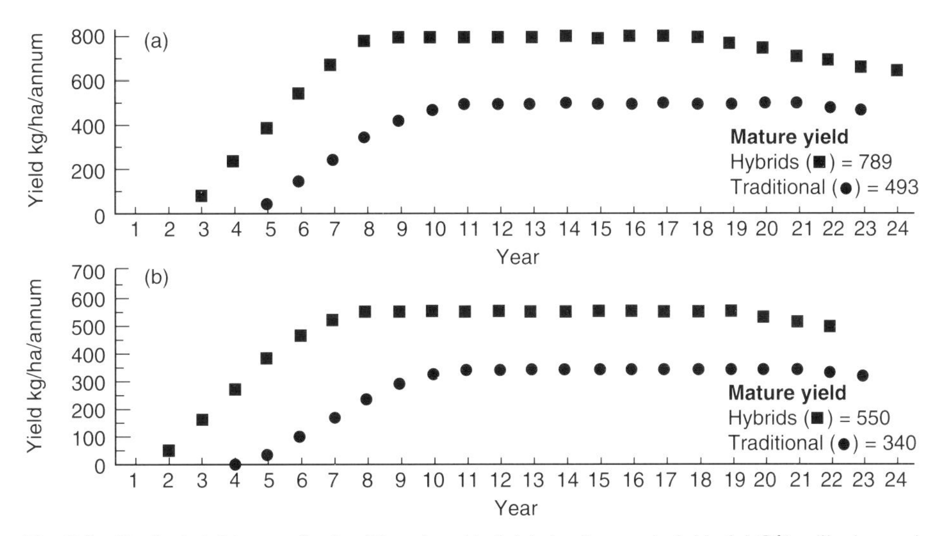

Fig. 2.6 Typical yield curve for traditional and hybrid planting material in (a) Côte d'Ivoire and (b) Ghana. *Source:* after Anon (13). Copyright Cadbury Limited.

Availability of suitable land

Since the earliest years of planting cocoa as a crop, it has tended to be planted following economic timber extraction from primary forest. This trend started in Central America, moving to South America, to West Africa, back to South America and now to South-East Asia. It seems that only in a few isolated cases was planting of cocoa ever the prime reason for this clearance of primary forest; timber extraction was the real reason. Much of the best land in the cocoa producing areas of South and Central America, West Africa and South-East Asia has now been planted either with cocoa or with other crops. Techniques for the successful replanting of such lands with a second crop of cocoa have not yet been developed; these are now very urgently needed for a wide variety of cocoa growing environments.

Lack of credit availability for smallholders

The banking systems are simply unable to provide suitable seasonal or replanting credits for smallholder cocoa growers in any of the major growing areas. In many cases this is because growers do not have any, or adequate, collateral as they often do not have formal title to the land that they have been using for decades. This is a major constraint to efficient cocoa cultivation.

Pest and disease losses

Total losses of cocoa production due to pests and diseases are substantial, as are the costs of attempting to control them effectively. There are no recent, comprehensive and reliable estimates of these total losses, either on a tonnage or a percentage basis. It seems as if the last serious systematic attempt to do this was by Cramer, published in 1967 (14). At that time, he concluded that with world cocoa production at 1 528 000 tonnes, the potential loss from diseases was 588 000 tonnes and that from pests and weeds together amounted to a further 788 000 tonnes, or 20.2% and 27.0% respectively of the potential production per annum. With world cocoa production in 1997/98 estimated at 2 673 000 tonnes, a straight extrapolation of Cramer's figures (which may or may not still be valid) gives huge potential global losses. Empirical evidence suggests that actual losses are very high and may well still reach such levels. The distribution of the major cocoa pests and diseases is shown in Table 2.4. There is a real need for improved research into better cultivars and new techniques of environmentally friendly pest and disease control suited to the small cocoa farmers of the world, no doubt involving integrated pest management practices.

On-farm processing of cocoa

Harvesting

In common with many other tropical crops, the cocoa harvest is spread over several months, usually with a major peak and a minor peak of pod ripeness/harvesting. A maturing crop tends to suppress further flowering, and some varieties have sharper peaks than others, though the new hybrids tend to have a flatter harvest pattern. These factors, together with weather variations between seasons, make accurate predictions of crop timing and size difficult.

Ripe pods can be harvested over a 2-week time-frame (before, as, or after they start to change colour) usually with no yield loss and can be left on the tree for a further 2–3 weeks without a reduction in flavour quality; though leaving ripe pods on the trees for any time is not really to be recommended as theft and losses from rodent pests and diseases will increase. Rodents are attracted by the sweet mucilage in the ripe pod; any rodent damage breaks the pod wall, exposes the ripe beans to oxygen and levels of germinated beans can become significant. These are classified as a defect and are to be avoided – the germ is often broken off and moulds can enter through the resulting hole.

Table 2.4 Matrix of major pest and disease problems faced by cocoa growers.

Problem	Country							Comments
	Brazil	Cameroon	Côte d'Ivoire	Ghana	Indonesia	Malaysia	Nigeria	
Capsid	X	X	XX	XX	X		X	
P. palmivora	X	XX	X	X			XX	Ubiquitous though not always serious problem
P. megakarya		XX		XX			XX	Spreading in Ghana; risk of spread to Côte d'Ivoire
Swollen shoot virus				X				
Witches' broom	XX							
Vascular streak dieback (VSD)					X	X		
Cocoa pod borer					XX	XX		
Average yields	L	L	M	L	H	H	M/L	

Source: Lass (15).

Careful removal of the pods from the trees with a knife is required to avoid damage to the flower cushions; in South America and West Africa pods are usually left on the ground to be gathered together for breaking by a team of workers after a few days' harvesting activity. It appears that this act of so-called 'pod storage' can have a highly beneficial effect on the subsequent development of chocolate flavour, though the precise biochemical pathways, conditions and processes involved are still poorly understood.

It seems that most, if not all, Ghana farmers have unknowingly adopted this technique of pod storage simply by their practice of using family labour to collect the harvested pods into a pile before organizing friends and neighbours to help break open the pods (16). Farmers clearly have to be careful to avoid loss from diseases in these pod piles, which can be a danger if they are left for much more than 10 days.

Research into the poor flavour quality characteristics of much South-East Asian cocoa identified that several days' pod storage could give significant flavour improvement to, for example, Malaysian estate cocoas (17). Unfortunately, there turned out to be insurmountable logistical difficulties in undertaking several days' pod storage on an estate basis, when several hundred tonnes of pods per day could often require to be split while hundreds more were still being stored.

Furthermore an insect pest – the devastating cocoa pod borer (see Table 2.4) which burrows around in the pod walls – requires that pods in areas affected by the pest (which now include a large percentage of the cocoa growing locations in South-East Asia) are harvested at the latest when pods are just showing a tinge of colour. In the foreseeable future, cocoas from South-East Asia seem set to have poorer flavour quality than West African beans.

Despite a number of attempts, the mechanical removal of cocoa pods from the tree has so far proved impossible, while mechanical pod breaking and extraction of wet beans, despite the construction of a number of prototype devices, still requires to be fully developed. The separation of the beans from the pieces of pod husk of about the same size when both are covered with mucilage has, so far, proved to be a major constraint to success.

Fermentation

Chocolate flavour is developed in two parts: the first on the farm by correct fermentation of the wet beans by the grower, and the second by the processor in the factory at the roasting step. Good chocolate flavour cannot be produced by only one of these stages. In the initial stages of fermentation, much of the pulp drains away and some time between 36 and 72 hours the beans are killed. The processes of flavour development are complex, and still quite poorly understood, though good progress has been made recently through the use of expert analytical and sensory evaluation (flavour profiling) techniques.

Very nearly all cocoa growers (with the exception of those owning some recent plantings in Indonesia, in particular in Sulawesi) fully ferment their cocoa. Unfermented or insufficiently fermented beans, such as those from Indonesia, have poor chocolate flavour and so are destined for cocoa butter extraction rather than for chocolate manufacture. Fermentation methods vary considerably from country to country and even from grower to grower, but there are basically two approaches: fermentation in a wooden box – often adopted by larger growers – and fermentation in a heap on the ground covered with banana leaves – more frequently used by the smaller grower (see Figs 2.7 and 2.8). Season, ambient temperature, duration, size of boxes/heaps, amount of mixing (turning), presence of a cover or not, presence of drain-holes, etc. all require careful optimization; all are variable, subject to local influences and affect the final flavour quality of the dry cocoa.

Fig. 2.7 A cascade of fermentation boxes with moveable side walls. *Source:* Wood (3).

In general, Criollo beans require much shorter fermentations: 1 or 2 days as opposed to 5–6 days (or sometimes more) for Forastero types. Beans from South-East Asia almost always have a flavour with a high degree of acidity, at a level which is unacceptable to many chocolate manufacturers and can even require cocoa processors to include additional processing steps. This acidity in Malaysian cocoa (and others from South-East Asia) seems to be due to the presence of larger amounts of acetic and lactic acids, in turn due to the condition of the mucilage (in particular the quantity) present when the beans are fermented,

Fig. 2.8 The start of a heap fermentation in Ghana. *Source:* Wood (3).

though the reasons for this are not at all clear. This acidity is a defect, which has proved very difficult to eliminate from such cocoas, despite extensive, recent research on improved fermentation techniques.

Drying

After fermentation, the moisture content of the beans needs to be reduced from 55% to 7.5% – an appropriate moisture content for secure storage of cocoa for a couple of months in the tropics. Smallholders lay the wet beans on raised bamboo mats or, less satisfactorily (for hygiene considerations), on concrete platforms on the ground in the villages. The drying beans would be covered at night or in the event of rain, watched for theft and be regularly turned during the day. The duration of drying depends on the weather, but it is unusual for sun-drying in West Africa to be completed in less than a week. Dull weather may extend it to 2 weeks or very occasionally more; after that length of time there would be a high risk of both surface and internal moulds developing on the beans. The care and sorting of the beans while they are drying tends to be the responsibility of the old and the young who remain in the villages during the day.

The best quality cocoa is always that produced when the beans have been fully dried (to about 7.5% moisture) in the sun, as happens in Ghana and most of the cocoa areas of Côte d'Ivoire and Nigeria. However, in West Cameroon much of the crop is harvested in the middle of the wet season and artificial drying, by hot air, is essential. In Brazil and several other countries, the climate is such that artificial dryers are required for at least part of the crop. Smallholders in Malaysia and Indonesia are mostly able to dry their cocoa effectively in the sun, though the drying is also often undertaken by the local bean collectors who can use artificial dryers. However, on the estates in both countries, cocoa fermentation and drying tend to be a highly mechanized, rapid and very efficient operation producing a uniform product at a low cost per kilogram, though the produce is often of a flavour which is not especially sought after.

Storage

Storage of cocoa in the tropics for over 2–3 months, even of fully dry cocoa (at 7.5% moisture) under good storage conditions, risks the development of mould and the spread of stored product pests and is not to be recommended.

Storage at these tropical temperatures and/or in damp humid conditions will lead to a rise in the free fatty acid levels after only a few weeks. All cocoa stores should be purpose built, designed to minimize these risks, while being cleaned and fumigated regularly. Bags should always be stored on pallets, never directly on the ground, and there should be free air movement around the stacks. Finally, storing cocoa in the tropics should be kept to the minimum duration possible.

Transport

Vessels, but in particular containers, used to transport cocoa beans should normally only be used for the transport of foodstuffs and never be used for the transport of toxic or noxious substances. They must be scrupulously cleaned prior to filling and treated to ensure they are free from infestations. Substances which might impart a flavour taint to cocoa beans should not be transported in the same hold.

Fine or flavour cocoas

The cocoa market distinguishes two main types of cocoa beans: the great majority being of the Forastero type, or so-called *bulk* cocoas comprising 93.5% of world cocoa production; the minority being the so-called *fine or flavour* cocoas – the specialist growths, often originating from white-seeded Criollo planting materials (see Table 2.2). Other than by reputation, there is no universally accepted criterion for distinguishing between fine or flavour cocoas and the bulk (or *ordinary* cocoas), though as sensory evaluation techniques advance, this may become possible. Planting material is important, though it cannot alone be used to distinguish between fine or flavour cocoas and bulk types (18).

The global demand for fine or flavour cocoas is static or falling gradually, but they are sold, often as particular estate marks, on the basis of specific flavour characters to specific buyers for speciality chocolates. If well prepared they can attract considerable premia over bulk cocoas, though fine or flavour trees (i.e. of Criollo planting material) tend to yield fewer beans, be more susceptible to pests and diseases and their preparation requires more care; in addition, they are more difficult and expensive to market (as they are usually sold on the basis that the buyer can examine a pre-shipment sample) and this all means that the premia paid may not be that rewarding to the growers. No grower should plant fine or flavour cocoas in the present market circumstances unless they already have a reliable outlet for the increased tonnage of flavour cocoa at a suitable premium price (18).

None of the seven major world cocoa producing countries are producers of fine or flavour cocoas, though 5000 tonnes are currently produced on government-owned plantations (or PTPs) in East Java, Indonesia, as a historic relic from Dutch colonial times. This represents an increasingly small percentage of total Indonesian cocoa production, which is likely to be some 345 000 tonnes in 1997/98. No further discussion on fine or flavour cocoas is included here[1].

Cocoa quality and its assessment

It is proposed that in practice, there are four components of cocoa quality: (1) physical bean quality, (2) yield of edible material, in particular the theoretical fat yield, (3) physical characteristics of the fat, and (4) flavour quality. A purchasing decision on cocoa for use by a cocoa processor for cocoa butter extraction is likely to pay most attention to the theoretical fat yield and the physical characteristics of the fat, rather less to the physical bean quality and little to the flavour quality. On the other hand, chocolate manufacturers are likely to pay most attention to flavour quality, as this will decide if they can use those beans in their products at all, or whether blending may be necessary or desirable. For example, a buyer of a highly prized (and thus expensive) fine or flavour cocoa may 'need' to blend such beans with those of a lower price in order to moderate the price of the product to the consumer.

There are a number of standards for measurement of physical bean quality, all of which would be measured by the traditional *cut test* of 300 beans and a bean count. These are: the International Cocoa Standards recommended by the Food and Agriculture Organization (FAO) (20); national standards of cocoa producing countries and some national standards for import into cocoa consuming countries (for instance, the Food and Drug Administration (FDA) have standards below which cocoa beans cannot be imported into the USA). However, most importantly in commercial terms, there are the contractual standards of the international cocoa trade which are drawn up and monitored by the trade regulatory bodies in London, Paris and New York. These latter standards are the basis on which physical cocoa is bought, sold and delivered.

Cocoa quality is presently measured for all these purposes by the somewhat primitive, but well-established, sampling and assessment technique known as the *cut test*. This is a visual assessment of the levels of defective beans in the official

[1] For further information see Fowler (18) and Anon (19). These contain explanations of the increasingly small market for these specialist cocoas.

sample of that cocoa, and the maximum acceptable levels of defects would be specified in every sales contract. These various standards differ in the level of defective beans which are permitted in their various quality grades in the cut test. The arrangements for taking the contract samples, their labelling, storage, the grades and any subsequent arbitration actions are clearly stated in the rules of the relevant contracts – the Cocoa Association of London (CAL), the Association Française Cafet Cacao (for the Paris contract) and the Cocoa Merchants Association of America (for the New York contract)[2].

Fundamentally, all the operations of the international cocoa trade, regardless of whether it is a trade using the contracts of London, Paris or New York, are based on mutual trust between buyer and seller and both parties will follow the specific rules of the body under whose aegis that particular contract has been written. Part of this trust is the expectation that the delivered cocoa will be of the quality specified in the contract, at the agreed price, in the contracted quantity and shipped/delivered in the specified shipment/delivery period.

There are no standard clauses in contracts of the three markets which cover fat yield, physical characteristics of that fat or flavour quality; however, such standards can be added by specific agreement between buyer and seller. Crespo (8) reports the results of defects as follows.

- A large percentage of unfermented beans produces an unflavoured, somewhat astringent flavour, lacking good chocolate characteristics.
- Mouldy beans bring an undesirable taste which cannot be masked out.
- Insect infested beans produce a lower fat yield and most likely a product with high count of insect fragments.
- Flat beans also have a low fat yield.

The theoretical fat yield of cocoa beans is the maximum quantity of usable fat which can theoretically be extracted from them, expressed as a percentage of the whole bean. It is calculated by measurement of percentage shell, percentage germ, percentage nib moisture and the percentage fat in dry nib using the following equation:

% Theoretical fat yield = (100 – % shell – % germ) 100 × (100 – % nib H_2O) 100 × % dry nib fat

[2] For example, in CAL contracts, *good fermented* beans have a defect level below 5% mouldy and 5% other defects and *fair fermented* beans below 10% of each. In summary, the International Cocoa Standards identify defects as follows: a *broken bean* is one in which a fragment is missing, though the missing part is less than half the bean; a *flat bean* is one which is too thin to cut; a *germinated bean* is one where the shell has been pierced, slit or broken by the germ; an *insect damaged* bean or a *mouldy* bean respectively show damage from, or presence of, insects or moulds visible to the naked eye; a *slaty* bean shows a slaty colour over half or more of the cut bean surface and thus demonstrates inadequate fermentation.

The physical characteristics of the fat are assessed by measurement of the fat hardness and the free fatty acid content of the fat. Today, fat hardness is typically measured by nuclear magnetic resonance (NMR) and is reported as the amount of solid fat present in the sample at a specified temperature (BS 684 method 2). To achieve meaningful and reproducible results, the fat has to be thoroughly melted to remove its crystalline history and then conditioned for specific times at different temperatures in order to crystallize it in a controlled way before the NMR measurement is made. Prior to measurement of hardness by NMR, hardness was estimated from melting and cooling curves or from the degree of unsaturation in the fat (*iodine value*).

Cocoa butter from South-East Asia seems to be harder than the West African which in turn is harder than the Brazilian; that of the Brazilian Temporo crop being the softest as it develops at a time of year when temperatures are lowest; temperature during pod development does influence the hardness. The free fatty acid (FFA) content of a fat indicates its 'freshness' or how well the cocoa beans have been prepared and handled in the supply chain. It is measured by titration of the fat with a standardized potassium hydroxide solution. The higher the FFA content of a fat, the more likely that the beans have been either poorly prepared and handled at origin or that they were stocked at origin for prolonged periods before export. When the FFA content of a fat is high it is also likely that the fat will be softer than otherwise. The FFA itself does not cause the softening but as the fat degrades to produce FFA, diglycerides are also produced. These interfere with the tight packing of the fat during crystallization, causing the final crystal packing to be unstable and thus to melt at a lower temperature.

The flavour quality can only be realistically assessed by making up samples of liquor or chocolate and submitting them to a trained taste panel for comparison with existing standards. Flavour issues are discussed in the next paragraph.

The origins of cocoa flavour characteristics

Almost all of the recent increase in cocoa production has come from Côte d'Ivoire, Indonesia and Malaysia. Less than 5% of the increase has come from Ghana, which earlier had set the industrial flavour standard for bulk cocoa. The result has been a marked divergence in flavour away from Ghanaian quality to supplies with less cocoa flavour and more pronounced acidity (23). Studies have been done in Ghana, Malaysia and Côte d'Ivoire to discover why the flavours differed and how to bring them closer to the Ghanaian standard. A project was undertaken by chocolate manufacturers with the local authorities in Côte d'Ivoire from 1982–86. It marked an important change in direction of the research from detailed analyses of the composition of the products to studies of field practices which producers were able to test and apply, and has led to a much better understanding of the origins of cocoa flavour characteristics in terms of the effects of planting materials and methods of primary processing. A key development has been the use of analytical and expert sensory evaluation.

Davies *et al.* (24) studied cocoas from Brazil, Malaysia and West Africa tasted as plain chocolates by panels of up to 20 trained assessors using a profiling method to characterize the flavours. Maps of the flavours (see Fig. 2.9) were

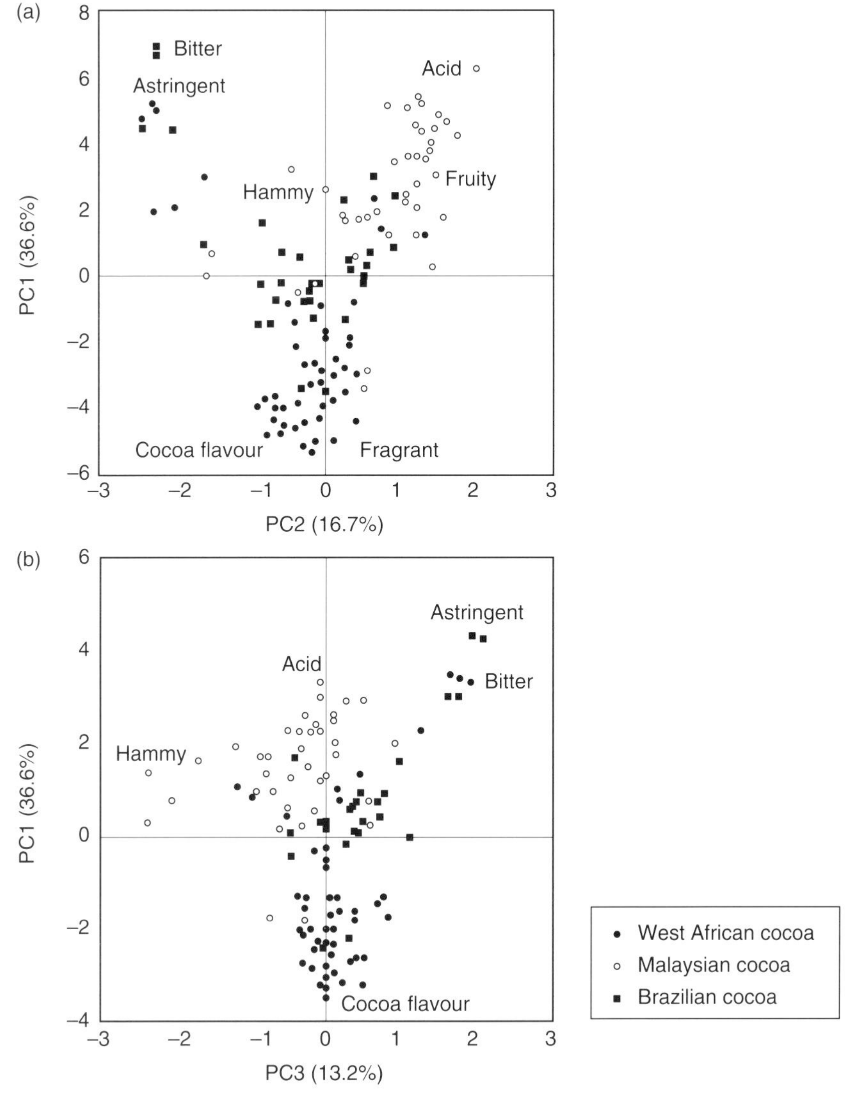

Fig. 2.9 Principal components analysis of flavour characterization data. (a) PC1 versus PC2 and (b) PC1 versus PC3. *Source:* Davies *et al.* (24) quoted by Clapperton (23). Copyright Cadbury Limited.

made using the results of principal component analysis of the characterization data. The first component (PC1) used to make both maps separated cocoas with well-developed cocoa flavour from others with either bitter and astringent tastes, typical of under-fermentation, or acid tastes. Fruity flavours were associated with acid taste. The second component (PC2) separated cocoas with acidic and fruity flavours from those with either bitter and astringent tastes or hammy flavours. The third component (PC3) contrasted bitter and astringent tastes with hammy or spoiled flavours caused by over-fermentation or smoke contamination.

Cocoas with known processing histories were used to interpret the flavour differences. The four samples of West African cocoa which were plotted among the Malaysian samples were prepared using Malaysian conditions of post-harvest processing, i.e. deep box fermentation with a four-step cascade, followed by fast drying at 60°C. Corresponding samples from the same farm, which were fermented and dried according to normal West African practices, were centred in the group of West African cocoas.

The two samples of Malaysian cocoa that are close to the main group of West African cocoas were from trials carried out by Lewis and Lee (25) in which Malaysian cocoa was processed using pod storage for 7–10 days followed by shallow box fermentation for 5 days with a single turn at the end of the second day. After fermentation these cocoas were sun-dried. The flavours of West African and Malaysian cocoas could therefore be moved in either direction depending on the conditions of post-harvest processing. Flavour appeared, therefore, to be process driven rather than geographical in origin.

Pulp reduction by pressing prior to fermentation, which had been applied to a number of the cocoas from Malaysia, was less effective in changing the flavour. Cocoas which were known to have received that treatment could not be separated from the rest of the Malaysian samples, other than those already discussed.

Studies have shown that flavours developed from different planting materials are clearly distinguishable and differ in cocoa flavour intensity, acid taste, bitterness, astringency and fruity/floral notes (26, 27). The planting materials developed for commercial production in Malaysia are by no means inferior to those grown in Ghana, but they are different and produce distinctly different flavours. The flavour differences between planting materials reside in the different compositions of the cotyledons and are unaffected by gross differences in pulp composition; effects of pod storage on flavour differ markedly between genotypes.

Storing pods for up to 12 days before breaking them for fermentation has for some time been part of post-harvest treatments recommended to improve the flavour of Malaysian cocoa. Shorter periods of post-harvest storage followed by partial removal of pulp by spreading the wet beans to dry in the sun, or by air-blasting, have also been recommended. Organic acids formed during pulp fermentation are clearly the source of the acidity. When beans are dried slowly after fermentation, ideally in the sun, the acids migrate from the cotyledons to the shells where they are lost by evaporation or decomposition while the beans remain moist and permeable. Under conditions of fast drying at higher temperatures the acids are concentrated in the cotyledons causing sharp and acid off-flavours. Trials in Brazil showed that acidity was reduced by sun drying and by slower mechanical drying.

Manufacturers agree about the off-flavours, which are unwelcome and are best avoided (28). However, characteristics of desirable cocoa flavours differ between manufacturers and specific flavour will be liked by some manufacturers and disliked by others depending on their traditional markets for finished goods. In the future, it is likely to become more important to identify and deliver the specific flavour qualities that manufacturers require, and careful checking and cross-checking of the flavours among researchers, producers and manufacturers will be necessary. Over recent decades, there have been contradictory methods for flavour assessment and a lack of consistency between manufacturers. Recent work has produced broad agreement within the industry and protocols are now available; a better understanding of the origins of cocoa flavour quality has also developed. Planting materials and the use of appropriate conditions of post-harvest processing determine the characteristics of normal flavour development, as opposed to off-flavours which result from defective processing.

Cocoa marketing

International cocoa trade

The major buyers of West African cocoas are in Western Europe, with the United States being a major buyer of South American and Asian cocoas. Cocoa is usually purchased from origin by international dealers (or traders) who subsequently sell it to final users (usually chocolate companies) as beans, or process it in their own plants and sell it to the final users as products (liquor, butter or powder). There are only a few large international dealers. Four of these may account for over two-thirds of total worldwide cocoa purchases from origin, though one of these companies is expected to be shortly leaving the cocoa trade. Chocolate companies and cocoa processors prefer to purchase through dealers because they bring the cocoa to a port close to the buyer, often pass it through customs, and the buyer can inspect the shipment and, if necessary, be compensated for any quality shortfall via arbitration, allowed for in the contracts. There is also an active international trade in cocoa products, through these same dealers, together with some smaller ones who specialize in certain products. The prices at any time will be determined by the cocoa futures (or terminal) markets in London (International Finance Futures Exchange (LIFFE)) and New York (Coffee Sugar and Cocoa Exchange (CSCE)) in response to the relative supply and demand balance at any given time.

Quality premia and discounts

The price levels of any physical cocoa traded will be related to the relevant cocoa terminal market price being traded on that market at the time. For example, Ghana cocoa for delivery in March would be expected to be priced on the basis of the London terminal price for March on the day in question, plus a certain premium per tonne called the *differential*. Ghana cocoa is considered by the market as the 'gold standard cocoa' against which all others are judged and so the differentials paid for Ghana cocoa are almost always the highest in the market.

The only exception to this could be the case of the small tonnage of speciality *fine or flavour* cocoas for which, on rare occasions, very much higher premiums can be achieved than those normally paid for Ghana cocoa. Unfermented Sulawesi cocoas tend to be sold at a significant discount as their fat value is quite low and their flavour quality is generally perceived as poor. Differentials vary over time with the market determining the appropriate differential for a particular cocoa at any given time on the basis of supply and demand for that type. Differentials will be derived from extrinsic factors (such as the overall cocoa market situation and the supply/demand relationship for that particular cocoa) and from intrinsic factors (due to the inherent quality/circumstances/reputation of that particular cocoa at that time and in the recent past). Among the intrinsic factors could be included:

(1) The recent reputation for the origin, and/or that particular seller for:

- (a) quality of service (shipment to contract terms, to time, dispatch of a full set of documents, quality of ships, bags, etc.);
- (b) responsiveness in case of problems;
- (c) insurance terms;
- (d) tonnage available (unless speciality cocoa is involved, small producing areas will tend to be at a disadvantage in contract negotiations).

(2) Components of cocoa quality as discussed earlier:

- (a) physical bean quality
- (b) theoretical fat yield
- (c) physical characteristics of the fat
- (d) flavour quality.

All cocoa buyers have anecdotal information, historic experience and probably a range of analytical data on which to base their purchasing decisions and make the selection as to which beans are most suited to their needs. Some quality factors of some origins are mentioned below; these are broad general views derived from discussions with a handful of experts in the world cocoa trade.

Cocoa quality in South and Central America and the Caribbean

As seen from the data in Table 2.1, this region is now a modest cocoa producing area involving cocoa growing across a vast geographical area of varying climatic conditions, with a number of very small cocoa producing countries, including a selection of fine or flavour producers. In many of these cases, their supply chains are very long and thus they are at something of a competitive disadvantage to the larger producers of West Africa who can offer parcels of several thousand tonnes of cocoa at a time.

The diversity of the types of cocoa from this area makes quality generalizations difficult, save to point out the very considerable disease problems faced by these cocoa growers; many are afflicted with the devastating witches' broom disease (Brazil, parts of Colombia, much of Ecuador, Grenada, Guyana, Suriname, Trinidad, parts of Venezuela) and some are also affected by moniliophthora as well (Costa Rica, Ecuador, Panama), a cocoa disease which is still spreading and, as yet, has no economic method of control.

These problems do have an effect on buyers' perceptions of cocoa quality and their plans to do business with exporters in these countries. There is an increasingly large domestic chocolate market developing (Brazil, Colombia) using locally grown cocoa beans and this trend is expected to continue. In fact, the ravages of witches' broom disease on Brazilian cocoa production have led to recent importing of beans (from Côte d'Ivoire) to satisfy the local market for chocolate. The marketing systems are generally liberal ones, with only limited government involvement in quality control in some countries in the regions.

Cocoa quality in West Africa

The majority of Cameroon cocoa production is from Trinitario planting material which produces beans with a high fat content in comparison to the other cocoas from West Africa and a powder with a specific red coloration. This ability to provide red colour is much sought after by the specialist cocoa powder vendors (many of whom are located in the Netherlands or Germany) who, with considerable skill, manufacture a wide variety of cocoa powders of various colours and fineness for speciality cakes, biscuits and confections and sales are often at premium prices. Buyers of Cameroon cocoa are only concerned about quality if the cocoa does not achieve a *fair fermented* grade (less than 10% defective beans). At any higher level of defects, there is a real risk that the FFA level will be high and so the fat content and quality of the beans may be reduced.

Cocoa in Côte d'Ivoire is moved very quickly from the farm gate to export warehouse and thus in an economic sense the cocoa industry in Côte d'Ivoire is very efficient. The first purchase is undertaken by itinerant buyers who visit villages during the cocoa buying season. In practice, only modest attention is paid to the levels of defects along the marketing chain and a considerable degree of blending of good quality and poor quality cocoa takes place at almost every stage. Cocoa from Côte d'Ivoire is clearly of better flavour quality than most cocoas from Asia (which are affected by the problem of acidic off-flavours and it appears that a high percentage is presently destined for cocoa butter extraction, rather than for chocolate production directly; this seems set to continue).

Ghana cocoa normally sells at a significant premium compared to all other bulk cocoas as on pure flavour grounds, cocoa from Ghana is much in demand. However, Ghana also has an excellent reputation for all the other intrinsic factors mentioned earlier, contributing to premia. Some of these matters are handled with rather less care by other cocoa origins and their prices would tend to be discounted versus those of Ghana because of that. The excellent physical quality and the almost certain absence of any off-flavours in Ghana cocoa make it especially well suited to chocolate production, in particular for the manufacture of milk chocolate, which tends have a very mild flavour. Many chocolate manufacturers in Japan, UK and USA rely on Ghana cocoa to impart its characteristic flavour to their chocolate.

Since colonial times in Ghana, cocoa has always been purchased through a system of fixed buying centres open at regular hours and located in villages. The cocoa farmers in Ghana can thus conveniently sell their cocoa when it suits them and when it is correctly dried to 7.5% moisture, thus removing a pressure seen elsewhere. This is a major reason why very little Ghana cocoa of poor or even marginal quality is produced. Elsewhere at the time of visits of itinerant buyers, farmers sometimes have to sell cocoa damp, as they need cash but are concerned that the buyer may not return for a few more days. In Ghana, the quality is officially checked in the centres (and each bag is sealed) before being moved and the quality is reconfirmed as it moves to the port and prior to loading for export, all of which is under Ghana Government control[3].

Until 1986, Nigerian cocoa had a quality reputation equivalent to that of Ghana, assured by a very similar system of cocoa quality control, then operated by a similar government monopoly. Hasty liberalisation of cocoa marketing (under World Bank pressure), involvement of hundreds of inexperienced cocoa buyers and inadequate attention to the continued need for a system of cocoa

[3] Quality control of Ghana cocoa is taken seriously. Firstly, there is the quality examination carried out by the buying clerk at the time of first purchase of the cocoa from the farmer, then there is an examination in the buying centre by the produce inspector from the Quality Control Division (a wholly owned subsidiary of the Cocoa Marketing Company) who seals the bag after a quality check. This seal enables anyone to identify the inspector who examined that cocoa, the date and the buying centre in which the bag was first sealed. As the cocoa moves into, or out of, the regional up-country store and the port warehouse, it will be given a confirmatory quality check.

quality control, led to an almost instant, disastrous reduction in quality and in reliability of cocoa exports. More than a decade later, the cocoa quality and the reliability of exporters has improved, though much Nigerian cocoa is still sold on a sample basis in stores in Europe or North America. Due to the low level of buyers' confidence, prices are now at a substantial discount compared to those of Ghana and the country has lost substantial sums in foreign exchange due to the too-rapid process of liberalization.

The quality of Indonesian cocoa is variable depending on the location in which it is grown. A large part of the expanding quantity of smallholder production from the various provinces of Indonesia tends to be insufficiently fermented according to the standards described previously, though fermentation is the norm in some small, specific locations. Larger growers tend to be more conscious of the requirements of careful on-farm processing to achieve the best prices and a few of them produce cocoa of excellent flavour quality, though the theoretical fat yield tends to be low (see Table 2.5). For the larger growers (using straightforward rapid drying on an industrial scale), acidic off-flavours are often observed. Indonesian cocoa quality standards (especially as regards bean size) are not yet applied with equal vigour at all export ports.

Table 2.5 Some quantitative parameters of cocoas from various origins.

Origin	Bean count[1]	Shell (%)	Fat in nib (%)	Theoretical fat yield[2] (%)
Côte d'Ivoire main crop	92–105	11.9	56.5	46.3
Ghana main crop	90–95	11.3	57.3	47.2
Nigerian (pre-1986)	90–95	11.7	56.8	46.7
West African cocoa[3]	93.7	11.8	57.3	47.5
Peninsular Malaysia[4]	82–125	16.0	57.0	44.5

Source: adapted from Wood and Lass (21).
[1] Number of beans per 100 g.
[2] Assumes moisture at 6.0%, germ at 0.9%; illustrative only, assumes data from Wood and Lass (21) calculated as per formula above.
[3] Data from Duncan and Veldsman (22) quoting major users' experience with Ghana cocoa over 5 years to 1993.
[4] Similar levels are also noted for Indonesian cocoa.

Malaysian cocoa has been plagued by acidic off-flavours and this has limited both the use and the price of Malaysian cocoa since the surge of plantings from the 1980s. The Malaysian authorities pay great attention to the application of their grading standards for both internal and external processing, with especial attention to their bean size standards. There seems to be particular variability in the bean size of Malaysian cocoa, perhaps due to climatic and planting material variations. The issue of acidity is discussed in some detail previously in this chapter.

Economics of cocoa cultivation

For studies of profitability to be of any value in between country comparisons, they need to be conducted with the same methodology and require substantial resources. Because of these demands, very few comprehensive, comparable studies on the economics, or profitability, of cocoa cultivation have been undertaken. In any event, such work is greatly complicated by the devaluation of the CFA franc in January 1994 which dramatically enhanced the profitability of cocoa cultivation in the French-speaking cocoa producing countries of West Africa – Cameroon, Côte d'Ivoire and Togo.

However, some work over the period 1992/93 to 1995/96 on the basis of average yields and average costs indicates that total profitability of cocoa cultivation in Brazil is very substantially negative, in Malaysia is negative, in Cameroon is slightly negative while in Ghana, Côte d'Ivoire, Indonesia and Nigeria it is increasingly positive. It is interesting to note that cocoa is actually being planted (and/or rehabilitated) in Ghana, Côte d'Ivoire and Indonesia and being planned actively in Nigeria. The disinterest in cocoa cultivation in Brazil (due to witches' broom disease) and in Malaysia (due to difficulties of management and labour shortages) is well known. Planning on a local basis continues in the western part of Cameroon but not in the oldest and low-yielding plantings in the east of the country. Predictably, the level of yield is a major factor in profitability.

Conclusion

Chocolate is much appreciated by many, though there remains a high percentage of the world's population who have still never tried a bar! Demand for cocoa from both existing and new markets is expected to continue offering good prospects for cocoa growers, especially the more efficient ones.

It is important that cocoa is only grown in the most suitable locations. In the early 1980s (in response to the very high prices of the late 1970s), substantial areas with less than suitable conditions were planted to the crop. Many of these have now died off or been grubbed. The profitability of cocoa cultivation is significantly influenced by the suitability of the conditions in which it is planted and the quality of the planting material. It can be a highly rewarding crop and is one well suited to smallholder cultivation.

Growers face many problems including high losses from a number of pests and diseases and inadequate systems of controlling them. Improved pest management strategies are urgently needed, as is improved cocoa planting material for many cocoa growing environments to enable significant improvements in productivity by growers, to match those seen in so many other tropical (and temperate) crops over recent decades.

The large number of Malaysian estates who planted cocoa – following the high prices in the late 1970s – have found that it is not an easy crop to manage on a large-scale basis. At the moment, the labour scarcity in Malaysia with the high labour needs and the significant peak harvest requirement for labour on cocoa are causing very real difficulty to the few estates who have not already uprooted their cocoa. There seems little comparative advantage to large, as opposed to smallholder, cultivation. It can be concluded that cocoa is not an easy crop to manage on an estate basis, but has a number of positive advantages for smallholders who now cultivate some 85% of world cocoa production.

Attention to the quality of raw cocoa beans will be increasingly important, as global food standards become increasingly strict.

Much of the world's cocoa is grown in small plots with a mixture of forest or planted trees as shade, while smallholders also often leave other economic trees amongst their cocoa. This gives a good level of biodiversity and makes cocoa a potentially very useful crop for inclusion in a sustainable farming system. Further research is needed to quantify these benefits.

References

1. Wood, G.A.R. (1991) A history of early cocoa introductions. *Cocoa Growers' Bull.* **44**, 7–12.
2. Anon (1998) *World Agricultural Production*, Circular Series, WAP-01-98. USDA, Washington DC.
3. Wood, G.A.R. (1985) *Cocoa*, 4th edn (ed. by Wood, G.A.R. and Lass, R.A.), Tropical Agricultural Series. Longman Group, London.
4. Anon (1998) *Review of Annual Forecasts of World Production and Consumption*, International Cocoa Organization, ICC/57/5. ICCO, London.
5. Laker, L.A. and Trevisan, O. (1992) The increasing importance of Cupuassu (*Theobroma grandiflorum*) in the Amazon region of Brazil. *Cocoa Growers' Bull.* **45**, 45–52.
6. Morris, D. (1882) *Cacao: How to Grow and How to Cure It*. Jamaica.
7. Cheeseman, E.E. (1944) Notes on the nomenclature, classification and possible relationships of cocoa populations. *Trop. Agric. Trin.* **27**, 144–159. Reprinted 1982 in *Arch. Cocoa Res.* **1**, 98–116.
8. Crespo, S. (1986) *Cacao Beans Today*. Silvio Crespo, Lititz, Pennsylvania.
9. Toxopeus, H. (1985) Botany, types and populations. *Cocoa*, 4th edn (ed. by Wood, G.A.R. and Lass, R.A.). Longman Group, London.
10. Cuatrecasas, J. (1964) Cacao and its allies. A taxonomic revision of the genus *Theobroma*. *Contrib. US Nat. Herb.* **35** (6), 379–614.
11. Kennedy, A.J., Lockwood, G.R., Mossu, G., Simmonds, N.W. and Tan, G.Y. (1987) Cocoa breeding: past, present and future. *Cocoa Growers' Bull.* **38**, 5–22.
12. Corley, R.H.V. (1985) Yield potentials of plantation crops. In *Proc. 19th Coll. Int. Potash Inst.*, Bangkok. International Potash Institute, Bern.
13. Anon (1993) *The World Cocoa Market – an Analysis of Recent Trends and of Prospects to the Year 2000*. ICCO, London.

14. Cramer, H.H. (1967) Plant protection and world crop production. *Pflanzenschutz Nachtrichter 'Bayer'* **20**, 1.
15. Lass, R.A. (1998) Cocoa – a promising crop for the new millennium. Seminar: trees as crops (Oxford, September 1997). *Trop. Agric. Newsletter* **18** (1), 25–27.
16. Duncan, R.J.E. (1984) A survey of Ghanaian cocoa farmers harvesting, fermenting and drying practices and their implications for Malaysian practices. *Proc. Int. Conf. Cocoa and Coconuts: Progress and Outlook* (Inc. Soc. Planters, Kuala Lumpur), 509–516.
17. Duncan, R.J.E., Godfrey, G., Yap, T.N. and Pettipher, G.L. (1989) Improvement of Malaysian cocoa bean flavour by modification of harvesting, fermentation and drying methods – the Sime–Cadbury process. *The Planter* **65**, 157–173.
18. Fowler, M.S. (1994) Fine or flavour cocoas: current position and prospects. *Cocoa Growers' Bull.* **48**, 17–23.
19. Anon (1991) *Fine or Flavour Cocoa, an Overview of World Production and Trade.* International Trade Centre, UNCTAD/GATT, Geneva.
20. Anon (1970) *International Cocoa Standards.* FAO, Rome.
21. Wood, G.A.R. and Lass, R.A. (1985) *Cocoa*, 4th edn. Tropical Agricultural Series. Longman Group, London.
22. Duncan, R.J.E. and Veldsman, I. (1994) A European chocolate manufacturer's experience of fermented Sulawesi cocoa. *Cocoa Growers' Bull.* **48**, 24–35.
23. Clapperton, J.F. (1994) A review of research to identify the origins of cocoa flavour characteristics. *Cocoa Growers' Bull.* **48**, 7–16.
24. Davies, A.M.C., Franklin, J.G., Grant, A., Griffiths, N.M., Sheperd, R. and Fenwick, G.R. (1991) Prediction of chocolate quality from near-infrared spectroscopic measurements of the raw cocoa beans. *Vibrational Spectroscopy* **2**, 161–172.
25. Lewis, J.F. and Lee, M.T. (1986) The influence of harvesting, fermentation and drying on cocoa flavour quality. *The Planter* (Kuala Lumpur) **62**, 134–140.
26. Clapperton, J.F., Yow, S.T.K., Chan, J., *et al.* (1991) The effect of cocoa genotype on flavour. *Proc. Int. Cocoa Conf.* (Challenges in the 1990s, Kuala Lumpur). Malaysian Cocoa Board, Kota Kinabalu.
27. Clapperton, J.F., Yow, S.T.K., Chan, J., *et al.* (1994) The contribution of genotype to cocoa (*Theobroma cacao* L.) flavour. *Trop. Agric. Trin.* **71** (4), 303–308.
28. Anon (1990) CAOBISCO: *Position Statement on Cocoa Quality.* Association of the Chocolate, Biscuit and Confectionery Industries of the EU (CAOBISCO), Brussels.

Chapter 3

Cacao Bean and Chocolate Processing

Ronald G. Bixler and Jeffrey N. Morgan

This chapter discusses the basic processes involved in converting cocoa beans into chocolate, and the major steps involved in the manufacturing of chocolate. It is not within the scope of this chapter to provide extensive detail on cocoa bean and chocolate processing. Rather, the intent of this chapter is to provide an overview of basic processes so that the reader gains an appreciation of how chocolate is manufactured and the types of chocolate produced. Fig. 3.1 shows the process diagrammatically.

Cocoa bean processing

The unit operations utilized in converting cocoa beans to chocolate liquor, cocoa powder and cocoa butter have changed little since the beginning of the 20th century. The changes have occurred in increased equipment efficiencies, controls have become very sophisticated and the raw material sources have changed significantly. For example, the table of contents in a book (1) published shortly after World War I lists topics very similar to those found in recent publications:

Chapter X	Cleaning, Sorting and Grading Cacao Prior to Roasting
Chapter XI	Roasting of Cacao
Chapter XII	Nibbing, Husking and Winnowing the Roasted Cacao
Chapter XIII	Milling–Preparation of Cocoa Powder–Expression of Cacao Butter

Each of the operations will be described in the following discussion.

Cleaning

After inspecting and conducting the necessary tests to insure compliance with internal standards and regulatory requirements, cocoa beans move to the next

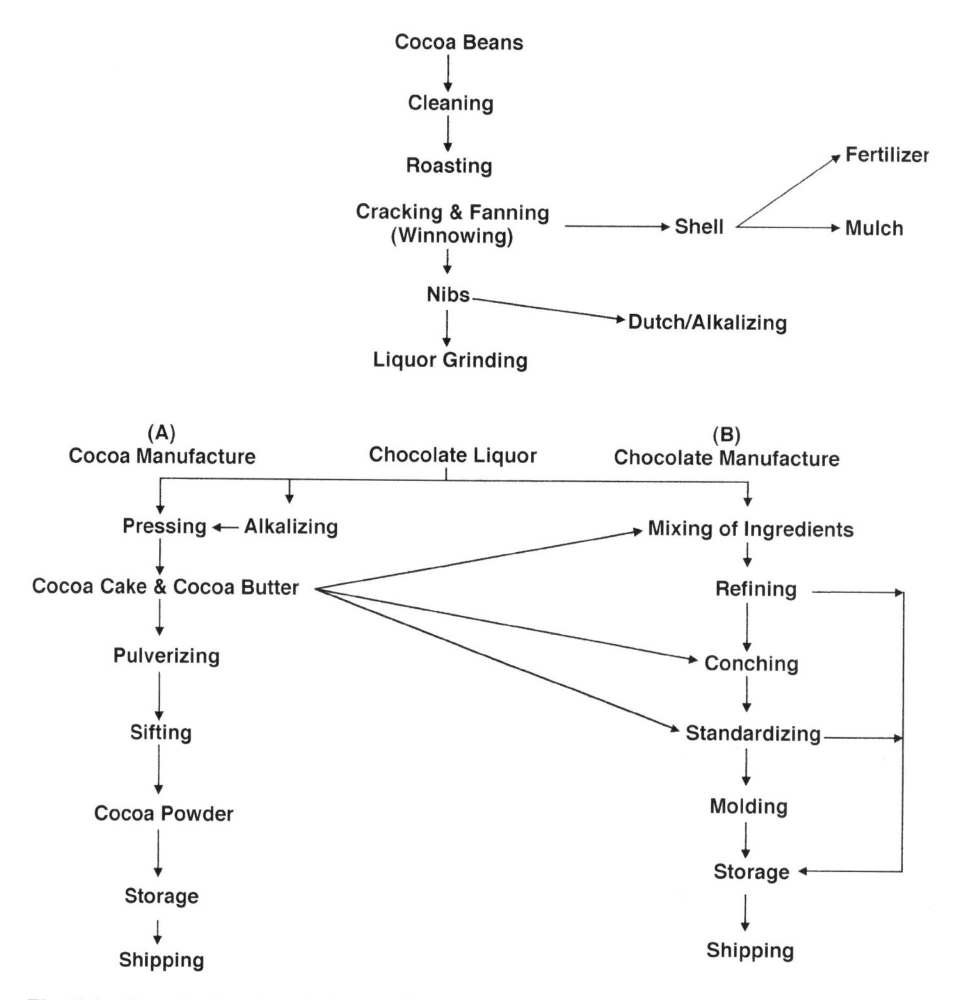

Fig. 3.1 Flowchart of chocolate manufacture.

part of the operation. Although it is mentioned above, grading of the beans at a processing facility into two or three grades is not as common as it was in the early part of the 20th century.

Cleaning is typically the next step and is shown in Fig. 3.2. Foreign material ranging from machete blades to shotgun shells can be found in shipments and this material must be removed both to minimize damage to downstream equipment and to maintain product integrity. The inside of an undamaged cocoa bean is clean in all respects and will remain so as long as it is stored properly and is not mixed with contaminated materials not completely removed during cleaning or subsequent steps downstream.

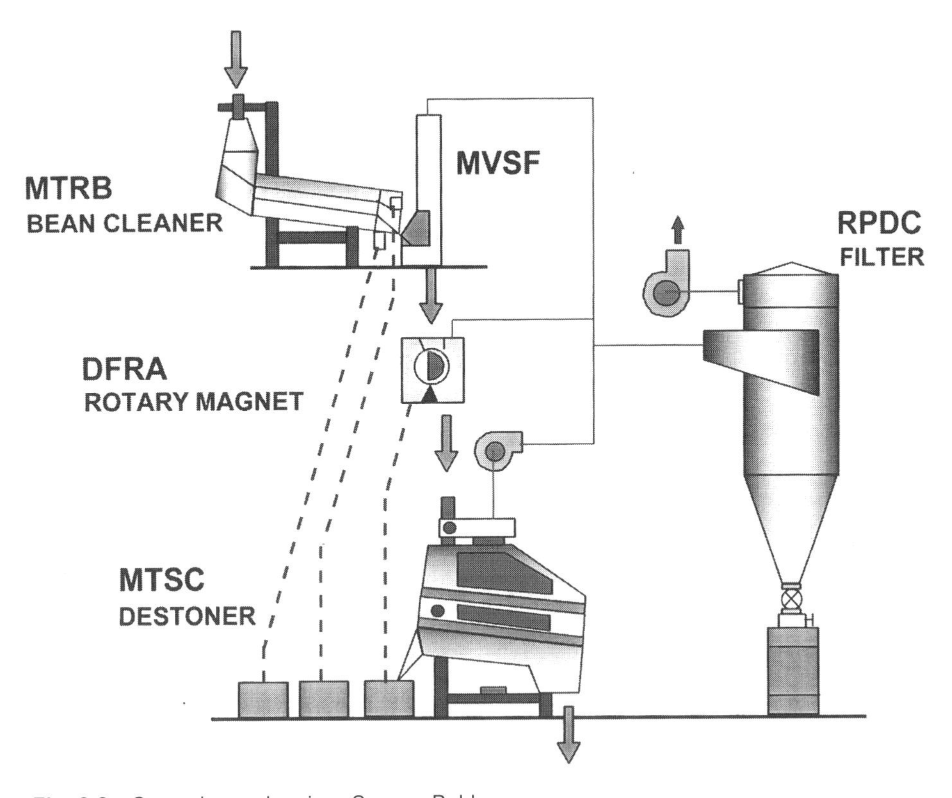

Fig. 3.2 Cocoa bean cleaning. *Source:* Buhler.

Roasting

Roasting cocoa beans can be described as an individual process. While all manufacturers have a similar goal of producing products efficiently, the flavor objectives for chocolate liquors usually differ from company to company and from country to country.

Therefore, the flavor target is a key factor in determining the type and blend of cocoa beans to be processed, whether to roast whole beans or nibs, type of roasting equipment and the roasting parameters employed. While it will not be discussed in detail, sufficient lowering of the microbiological counts during the roasting process must also be a major determinant in selecting a roasting system.

One of the classic debates in the roasting process has been which roasting technique produces the chocolate liquor in the most efficient manner – the roasting of whole beans or the roasting of nibs? While actual production figures have not been located, nib roasting appears to be winning the debate. Another aspect of roasting which continues to be studied and discussed is the type of roaster to be used. Available types range from rotating drums to fluidized beds with all types being operated successfully around the world.

Regardless of the process employed, a broad flavor generalization can be made about the chocolate liquor flavors favored by the European roasters and the US roasters. Europe tends to favor a light roast and the American processors favor a dark roast. This has led to European chocolate liquor (cocoa mass) being described as green in the USA and the US material described as burnt by European producers.

In summary, there is not a right or wrong roast level nor is there a correct way or an incorrect way to obtain the target roast level. The correct way to roast and the proper roast level is that process which provides nibs in an efficient and cost-effective manner with the flavor system and produces the products meeting the consumers' needs in a specified market.

Winnowing

Winnowing, cracking, fanning and hulling are some of the terms and phrases which describe separation of shell (hull) and meat of the bean (nib). It is a process where obtaining a clean separation of the two components is driven by economics, product integrity and, in many countries, government regulation. A winnower is shown in Fig. 3.3.

Food regulations usually specify a maximum level of shell in nib. For example, in

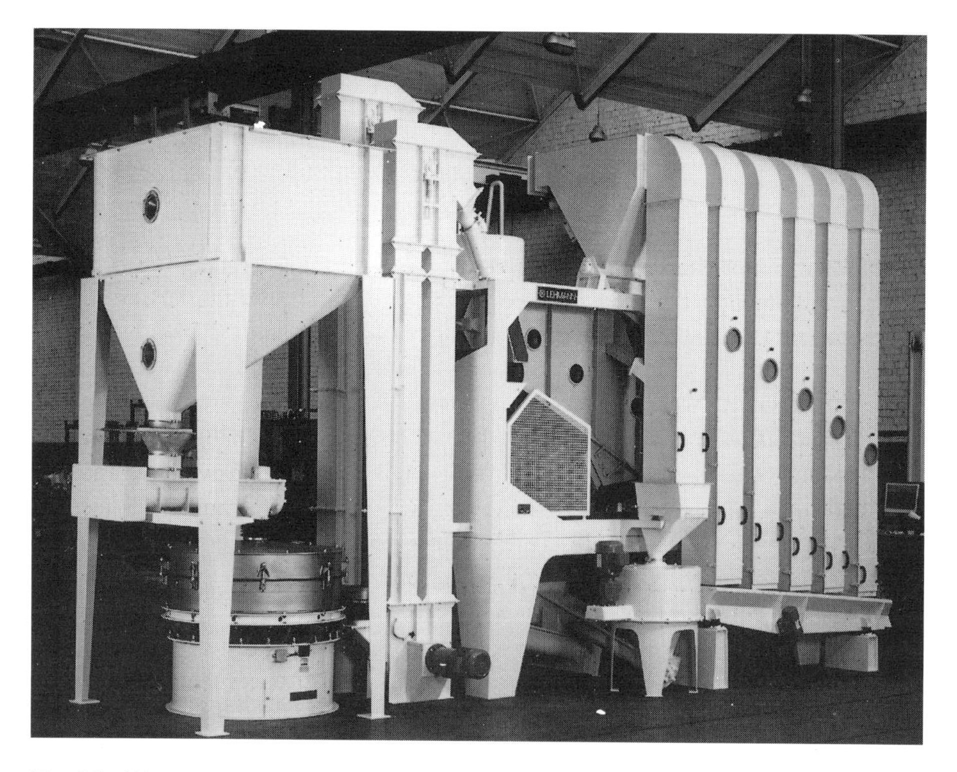

Fig. 3.3 Winnower. *Source:* Carlo and Montanari.

the USA this level is a maximum of 1.75%. The level of nib in shell is of no concern to anyone other than the manufacturers and their concern is one of economics.

Having excessive levels of nib in shell is one way for a manufacturer to lose large amounts of money quickly, because nib is typically worth in excess of $2000/tonne, but as part of the shell stream it essentially becomes worthless.

The amount of shell in nib can also impact on the life of the processing equipment. Cocoa shell is very abrasive and can accelerate equipment wear resulting in increased maintenance costs.

Finally, the level of shell in nib can affect the quality level of the resulting chocolate liquor. Excessive levels of shell can alter the flavor of the finished product by contributing off-flavors. Examples of off-flavors are fiber-like notes and, occasionally, moldy notes. Of major concern is production of chocolate liquor that contains pathogenic organisms, unacceptably high microbiological counts or high levels of extraneous materials. As stated earlier, the nib is clean. Bacteria are found on the shell and other foreign material and it is the incomplete separation of the shell and nib as well as inadequate roasting that can lead to chocolate liquor with the previously mentioned defects.

Nib grinding

Nib grinding has seen many advances in the last half of the 20th century. Before liquor mills, one method of grinding nibs was mixing nibs with granulated sugar and placing the mixture in a mélangeur. This process yielded a material with a consistency ranging from a paste to a fluid.

It should be stated that a modern nib grinding system consists of a pre-grinder (e.g. Fig. 3.4) and a finish grinder. Pre-grinding of nibs results in a fluid material with a particle size normally exceeding 100 μm. Finish grinding reduces the particle size of the chocolate liquor to a value dependent on the planned use of the material. The particle size and particle size distribution values are critical if chocolate liquor is to be used in operations such as cocoa pressing and chocolate making.

The early mill of choice appears to have been the stone mill. This is one type of three basic mills which will be reviewed. Two others are the ball mill and roller refiner. Many other pre-grinding and finish grinding devices are available and are described in the literature. When selecting a grinding device, numerous factors need to be considered. Examples are the initial cost, operating costs, floor space required, control capability of particle size and particle size distribution and flavor impact on finished product (2).

The ball mill (as shown in Fig. 3.5) is a grinding system which has gained wide use in recent years. Both continuous and batch types are available. The ball mill requires an upstream nib-grinding device, which will allow the liquor to be pumped into the grinding chamber for exposure to the grinding action of moving balls. The grinding chamber is filled with steel or ceramic balls with diameters

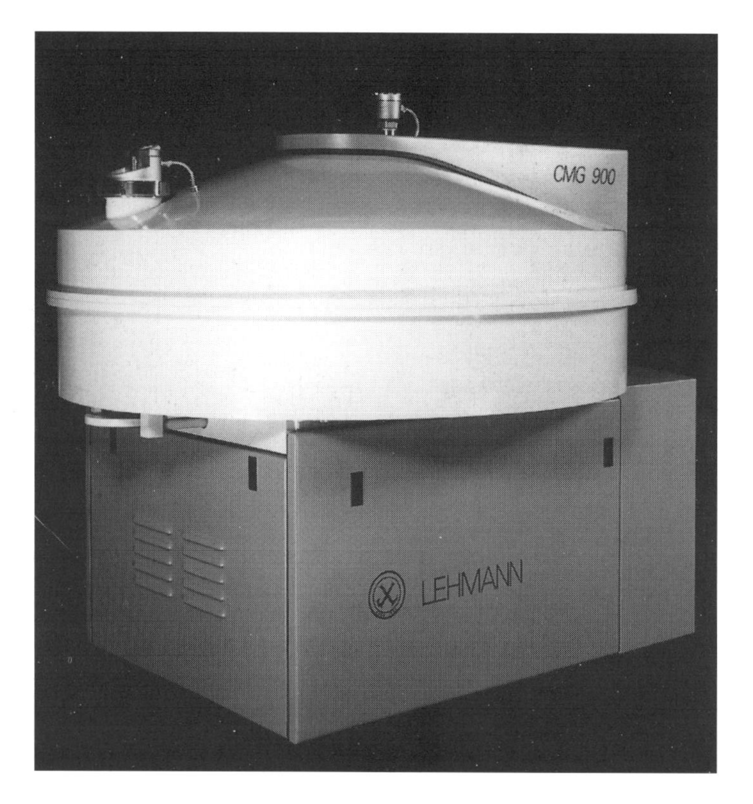

Fig. 3.4 Cocoa nib pre-mill. *Source:* Carlo and Montanari.

ranging from 3–8 mm and they are set in motion by rotation of the shaft to which blades are attached. It has been stated in the literature that chocolate containing ball mill-processed chocolate liquor requires shorter conche times; a desirable end result if applicable to the chocolate recipe being produced.

A second mill type, in use for many years, is the triple stone mill as shown in Fig. 3.6. A common configuration is three pairs of circular stones arranged in a vertical manner. Each stone is grooved and liquor moves from the center to the outside as grinding occurs. It can now move to the next pair of stones for further grinding or be pumped downstream for use.

A third mill type is the roller refiner which is similar to the five-roll refiner common in the manufacture of chocolate. It usually consists of three steel rolls. The nibs are pre-ground with a pin mill or a hammer mill although having grooved rolls as the first pair of the series can serve as a pre-grinder. A roller refiner produces flat particles while ball mills and stone mills produce particles more spherical in shape. This can be a negative since the increased surface area of the flat particle can require additional cocoa butter to reach a specified viscosity, which is a point to be considered during mill selection should this mill type be under consideration for installation.

Fig. 3.5 High-capacity ball mill refiner. *Source:* Wiener and Co. (Amsterdam).

The above efforts result in a product we have been referring to as chocolate liquor, the common name in the USA; to the rest of the world, it is cocoa mass. Other names are baking chocolate, cooking chocolate and chocolate. Chocolate liquor processed in the above manner is described as natural liquor. The next section describes a process producing a second liquor type.

Alkalization

The alkalizing process is applied to modify the flavor and color of chocolate liquors and cocoa powders. It is also known as *Dutch processing*. The starting material can be cocoa cake, nibs or chocolate liquor. Although confirming figures have not been located, nibs and cake are the most common starting materials

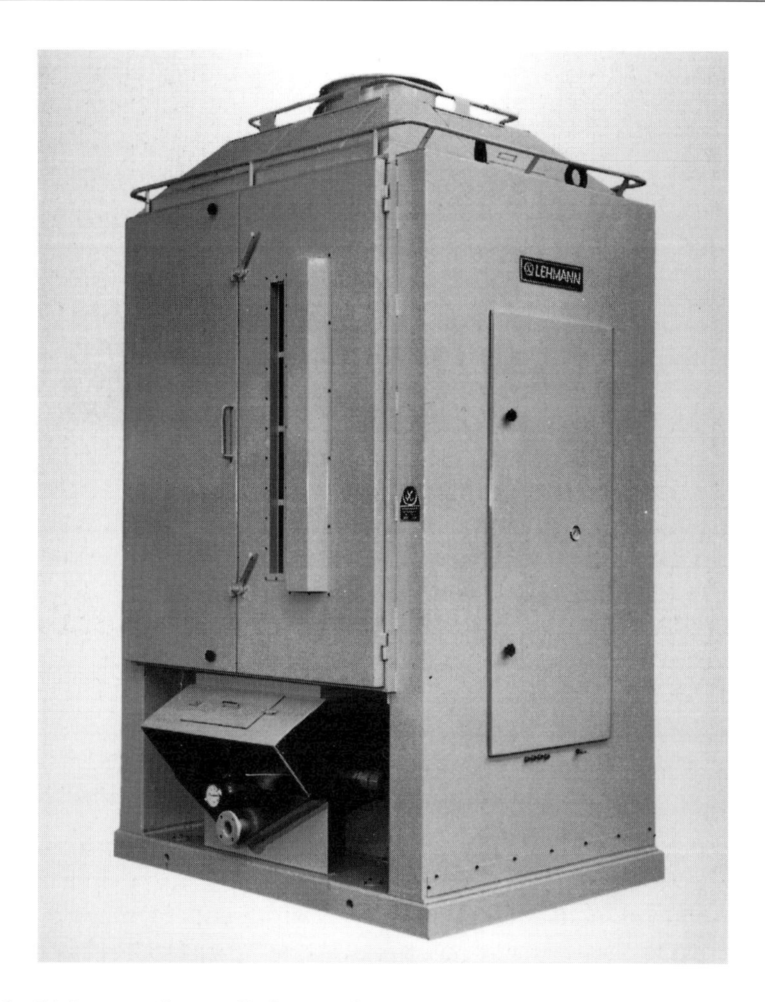

Fig. 3.6 Triple cocoa liquor mill. *Source:* Carlo and Montanari

with nibs being the most prevalent. Simply stated, the process consists of mixing the selected cacao material with an aqueous solution of the specified alkaline compound and mixing at elevated temperatures and possibly increased pressures. Fig. 3.7 shows an example of an alkalizer.

There are process schemes where the mixture is allowed to set for several hours without agitation. Common alkaline compounds are potassium carbonate, calcium carbonate and sodium hydroxide. The resulting product colors range from a light red to a charcoal black. Flavors vary over a wide range and the only general statement to be made is that alkalized cocoa powder or chocolate liquor tends to be less acidic. It must also be stated that the flavor of an alkalized powder varies with the starting material – nib or cocoa cake.

As flavor is subjective and personal, it is difficult to say which yields the *best* product. Excellent products are available from both starting materials, and the

Fig. 3.7 Cocoa liquor alkalizer. *Source:* A.A. Martin Lloveras (Spain).

author's experience has been that powders produced by alkalizing nibs lack harsh notes and exhibit a more balanced flavor profile. They also tend to be more expensive and the product development specialist must determine if a product difference exists and warrants the increased cost. The pH of finished products also changes during *Dutch processing* and can range from 6.5 to 9.0 but can be adjusted to neutrality, if needed. Suffice it to say, the final colors and flavor profiles that are available vary as much as the numerous combinations of starting materials, process schemes and process conditions practiced by this phase of the industry (3).

All of the above efforts lead to one of the key ingredients in chocolate. It is chocolate liquor that can be natural or alkalized. Generally, the liquor will now move in one of three directions. It can be used as a source of flavor and color in a wide array of foods other than chocolate; it can go through a separation step which yields cocoa powder and cocoa butter; or it can be moved to the chocolate plant for producing chocolates and related products. Both natural and alkalized chocolate liquors are used in all these applications.

The use of chocolate liquor as a flavoring and coloring material in foods other than chocolate is not discussed further here. The separation step is covered here briefly since it does provide cocoa butter, an ingredient essential in chocolate making.

Chocolate liquor pressing

Typically, the separation or pressing process begins by pumping hot chocolate liquor (200°C) into a horizontal hydraulic press with operating pressures of up to 550 bars. Screw presses or expellers are in limited use and are not discussed here in any detail. They are usually used to remove cocoa butter from whole beans and other cacao-containing materials. Both natural and alkalized chocolate liquors can be pressed to separate the cocoa butter from the solid cocoa cake. Pressing times are determined by the final fat content of cocoa cake, with 8.0% being the lowest practical value obtainable with the hydraulic press. Typically, a cycle time of 15 min is required to produce cocoa powder with fat content of 22–24%. On the low end, 25–30 min are required to reach a fat content of 10–12%.

In the USA, cocoa powder is usually available in several fat content ranges including 22–24%, 16–18% and 10–12%. Fat-free powders, which are now available, are possible by additional fat removal through application of critical fluid extraction or solvent extraction. The 10–12% fat content powder is the most widely used for numerous reasons with economics probably being most important. Product developers should consider the higher fat content powders for the improved flavor they can contribute. If alkalized liquors are pressed the colors and flavors available increase and can best be described by evaluating various manufacturers' offerings. The cocoa cake formed from the pressing operation described above is then broken and milled to specified particle sizes as determined by end use.

Cocoa grinding

Before grinding of this pressed cocoa cake can begin, it must go through a series of cooling steps. The hammer mill and disc mill are the more common mills being used by industry and a classifier is usually included as part of the system to improve grinding efficiency. In reality, the term grinding is misleading. The particle size of the finished powder is dependent upon particle size of the chocolate liquor pressed. Commonly used mills do not reduce the particle size of the cake; instead they tend to break up agglomerates. Figure 3.8 shows an example of a pulverizer used for this duty.

It must also be mentioned that cocoa powder must be tempered. This is a sort of controlled cooling operation during which the powder is held at specified temperatures for predetermined times to allow for the cocoa butter to form into a stable crystal configuration. The powder should then be stored properly if it is to maintain its color and remain soft and in a flowable state. Cocoa powder is susceptible to fat bloom as is found in chocolate. The powder will turn gray in color and will became rock hard if not tempered during the grinding process.

Fig. 3.8 Cocoa cake pulverizer. *Source:* S.A. Martin Lloveras (Spain).

Strong measures will be required to remove it from a bag and get it to flow as part of a process stream.

The manufacture of chocolate confectionery, particularly milk chocolate, requires the addition of more cocoa butter; this can be as much as 20%. However, the ratio of cocoa butter to cocoa powder separated from the raw material is approximately 1:1. Consequently, an imbalance is created; one that industry needs to address. It is not unusual to encounter problems such as producers having warehouses filled with cocoa cake generated to satisfy cocoa butter needs but with no outlets for the pressed cake. Such imbalances can lead to serious financial losses, so the need to find new uses for cocoa powder is urgent.

Cocoa butter

The other product obtained from pressing chocolate liquor is cocoa butter. This is the most expensive of the major ingredients in a chocolate recipe. Historically, in the USA, cocoa butter is only filtered before it is used in the manufacture of chocolate. This is known as natural cocoa butter. In Europe, much of the cocoa butter is deodorized, a process for removing flavor components by treating cocoa butter with superheated steam under vacuum.

Cocoa butter is available from a large number of producers worldwide and can be derived from raw materials varying widely in type and quality level. It is for these reasons that the material inside a container labeled cocoa butter can exhibit significant variation with respect to flavor, hardness and color. The most extreme flavor variation is usually found in natural cocoa butters, but unacceptable flavors have been frequently detected in shipments labeled deodorized cocoa butter.

Cocoa butter color can be an excellent indicator of potential problems. For example, shipments that are dark brown rather than golden in color warrant further testing. It can indicate improper handling during processing and storage.

The determination of cocoa butter's melting profile is more difficult and requires sophisticated instrumentation. When cocoa butter is a component in a chocolate recipe, its melting characteristics can impact significantly on the downstream processing of chocolate confections. For example, an understanding of this characteristic is important in efficient operation of a high-speed chocolate bar moulding plant or an enrobing line. Should either type of line be operating at full capacity, having a soft cocoa butter as part of the chocolate recipe can cause enrobed products to stick to cooling belts or moulded items not releasing from moulds. One possible solution, depending on the severity of the problem, is slowing of the line but the softness of the cocoa butter can reach a point where nothing can be done on-line to allow production of an acceptable finished product (4).

Chocolate manufacturing

The above operations produce two of the major ingredients required to produce the various types of chocolate found throughout the world: chocolate liquor and cocoa butter. Before discussing specific types of chocolate and their variations, the manufacturing of chocolate will be divided into four areas:

(1) Batching
(2) Particle reduction
(3) Conching
(4) Standardization

Much research has been done and work continues in an effort to 'streamline' the manufacture of chocolate with the ultimate goal being a low asset base continuous chocolate system. How successful the efforts have been to date and will be in the future is a subject that will be hotly debated in technical meetings for many years. For this publication, we review the classic chocolate making systems by discussing the four previously mentioned areas.

Chocolate batching

Batching, as the name indicates, is the combining of chocolate liquor, sweeteners, milk solids (if a milk chocolate), cocoa butter and non-volatile flavoring materials as specified by a recipe or formula. Batch mixing times can range from 3–20 min with development efforts having been directed towards making this part of the process continuous. Controversy over batch versus continuous does exist, although it is diminishing as continuous batching is becoming accepted by more manufacturers.

Particle reduction

The particle reduction portion of the chocolate manufacturing process has received and continues to receive much research and development attention. When one considers that about 75 years ago, a batch of chocolate was refined up to eight times, progress has been made (1). It must be stated that this large number of refinings was driven as much by desired flavor changes as particle reduction needs. In fact, it was suggested that refinings be alternated between steel and granite refiners, as it was believed that steel refiners were inadmissible in preparation of the best chocolates because of metallic flavors.

Today, many chocolate paste-grinding techniques exist. In fact, some manufacturers choose the process based on chocolate type and, consequently, will employ several types of grinding devices and process schemes throughout their organization. It may be a single five-roll steel refiner grinding the complete batch or a two-roll pre-refiner may be placed before a five-roll refiner to improve the grinding process. In some cases, the ingredients are pre-milled and some may bypass the refiner completely in an effort to increase capacity or minimize assets. Finally, various types of grinding devices are available which do not use steel rolls. Examples are ball mills, hammer mills and refiner conches. Refiner conches have the grinding equipment inside a steel tank and the batch is said to be conched as it is refined during a 12–24 hour cycle depending on the target particle size.

One could continue to describe chocolate grinding schemes, but we can summarize by stating that particle reduction of chocolate ingredients is required to produce the smooth product most consumers expect. The debate as to the 'best' equipment and the 'best' process scheme will continue as long as two chocolate makers exist. As it should be, the consumer will decide. Reduced asset costs, reduced floor space requirements, increased tonnage per hour and flexibility are meaningless if an accurate consumer measurement indicates the chocolate is not highly acceptable and results in lower consumption.

Conching

Conching is a complex process that has been, and continues to be studied in great detail in an effort to fully understand changes that occur in chocolate during this process. Most of the changes are noted in the flavor of chocolate, although depending on the recipe and type of machine employed, textural changes have been reported. Typically, conched chocolates are described as being less harsh and exhibiting a more balanced flavor profile. While not fully understood, the chemical reactions and losses which occur during conching are important factors in the reported flavor changes and are widely reported in the literature. Pyrazines, sulfur containing compounds and the loss of volatiles are examples of chemicals that have been studied by industry and by several universities.

Another area of activity is development of equipment and processes which will improve the efficiency of conching with continuous conching being the major target. While chocolates are already being produced using continuous conching, batch conching continues to be practiced widely because producers have not been able to duplicate the flavors and flow properties typical of their batch conched chocolates with continuous devices (5).

What actually happens to a chocolate which makes the conching process so important to many chocolate makers? Properly practiced, conching will:

- drive moisture from the refinings;
- remove volatile components from the chocolate;
- modify the chocolate's flow properties.

A fourth benefit of conching has been reported and was discussed widely in early technical gatherings: particle size reduction was attributed to the conche. Research sponsored by the Chocolate Manufacturers Association of the USA in co-operation with Franklin and Marshall College investigated particle reduction during conching. William Duck, in his project report of 1962, stated that only in the longitudinal conche was particle reduction noted but he found this reduction to be too small to measure accurately. If particle reduction does not occur during conching, what then is responsible for textural differences noted before and after conching by trained sensory panels? One answer is a rounding of the rough edges on some of the particles and, to a lesser extent, the breaking up of agglomerates. There may be other changes that occur in chocolate during the conching process, and they will become evident as the research continues. How, then, is chocolate conched to obtain the benefits described?

A search of the indexes of books cited at the end of this chapter and the author's experience have yielded 12 different manufacturers of batch and continuous conches. One example is shown in Fig. 3.9; however, space does not allow detailed coverage of each of the conche types and each of the manufacturers' variations. It is suggested that the reader review information in the references and contact manufacturers if further details are required.

There exists a school of thought that believes conching, in the classic sense, is unnecessary. By pre-treating the component ingredients to remove any harsh and

Fig. 3.9 Frisse conche. *Source:* Richard Frisse, Salzuflen, Germany.

other undesirable flavors, one will have a chocolate with a mellow and well-balanced flavor profile. Exposing chocolate to high rates of shear so that each particle is coated with fat will stabilize the flow properties of chocolate. The moisture will be removed by pre-treatment processes and/or the shearing process. Therefore, all three elements provided by the batch conching process are obtainable without any type of conche. In reality, it appears that raw material pre-treatment and additional shear can complement the conching process resulting in decreased conche cycles and a more efficient use of assets.

As was noted when discussing cocoa bean roasting, the number of 'right' ways to conche chocolate is almost equal to the number of chocolate manufacturers. Not only is there disagreement in how to conche, questions have been raised as to the need to conche chocolate if some of the newer practices are adopted. Our knowledge base and understanding of the flavor development process are increasing as our quest for greater efficiencies continues. We have come from processing chocolate in longitudinal conches for days and weeks to cycle times of hours in modern conches, to continuous conching and, finally, no conching in the classic sense. Another debate which will continue is determining the impact the changes in conching will have on flavor, texture and flow properties of finished chocolate. Most important is the need to understand the impact these changes may have on consumers' acceptance of chocolates. This is an international issue and much work needs to be done.

Standardization

Conching is usually considered the final step in the manufacture of chocolate. In these days of high-speed moulding and enrobing lines, another step, which has been identified as standardization, has become very important and must be added to the chocolate manufacturing scheme. This is the step where the viscosity of chocolate is adjusted with addition of fats, usually cocoa butter, and emulsifiers. Volatile flavoring materials may also be added at this point. It is well understood that delivery of chocolates with improper viscosities, or with a wide range of viscosities, to moulding or enrobing lines will make product weight control extremely difficult, and will also increase the level of visual defects. Depending on the severity of defects, product rejection can result with attendant economic loss. Another consequence of poor viscosity control is having the line operator modify the chocolate temper level to adjust its viscosity. Chocolate tempering is the controlled cooling which is necessary to ensure that stable, fluffy cocoa butter crystals are formed. Adjusting the temper level to compensate for poor viscosity control usually results in a less than optimal temper level and a finished product that lacks gloss and may eventually bloom, resulting in consumer dissatisfaction.

Standardization seems to be a simple and minor step in producing chocolate, but time is spent discussing it here because its importance is not appreciated. When a chocolate plant is being designed and budget cuts are needed, the level of mixing assets in the standardization area is one of the first areas to be reduced. When pressure is applied to the chocolate plant to increase output, mixing times during standardization are usually decreased. Both these actions can have a negative impact on the flow properties of the chocolate being produced. In addition to the impact on coverage quality and weights of finished confections, an increase in cocoa butter use is a distinct possibility. All are familiar with the economic consequences of this situation.

Chocolate types

The manufacturing of chocolate has been discussed with no reference to chocolate type. This has been done because the variations required to produce the various types of chocolates manufactured worldwide involve changes in processing protocols rather than major equipment changes, the latter being beyond the scope of this chapter. Detailed reviews of specific processes required to produce various chocolate types are available in the cited references.

To begin the discussion of chocolate types, we need to state that no international classification system has been developed, although most producers and geographical areas generally agree to two basic types – milk and dark. In many markets, a white chocolate is also recognized. In most countries, chocolate types are defined through their food regulations and many definitions for a given type exist, which results in trade and other difficulties. The food regulations of the USA, for example, define four milk chocolates and two sweet chocolates. They are milk, skim milk, buttermilk mixed dairy, sweet, semisweet and bittersweet chocolates. In Japan, standards define a 'pure' milk chocolate, milk chocolate and semi-milk chocolate. At least two definitions for couverture chocolate exist in Europe with one standard allowing a maximum of 5% non-cocoa vegetable fat while no non-cocoa vegetable fat is allowed by the other.

Listing the differences which exist in chocolate standards throughout the world is an exercise in futility because changes are being proposed on a continuous basis. The reason for most of the changes appears to be an effort to harmonize standards, which is a difficult but much needed effort as the industry continues to globalize. Two specific chocolate regulatory situations receiving or needing attention are mentioned here. The first is partial replacement of added cocoa butter with vegetable fat and the second is requiring of a specified minimum total fat content for finished chocolate.

The vegetable fat issue is very controversial and its resolution is most pressing in Western Europe. It will also become an issue in other parts of the world depending on the outcome in Europe. Discussions have already taken place in major producing areas outside of Europe showing the same high level of controversy. While the European chocolate industry will continue to produce chocolate while the vegetable fat issue is debated by many groups, its resolution is imperative, so the energy being expended can be directed to other needed areas within the industry. Finding a solution has become more difficult because the vegetable fat debate appears to have left the technical arena and has become a political issue. Include the economic factors involved and a solution does not appear to be on the immediate horizon, but discussions must and will continue.

The practice of specifying minimum total fat content in chocolate regulations also needs to be reviewed by the appropriate authorities. Chocolate has the reputation of being a high-fat food and many believe that this will continue to harm chocolate consumption if fat levels are not reduced. Technology is advancing to the point where acceptable chocolates can be produced with total fat contents below those specified by the food regulations of some of the major chocolate producing countries. The need for change is obvious. Let the consumer decide if technology advancements produce an acceptable product. It is not prudent practice to limit the development of new products by regulations with no technical or economic basis.

Summary

The basic protocol followed in the processing of cocoa beans is quite consistent around the world. It is the variation in the equipment used, the raw material sources, the cocoa bean blends and operating conditions which provide the noted product differences. The same situation exists in the production of chocolate. Typically, a manufacturer will have a process scheme which consists of batching the materials, particle reduction, conching and standardization. Again, it is the type of equipment that is selected, processing conditions, raw materials and the recipes which provide the flavor and textural differences noted in chocolates from around the world.

What are the differences noted in chocolates produced around the world? Usually, one hears statements that European chocolates are mellow and well balanced with a texture described as smooth and creamy. It is said that typical American chocolates are the opposite since they tend to be harsh, with a hard bite and a coarse texture. This is another chocolate debate which will never end, so let it be said that the best flavor and texture is that which is significantly preferred over whatever competition exists in a market area. As there is no best or perfect chocolate flavor, there is no best or perfect manufacturing facility. The choices in both cases are almost unlimited. Develop clear and concise product objectives and produce them on a system that is affordable and efficient in manufacturing good quality products at competitive costs.

References

1. Whymper, R.A. (1921) *Cocoa and Chocolate, Their Chemistry and Manufacture*, 2nd edn. P. Blakiston's, Philadelphia.
2. Minifie, B.W. (1989) *Chocolate, Cocoa and Confectionery: Science and Technology*, 3rd edn. Van Nostrand Reinhold, New York.
3. Anon (1993) *The Cocoa Manual – a DeZaan Publication*. Cacao DeZaan B.V.
4. Cook. L.R. and revised by Meursing, E.H. (1982) *Chocolate Production and Use*. Harcourt Brace Jovanovich, New York.
5. Beckett, S.T. (Ed.) (1994) *Industrial Chocolate Manufacture and Use*, 2nd edn. Blackie Academic & Professional, Glasgow.

Section III

Nutrient Information and Nutritional Aspects

Chapter 4

Analysis and Nutrient Databases

Susan L. Cheney

Nutrient databases are employed throughout the world for nutrition research, product development, prevention and control of nutrient deficiencies, labeling the nutrient content of foods, assessing and monitoring nutritional intake, and development of nutrition and agriculture properties. Most databases provide the nutrient composition for whole foods as they are consumed; few are designed to provide the nutrient composition of the ingredients that constitute the finished foods, with the exception of proprietary databases maintained by manufacturers. Most nutrient databases rely heavily on private and public sources for nutrient data, rather than on analytical results, making it difficult to evaluate the quality of reported data. It is also important to note that bioavailability of nutrients is only now beginning to be understood, and is not reflected in any of the nutrient databases shown here.

The Chocolate Manufacturers Association (CMA) in the USA has developed a *Nutrient Database for Three Selected Major Ingredients Used in the CMA Recipe Modeling Database: Chocolate Liquor, Cocoa Powder, and Cocoa Butter* (1), dedicated to providing analytical data on the nutrient composition of *cacao*-based ingredients. Other databases in use throughout the world provide information on cocoa powder and cocoa butter, and specific chocolate products, but they rely on industry information and printed sources for nutrient composition, or analyze store-bought samples for nutrient composition rather than obtaining nutrient data from the raw product.

This chapter briefly details the process of developing a nutrient database, and compares nutrient data on chocolate liquor, cocoa powder and cocoa butter from five sources: the CMA *Nutrient Database* and US Department of Agriculture (USDA) *Agriculture Handbook Nos 8–19* from the USA; *Miscellaneous Foods: Supplement to McCance and Widdowson's Composition of Foods* from the UK, the Netherlands' *NEVO Tables*; and Germany's *Food Composition and Nutrition Tables* (1–5).

Database development

The goal of developing ingredient-based databases is to provide valid estimates of nutrient content for use in labeling of products so they are in compliance with nutrition labeling regulations. In accordance with this, the US Food and Drug Administration (FDA) has developed a set of guidelines for use in developing ingredient databases (6). As the CMA database is the only one devoted to *cacao*-based ingredients, it will be the focus of this review.

Sampling plans for ingredient databases must provide a representative sampling of nutrient analysis of the basic product. The CMA database developed sampling plans for the following ingredients:

(1) *Chocolate liquor* conforming to US standard of identity (21 CFR 163.11) (7). Chocolate liquor was sampled by manufacturer's blend, which is proprietary with regard to country of origin and composition. Liquor producers measure fat content of blends with a high degree of accuracy, and provided the fat content of each blend sampled on an individual basis.
(2) *Cocoa powder* conforming to US standard of identity (21 CFR 163.113). Cocoa powders are pressed to very specific fat content levels, and powder producers provided the precise fat content for each pressing of cocoa powder. Only natural, non-alkalized cocoa powders were analyzed because of the variable potassium and sodium levels achieved with alkalization.
(3) *Cocoa butter blends*. Cocoa butters are the natural fat of the cocoa bean and are extremely consistent in that they are 100% fat, and the fatty acids vary only slightly. Cocoa butter samples included deodorized and non-deodorized blends.

The CMA database used a multi-stage sampling method to determine if manufacturer, bean blend and fat content variations occur in *cacao*-based ingredients. Sample size was estimated by developing typical recipes in which *cacao*-based ingredients are used and calculating the amount of *cacao*-based ingredients in the final foods. Then a nutrient literature search was conducted, calculating an estimated sample size from each nutrient literature population, using a 5% margin of error, and determining the contribution of nutrients to final product. This was done by converting nutrient values of *cacao*-based ingredients from 100 g amounts to the maximum amount used in the reference amount of the final product, and developing the sample size for the nutrient database research using estimated sample size and contribution of nutrients to the final product.

Methodology

The CMA database analyzed *cacao*-based ingredients using *Official Methods of Analysis*, the Association of Official Analytical Chemists (AOAC)-approved methodology (8), while the other sources gathered nutrient data primarily from industry, government and published sources, making it impossible to evaluate the appropriateness of the methodology used. It is important to note that even when appropriate analytical methods are used, the same sample analyzed in different laboratories may have a large variation in reported results (9). Regardless of the method used when compiling nutrient data, it is vital that the performance of all analytical procedures is validated using various tools and techniques, the main one being the control sample. Due to the complex matrix of *cacao*-based ingredients, control samples must be validated to determine specific precision parameters. Good laboratory practices also ensure the integrity of the nutrient data obtained.

For each nutrient analyzed, there may be several methods available for determining content. Fats and carbohydrates are particularly difficult classes of foods to analyze because of the complexity of their makeup, and the time-consuming sample preparations that they must undergo before they can be analyzed. For this reason, most database values are based on simple methods and the values obtained tend to be method dependent. Method documentation is extremely important to the integrity of the database.

Energy values for food nutrients can be measured using one of the following three methods:

(1) *General food factors* – adjust protein, fat and carbohydrate heats of combustion to allow for fecal losses by multiplying the average heats of combustion for mixed proteins, fats and carbohydrates by the digestibility of the three compounds. Results in energy values of 4 kcal/g for protein, 9 kcal/g for fat and 4 kcal/g for carbohydrates. This method is widely used throughout the world.

(2) *Specific Atwater factors* – use heats of combustion values specific for different foods or food groups rather than average values. Results in energy values of 1.83 kcal/g for protein, 8.37 kcal/g for fat and 1.33 kcal/g for carbohydrates. This method can result in slightly lower energy values, and databases using this system must document precisely which factors have been used. The CMA database employed this method when determining energy values.

(3) *Modified Atwater* – used in the UK. Measures carbohydrates directly rather than by difference. Available carbohydrates are given the value of 3.75 kcal/g and unavailable carbohydrates are given the value of 0 kcal/g. Energy values for protein and fat are 4 kcal/g and 9 kcal/g, respectively. This method can result in lower energy values for plant foods rich in plant cell wall material.

The differences between the methods do not result in significant differences between energy values (10). The general food factors method tends to overestimate metabolizable energy, while the modified Atwater method can underestimate energy values. It is important to remember that the differences between the three methods are much smaller than the errors resulting from ignoring the individual differences in digestibility and in measuring food intake.

Analysis of lipids and fatty acid components is dependent on the type of food, the number of samples and the laboratory facilities. The information required on lipids for a nutrient database include the total lipid content for calculation of energy content, as well as comparison between the foods, the fatty acid types and the amount of cholesterol in the food. *Cacao*-based ingredients are assumed to contain zero cholesterol because they are plant-based components.

Due to their lipid content and makeup, *cacao*-based ingredients require a complex solvent extraction procedure followed by chromatographic and/or spectrophotometric analysis of all lipid classes. Extraction methods involve denaturing lipoproteins and enzymes in alcohol followed by an extraction into an organic solvent. The type of food will determine the nature of the organic solvent used. Once lipids have been extracted, it is necessary to determine the fatty acid content and composition where gas chromatography methods are preferred. In the USA, standard of identity 21 CFR 163.5 cites the extraction method that must be used to determine fat content of chocolate liquor and cocoa powder (7).

Protein is typically calculated from total nitrogen (N) values as measured by Kjeldahl. This result is a measure of the crude protein and is only an approximation. A true measure of protein is obtained by using the conversion factor of $6.25 \times N$. *Cacao*-based ingredients are a challenge in that they contain a large percentage of the nitrogenous compounds theobromine and caffeine. To obtain true protein content, the conversion factor of 6.25 must be employed after removing the theobromine and caffeine components from the total protein content.

Carbohydrates are characterized by digestible portions (i.e. starch and sugars) and indigestible portions (i.e. dietary fiber). The European community defines digestible carbohydrates as ‘available’ and indigestible carbohydrates as ‘unavailable’. Databases typically calculate total carbohydrates by *difference* – the material remaining after subtracting moisture, ash, fat and protein. A major drawback of this method is that calculation by difference accumulates analytical errors from fat, protein, ash and moisture determinations. However, difference methods can and do provide reasonable estimates of total available carbohydrates. It should be noted that carbohydrate values obtained ‘by difference’ and those obtained by analysis are incompatible and attempts to compare the two should not be made. In foods that do not contain a large amount of non-carbohydrate components, the sum of the individual components will approximate to total carbohydrate ‘by difference’.

Dietary fiber can be measured either by a gravimetric method where the non-fiber components are removed and the residue is weighed, or by using enzymatic gravimetric methods that use purified enzymes to measure total dietary fiber or soluble and insoluble components. Colorimetric and chromatographic methods have also been developed. Enzymatic gravimetric methods provide specific and precise analysis of carbohydrates and are ideal for the analysis of food carbohydrates; however, results can be overestimated if the fiber components remain in the residue. Colorimetric methods can also overestimate fiber content, while chromatographic methods can underestimate fiber content. Collaborative studies have shown that estimates of total dietary fiber using either enzymatic gravimetric methods or chromatographic methods are not significantly different (10).

Validation

Before adding analytical data to a database, it should be manipulated using statistical software to determine, at a minimum, means, standard deviations and prediction intervals. Separation of means should occur at $P < 0.05$, and one-sided 95% prediction intervals are calculated to ensure label values will have a high probability of being in compliance with nutrition labeling regulations, and accurately report the nutrient content of *cacao*-based ingredients.

The CMA database employed labeling statistics such as means, standard deviations, coefficients of variation and prediction intervals to test results. The final summarized numbers were rounded for the purposes of nutrition labeling only for use on final products, such as retail bakers' chocolate and retail cocoa butter.

All data contained in nutrient databases should be validated to provide evidence of a functional database that effectively supplies nutrient information for its intended use. Validation requires reviewing the data for consistency and verifying any numbers that appear incorrect.

CMA validates its database by comparing the analytical values of nine *cacao*-based products to database-generated values. Currently, the CMA database is undergoing another validation so it will continue to receive FDA acceptance as an approved ingredient database.

The CMA database provides nutrient composition data for chocolate liquor (50–58% fat), natural cocoa powder (10–17% fat) and cocoa butter for use in a recipe modeling system. Because these items are strictly ingredients and are not sold at retail, serving size considerations were not addressed.

Comparison of databases

Data tables for chocolate liquor and cocoa powder were developed for the nutrition labeling of all mandatory nutrients (i.e. calories, calories from fat, total fat, saturated fat, cholesterol, sodium, total carbohydrate, dietary fiber, sugars, protein, vitamin A, vitamin C, calcium and iron) plus eight optional nutrients (polyunsaturated fat, monounsaturated fat, potassium, phosphorus, magnesium, zinc, manganese and copper).

Nine listings for chocolate liquor were developed for inclusion in the database (Table 4.1). Each is defined by fat content with an overall range from 50–58% fat. When using the database for nutrition labeling, manufacturers are able to select the listing that represents the fat content of the liquor they are adding to their chocolate and confectionery products.

Eight listings for cocoa powder were developed for inclusion in the database (Table 4.2). Each is defined by fat content with an overall range from 10–17% fat. The standard of identity for cocoa powder allows up to 22% fat, but it is rarely used at such high fat levels. Again, when using the recipe modeling database to determine label values, manufacturers will select the listing that represents the fat content of the cocoa powder that they are adding to their products.

The database for cocoa butter includes proximates and fatty acids only (Table 4.3). Fat and fatty acids are the only components present at significant levels.

Database values were selected using combined means and compliance calculations. Because of the significant amount of nitrogen-containing alkaloids (caffeine and theobromine) in *cacao* products, specific calculations were carried out to avoid overestimating or underestimating protein, carbohydrates and calories (11). Based on analytical results, further calculations were done to derive database values applicable to the various levels of fat found in *cacao* products. No changes were made to water, ash, protein, sugars, dietary fiber, vitamins or minerals, as these components do not show a proportional change with fat content. Fat and carbohydrates were adjusted so that an increase in fat yielded a decrease in carbohydrates and vice versa. Calories and fatty acids were adjusted in proportion to the recalculated levels of fat and carbohydrates.

Laboratory analysis of chocolate liquor, cocoa powder and cocoa butter samples indicate no significant variation due to bean blend, fat content or manufacturer. Due to wide variations in the iron data obtained from analysis of cocoa powder samples, the CMA database used the more conservative value taken from the USDA *Handbook 8–19*: cocoa powder, unsweetened. Laboratory results of the cocoa butter samples showed trace amounts of water, protein and carbohydrates, most likely due to the limitations of analytical methodology. In order to arrive at representative database values, the fat content was converted to 100% and fatty acids were adjusted based on their percentage of the total fat. Proximates do not equal 100% because not all fatty acid and carbohydrate components are reported.

Tables 4.4, 4.5 and 4.6 compare CMA database values with those found in the USDA *Handbook 8–19*; *Miscellaneous Foods* and *Dairy Ingredient Supplements to McCance and Widdowson's Composition of Foods*; the Netherlands' *NEVO Tables*; and Germany's *Food Composition and Nutrient Tables* (1–5, 12).

Chocolate liquor values are not available from any country other than the USA. The above databases provide data for finished chocolate products such

Table 4.1 CMA database values for chocolate liquor for varying fat contents per 100 g.

Nutrient	Percentage fat (%)								
	50	51	52	53	54	55	56	57	58
Calories (kcal)	485.09	492.13	499.17	506.21	513.25	520.29	527.33	534.37	541.41
Calories from fat (kcal)	418.50	426.87	435.24	443.61	451.98	460.35	468.72	477.09	485.46
Fat (g)	50.00	51.00	52.00	53.00	54.00	55.00	56.00	57.00	58.00
Saturated fat (g)	30.02	30.63	31.21	31.82	32.46	33.10	33.75	34.07	34.71
Polyunsaturated fat (g)	1.49	1.52	1.55	1.58	1.62	1.65	1.68	1.70	1.73
Monounsaturated fat (g)	16.36	16.70	17.01	17.34	17.70	18.05	18.40	18.57	18.92
Cholesterol (mg)	0[1]	0[1]	0[1]	0[1]	0[1]	0[1]	0[1]	0[1]	0[1]
Carbohydrates (g)	35.61	34.61	33.61	32.61	31.61	30.61	29.61	28.61	27.61
Dietary fiber (g)	15.75	15.75	15.75	15.75	15.75	15.75	15.75	15.75	15.75
Sugars (g)	1.01	1.01	1.01	1.01	1.01	1.01	1.01	1.01	1.01
Protein (g)	10.82	10.82	10.82	10.82	10.82	10.82	10.82	10.82	10.82
Potassium (mg)	1023.8	1023.8	1023.8	1023.8	1023.8	1023.8	1023.8	1023.8	1023.8
Sodium (mg)	3.24	3.24	3.24	3.24	3.24	3.24	3.24	3.24	3.24
Vitamin A (IU)	0[1]	0[1]	0[1]	0[1]	0[1]	0[1]	0[1]	0[1]	0[1]
Vitamin C (mg)	0[1]	0[1]	0[1]	0[1]	0[1]	0[1]	0[1]	0[1]	0[1]
Calcium (mg)	91.36	91.36	91.36	91.36	91.36	91.36	91.36	91.36	91.36
Iron (mg)	13.52	13.52	13.52	13.52	13.52	13.52	13.52	13.52	13.52
Phosphorus (mg)	432.88	432.88	432.88	432.88	432.88	432.88	432.88	432.88	432.88
Magnesium (mg)	314.17	314.17	314.17	314.17	314.17	314.17	314.17	314.17	314.17
Zinc (mg)	4.29	4.29	4.29	4.29	4.29	4.29	4.29	4.29	4.29
Copper (mg)	2.36	2.36	2.36	2.36	2.36	2.36	2.36	2.36	2.36
Manganese (mg)	2.57	2.57	2.57	2.57	2.57	2.57	2.57	2.57	2.57
Water (g)	0.63	0.63	0.63	0.63	0.63	0.63	0.63	0.63	0.63
Ash (g)	3.22	3.22	3.22	3.22	3.22	3.22	3.22	3.22	3.22

[1] These values were taken from the USDA database and are assumed values.

Table 4.2 CMA database values for cocoa powder for varying fat contents per 100 g.

Nutrient	Percentage fat (%)							
	10	11	12	13	14	15	16	17
Calories (kcal)	203.85	210.89	217.93	224.97	232.01	239.05	246.09	253.13
Calories from fat (kcal)	83.70	92.07	100.44	108.81	117.18	125.55	133.92	142.29
Fat (g)	10.00	11.00	12.00	13.00	14.00	15.00	16.00	17.00
Saturated fat (g)	5.93	6.52	7.11	7.69	8.29	8.89	9.49	10.08
Polyunsaturated fat (g)	0.30	0.33	0.36	0.39	0.42	0.45	0.48	0.51
Monounsaturated fat (g)	3.30	3.62	3.95	4.27	4.61	4.94	5.27	5.60
Cholesterol (mg)	0[1]	0[1]	0[1]	0[1]	0[1]	0[1]	0[1]	0[1]
Carbohydrates (g)	56.85	55.85	54.85	53.85	52.85	51.85	50.85	49.85
Dietary fiber (g)	27.90	27.90	27.90	27.90	27.90	27.90	27.90	27.90
Sugars (g)	1.66	1.66	1.66	1.66	1.66	1.66	1.66	1.66
Protein (g)	19.59	19.59	19.59	19.59	19.59	19.59	19.59	19.59
Potassium (mg)	1495.5	1495.5	1495.5	1495.5	1495.5	1495.5	1495.5	1495.5
Sodium (mg)	8.99	8.99	8.99	8.99	8.99	8.99	8.99	8.99
Vitamin A (IU)	0[1]	0[1]	0[1]	0[1]	0[1]	0[1]	0[1]	0[1]
Vitamin C (mg)	0[1]	0[1]	0[1]	0[1]	0[1]	0[1]	0[1]	0[1]
Calcium (mg)	169.45	169.45	169.45	169.45	169.45	169.45	169.45	169.45
Iron (mg)	13.86[2]	13.86	13.86	13.86	13.86	13.86	13.86	13.86
Phosphorus (mg)	795.27	795.27	795.27	795.27	795.27	795.27	795.27	795.27
Magnesium (mg)	593.64	593.64	593.64	593.64	593.64	593.64	593.64	593.64
Zinc (mg)	7.93	7.93	7.93	7.93	7.93	7.93	7.93	7.93
Copper (mg)	4.61	4.61	4.61	4.61	4.61	4.61	4.61	4.61
Manganese (mg)	4.73	4.73	4.73	4.73	4.73	4.73	4.73	4.73
Water (g)	2.58	2.58	2.58	2.58	2.58	2.58	2.58	2.58
Ash (g)	6.33	6.33	6.33	6.33	6.33	6.33	6.33	6.33

[1] These values were taken from the USDA database and are assumed values.
[2] USDA *Agriculture Handbook No. 8–19*: cocoa powder, unsweetened (2).

Table 4.3 CMA database for cocoa butter.

Nutrients	Per 100 g
Calories (kcal)	884.00
Calories from fat (kcal)	884.00
Fat (g)	98.45
Saturated fat (g)	59.63
Polyunsaturated fat (g)	3.88
Monounsaturated fat (g)	34.99
Carbohydrates (g)	0
Protein (g)	0
Water (g)	0
Ash (g)	0

Table 4.4 Nutrient database comparisons of chocolate liquor blends per 100 g.

Nutrient	USA[1]	
	CMA[2] Database	USDA Handbook 8–19[3]
Calories (kcal)	513.25	522.00
Calories from fat (kcal)	451.98	462.86
Fat (g)	54.00	55.30
Saturated fat (g)	32.42	32.60
Polyunsaturated fat (g)	1.61	1.76
Monounsaturated fat (g)	17.67	18.46
Cholesterol (mg)	0[4]	0
Carbohydrates (g)	31.61	28.3
Dietary fiber (g)	15.75	15.4
Sugars (g)	1.01	0.60
Protein (g)	10.82	10.30
Potassium (mg)	1023.8	833.0
Sodium (mg)	3.24	14
Vitamin A (IU)	0	0
Vitamin C (mg)	0	0
Calcium (mg)	91.36	74.00
Iron (mg)	13.52	6.32
Phosphorus (mg)	432.88	417.0
Magnesium (mg)	314.17	310.00
Zinc (mg)	4.29	4.01
Copper (mg)	2.36	2.17
Manganese (mg)	2.57	1.92
Water (g)	0.63	1.30
Ash (g)	3.22	3.00

[1] No data available from the UK, the Netherlands or Germany.
[2] Mean from analytical data.
[3] Mean obtained from published and unpublished sources.
[4] These were taken from the USDA database and are assumed values.

Table 4.5 Nutrient database comparisons on cocoa powders per 100 g.

Nutrient	USA		UK	Netherlands	Germany
	CMA Database[1]	USDA Handbook 8–19[2]	Miscellaneous Foods [3]	NEVO Tables[4]	Food Composition and Nutrition Tables[5]
Calories (kcal)	228.49	229.00	312.00	315.00	343.06
Calories from fat (kcal)	112.99	114.67	—	—	—
Fat (g)	13.50	13.70	21.70	21.70	24.50
Saturated fat (g)	8.00	8.07	12.80	12.90	—
Polyunsaturated fat (g)	0.41	0.44	0.60	0.60	—
Monounsaturated fat (g)	4.45	4.57	7.20	7.30	—
Cholesterol (mg)	0[6]	0	0	0	—
Carbohydrates (g)	53.35	54.30	11.50[7]	45.50	10.84[7]
Dietary fiber (g)	27.90	29.80	—	34.00	30.43
Sugars (g)	1.66	0.90	Trace	2.20	2.10
Protein (g)	19.59	19.60	18.50	18.50	19.80
Potassium (mg)	1495.50	1524.00	1500.00	4000.00	1920
Sodium (mg)	8.99	21.00	950.00	100.00	17.00
Vitamin A (IU)	0[6]	0	0	0	—
Vitamin C (mg)	0[6]	0	0	0	—
Calcium (mg)	169.45	128.00	130.00	150.00	114.00
Iron (mg)	13.86[8]	13.86	10.50	15.00	12.50
Phosphorus (mg)	795.27	734.00	660.00	600.00	656.00
Magnesium (mg)	593.64	499.00	520.00	525.00	414.00
Zinc (mg)	7.93	6.81	6.90	7.00	5.73
Copper (mg)	4.61	3.79	3.90	—	3.81
Manganese (mg)	4.73	3.84	—	—	—
Water (g)	2.58	3.00	3.40	—	5.60
Ash (g)	6.33	5.80	—	—	—

[1] Mean obtained from analytical data.
[2] Mean from published and unpublished sources.
[3] Mean obtained from analysis of 12 store-bought samples.
[4] Mean obtained from food composition tables and calculation factors.
[5] Mean obtained from analysis and literature sources.
[6] These were taken from the USDA database and are assumed values.
[7] Calculated as available carbohydrate.
[8] USDA *Agriculture Handbook 8–19*: cocoa powder, unsweetened.

as milk chocolate and baking chocolate rather than the ingredient chocolate liquor.

Both the USDA *Handbook 8–19* and the CMA database calculate energy values using the general food factors method. In both databases, carbohydrates are calculated by difference and protein values were calculated after adjustments were made for non-protein nitrogenous compounds. Differences between reported nutrient data in the two sources may be a result of the USDA obtaining analytical data from analysis, government agencies, literature reviews and manufacturers, where the CMA database relied solely on analytical data. The

Table 4.6 Nutrient database comparisons on cocoa butter blends per 100 g.

Nutrient	USA		UK	Netherlands	Germany
	CMA Database[1]	USDA Handbook 8–19[2]	Miscellaneous Foods [3]	NEVO Tables	Food Composition and Nutrition Tables[4]
Calories (kcal)	884.00	884.00	896.00	No data	900.00
Calories from fat (kcal)	884.00	884.00	—	—	—
Total fat (g)	98.45	100.00	99.50	—	100.00
Saturated fat (g)	59.63	59.70	59.00	—	—
Polyunsaturated fat (g)	3.88	3.00	3.30	—	—
Monounsaturated fat (g)	34.99	32.90	32.80	—	—
Cholesterol (mg)	0[5]	0	3.00	—	2.70
Total carbohydrate (g)	0	0	0	—	—
Protein (g)	0	0	0	—	—

[1] Mean obtained from analytical data.
[2] Mean obtained from published and unpublished sources.
[3] Mean obtained from analysis and literature sources.
[4] Mean obtained from analysis and literature.
[5] These were taken from the USDA database and were assumed values.

USDA database has been criticized as not accurately reporting the nutrient content of food. Concerns have also been raised regarding the accuracy of the data, the adequacy of analytical methods used to produce the data, the sufficiency of documentation related to the data and the adequacy of documentation on the criteria for acceptance of data (13). With this in mind, the USDA *Handbook 8–19* values should not be used for nutrition labeling purposes if analytical data are available.

Nutrient values for cocoa powder are reported in numerous databases. Table 4.5 compares five data sources: CMA, USDA, the UK, the Netherlands and Germany. Of these, the UK, Netherlands, Germany, and USDA all report energy values using the general food factors method. The UK and Germany calculate 'available' carbohydrates rather than reporting carbohydrates by difference. The differences in total fat are most likely due to methodology, but methods of analysis are not listed for the USDA, UK, Dutch, or German nutrient data. Another possible source of the variation is reliance on industry and published sources as several of the databases do. The quality of data obtained from industry and published sources is difficult, if not impossible, to evaluate.

The Netherlands data for cocoa powder was obtained almost exclusively from food composition tables, estimation or by calculation. The UK obtained nutrient data after analyzing ten samples obtained from two store-bought cocoa powders. Due to the extremely high levels of potassium and sodium reported by these two countries, one must assume that the reported nutrient values are for alkalized cocoa powders, since the alkalization process results in variable levels of these nutrients.

Comparisons of cocoa butters do not show any significant variation as expected since cocoa butter is 100% fat. Furthermore, since cocoa butter is of plant origin, it is expected that the concentration of cholesterol would be negligible and this was generally borne out by earlier analysis. Nonetheless, the UK and Germany do report small concentrations of cholesterol; however, this is thought to be most likely due to analytical methodology.

Summary

Ingredient databases provide significant benefits to both the industry and the consuming public by making accurate nutrient values readily available to the chocolate manufacturer and confectioner at a reasonable cost and with rapid timing. Before choosing a database, one must carefully look at the analytical methodology, validation procedures and reporting methods used in developing the database to ensure that the database provides a credible source of nutrient information. Reported nutrient data can vary due to the type of food examined, the nutrients analyzed and the method used to report data. Before using any nutrient data, it is important to understand these variables as they can result in large differences in reportable data.

Currently, the CMA database is the only non-manufacturer database devoted to providing nutrient composition data for all three of the *cacao*-based ingredients – chocolate liquor, cocoa powder and cocoa butter – based on direct analysis of the raw products. Other databases are generally constructed from industry data, printed material or from analysis of store-bought finished products. Consequently, when making use of these databases, their origins should be carefully considered against their purpose, as they can affect the quality of the data supplied.

References

1. National Confectioners Association (NCA) and CMA (1997) *Nutrient Database for Three Selected Major Ingredients Used in the NCA/CMA Recipe Modeling Database: Chocolate Liquor, Cocoa Powder, and Cocoa Butter*. Chocolate Manufacturers Association, McLean, VA.
2. USDA (1991) *Composition of Foods: Snacks and Sweets. Agriculture Handbook No. 8–19*. US Department of Agriculture, Hyattsville, MD.
3. Chan, W., Brown, J. and Buss, D.H. (1994) *Miscellaneous Foods: Supplement to McCance and Widdowson's Composition of Foods*. Royal Society of Chemistry and Ministry of Agriculture, Fisheries and Food, London.

4. NEVO Foundation (1996) *NEVO Tables – Dutch Food Composition Database*. NEVO Foundation, The Hague.
5. Souci, S.W., Fachmann, W. and Kraut, H. (1994) *Food Composition and Nutrition Tables*, 5th edn. CRC Press and Medpharm Scientific Publishers, Stuttgart.
6. US FDA (Ed.) (1993) *FDA Nutrition Labeling Manual: A Guide for Developing and Using Databases*. US Food and Drug Administration, Washington, DC.
7. US FDA (Code of Federal Regulations) (1997) 21 CFR 163.110–163.155. (Cacao products, Standards of Identity). US Food and Drug Administration, Washington, DC.
8. AOAC (1990) *Official Methods of Analysis*, 15th edn. Association of Official Analytical Chemists, Arlington, VA.
9. Rand, W.M., Pennington, J.T., Murphy, S.P. and Klensin, J.C. (1991) *Compiling Data for Food Composition Data Bases*. United Nations University Press, Tokyo.
10. Greenfield, H. (1995) *Quality and Accessibility of Food-Related Data*. AOAC, Arlington, VA.
11. Merrill, A.L. and Watt, B.K. (1955) *Energy Value of Foods – Basis and Derivation. USDA Handbook 74*. USDA, Hyattsville, MD.
12. Holland, B., Unwin, I.D. and Buss, D.H. (1989) *Milk Products and Eggs: Supplement to McCance and Widdowson's Composition of Foods*. Royal Society of Chemistry and Ministry of Agriculture, Fisheries and Food, London.
13. Katch, F.I. (1995) US Government raises serious questions about reliability of US Department of Agriculture's food composition database. *Int. J. Sport Nutr.* **5**, 62–67.

Macro-Nutrients

Chapter 5

Cocoa Butter and Constituent Fatty Acids

David Kritchevsky

Cocoa butter – the fat component of cocoa – comprises predominantly stearic and oleic acids, together with a smaller contribution from palmitic acid and only traces of several other fatty acids. The metabolic roles of stearic acid have not been studied as thoroughly as those of other fatty acids because it does not have a strong effect on human lipidemia (1, 2) and much of recent lipid research has been driven by effects on lipids, lipoproteins and atherosclerosis. However, stearic acid is a component of a number of dietary fats and its biologic and physiologic effects merit investigation. This is not to say that the biology of stearic acid has not been studied; it has, but not to the same extent as other long-chain fatty acids, namely, lauric, myristic and palmitic. There are differences in the specific utilization of all these fatty acids and eventually we will have to focus on individual effects rather than on the rough classification *long-chain saturated fatty acids*.

Although fatty acid metabolism has been studied using individual compounds, it is important to note that these fatty acids are present in our diet as components of triglycerides. Although the topic of triglyceride structure is not discussed here, it is important to note that the specific position of a fatty acid in a triglyceride may also influence its metabolic behavior (3, 4).

Digestibility of stearic acid

The absorbability of stearic acid was first studied by Arnschink (5) who reported that tristearin digestibility in dogs was 9–14%. Hoagland and Snider (6) found that in rats, tristearin and tripalmitin were 7% and 38% digestible, respectively. When the individual fatty acids were dissolved in olive oil, digestibility of palmitic acid/olive oil mix rose from 5–95% (39.6% digestible) to 15–85% (31.2% digestible). Stearic acid digestibility rose from 9.4% (5–95% mix of stearic acid and olive oil) to 21.0% (15–85% mix).

Carroll and Richards (7) found that the digestibility of fats in rats was influenced, in part, by the composition of the mineral mix present in the diet. Oleic acid was almost totally digestible when fed as part of a calcium-free diet, but digestibility fell in the presence of various salt mixes and seemed to depend on the calcium:phosphorus (Ca:P) ratio of the salt mix. Thus, digestibility of oleic acid was 78% in diets containing 5% of Phillips–Hart salt mix (Ca:P = 1.4) and 42% in diets containing 5% of the Hubbell–Mendel–Wakeman mix (Ca:P = 4.25).

Mattson *et al.* (8) examined the effects of calcium- and magnesium-containing diets on the absorption of various triglycerides, specifically OSO, SOO, OSS and SOS (O = oleic acid; S = stearic acid). They found that diets replete in calcium and magnesium affected absorption when stearic acid was in the 1 or 3 position (SOO, SOS) but not when it was in the 2 position (OSO or OSS). In a study of the digestibility of various natural and hydrogenated fats, Calloway *et al.* (9) concluded that digestibility depended primarily on chain length of the saturated fatty acids and their position in the triglyceride.

Mattson (10) studied the absorption of stearic acid using mixtures of hydrogenated linseed and safflower oils. The oils were mixed or randomized. Absorbability was a function of the total amount of stearic acid and the level of tristearin present in the test fat (Table 5.1).

Metabolism of stearic acid-rich fats

Bergstedt *et al.* compared the absorption of tristearin and triolein by the small intestine of rats bearing lymph fistulas (11) as well as the effects of triolein and

Table 5.1 Influence of fat composition on absorbability of stearic acid.

Mix*		% 18:0	% tristearin	Coefficient of absorption
% HLO	% SFO			
0	**100**	13	0	96.7
5	**95**			
Mix		16	4	93.8
Randomized		16	0	95.7
40	**60**			
Mix		45	36	73.2
Randomized		46	9	86.8
70	**30**			
Mix		72	63	50.0
Randomized		71	36	63.8
100	**0**	97	91	15.4

Source: after Mattson (10).

* HLO = hydrogenated linseed oil; SFO = safflower oil.

tripalmitin on absorption of tristearin (12). Lymph flow was slower in rats given tristearin and its triglyceride content was lower. The distribution of tristearin and triolein in lymph, mucosa, and lumen is given in Table 5.2 and distribution of labeled triglyceride in the intestine is given in Table 5.3. The absorptive index (100% – % dose recovered in the lumen) was 94.3 ± 1.0 for triolein and 56.7 ± 7.8 for tristearin. Lymphatic output of triglyceride was higher when tristearin was mixed with triolein than when it was mixed with tripalmitin (12). As in the earlier study (11), more tristearin remained in the lumen when it was undiluted by triolein or tristearin. The absorptive indices for the three fats were tristearin, 63.8 ± 5.1; tristearin + triolein, 94.7 ± 0.9; and tristearin + tripalmitin, 86.8 ± 6.3. Absorbability of tristearin was significantly ($P < 0.05$) lower than for the other two fats.

Chen *et al.* (13) compared the absorption of cocoa butter, corn oil and palm kernel oil in lymph fistula rats. Recovery of fatty acids was significantly lower in the rats fed cocoa butter. Cholesterol absorption was also lowest in the cocoa butter-fed rats (Table 5.4).

Table 5.2 Distribution of radioactive lipid in rats fed various triglycerides.

	Recovery (% of dose)	
	Triolein*	Tristearin*
Lymph	57.8 ± 4.5	20.4 ± 6.5
Mucosa	5.6 ± 0.4	3.0 ± 0.9
Lumen	5.7 ± 1.0	43.3 ± 7.8
Total	69.1 ± 5.0	66.7 ± 4.6

Source: after Bergstedt *et al.* (11).
* Means ± SE. Values obtained 8 hours after infusion of radioactive lipid. Six rats per group.

Table 5.3 Recovery of radioactive lipid (% of dose) remaining in lumen of small intestine after feeding rats triolein or tristearin.

	Recovery (% of dose)	
	Triolein*	Tristearin*
Stomach	0.63 ± 0.17	1.58 ± 0.42
Small intestine		
Segment 1	3.99 ± 1.02	2.40 ± 0.72
Segment 2	0.40 ± 0.09	4.38 ± 0.78
Segment 3	0.26 ± 0.05	10.46 ± 3.03
Segment 4	0.27 ± 0.02	17.42 ± 4.42
Cecum	0.02 ± 0.02	7.08 ± 2.31

Source: after Bergstedt *et al.* (11).
* Means ± SE. Values obtained 8 hours after infusion of radioactive lipid. Six rats per group.

Table 5.4 Recovery of absorbed fatty acids and cholesterol from rat lymph.

Dietary fat	% Recovery	
	Fatty acid	Cholesterol
Control	—	49.8 ± 9.3
Corn oil	100[1] a[2]	55.4 ± 5.8c
Palm kernel oil	82.3 b	53.8 ± 8.2d
Cocoa butter	63.0 ab	42.9 ± 13.5 cd

Source: after Chen *et al.* (13).
[1] Means ± SE. Corn oil recovery set at 100 – other values relative to it.
[2] Values bearing same letter are significantly ($P < 0.05$) different.

Apgar *et al.* (14) measured the digestibility of corn oil and the stearic acid-rich fat, cocoa butter, in rats. Digestibility coefficients for 5, 10 or 20% corn oil were 92.7 ± 0.9, 96.9 ± 0.1 and 96.3 ± 0.2, respectively. For cocoa butter, the digestibility coefficients (%) were 5%, 58.5 ± 0.5; 10%, 60.3 ± 1.4; and 20%, 71.7 ± 1.3. Body weight gain and food intakes were similar for all groups. Fecal excretion of palmitic acid was roughly similar for all groups (30–34%) but rats fed cocoa butter excreted significantly more stearic acid (61% versus 27%).

Elovson (15) injected carboxyl-labeled stearic acid into the jugular vein of rats and 5 min later recovered 8% of the radioactivity in liver oleic acid. When [3,4-^{3}H]-stearic acid was injected intravenously into rats, [^{3}H]-oleic acid was recovered from liver lipids. However, before postulating a major conversion of stearic to oleic acid, one must take into account the mode of administration (intravenous injection), the rapid reutilization of $^{14}CO_2$ obtained as a result of the metabolic decarboxylation of stearic acid. The lability of ^{3}H must also be taken into account. Emken (16) found that the conversion of deuterated stearic to oleic acid in five male subjects averaged 9.2%.

Leyton *et al.* (17) fed rats emulsions of carboxyl-labeled lauric, myristic, palmitic, stearic, oleic, linoleic, α-linolenic, dihomo α-linolenic or arachidonic acids. Recovery of $^{14}CO_2$ from rats given labeled stearic acid was 56% lower than from those given oleic acid and 22% lower than from rats given palmitic acid. The saturated fatty acids were incorporated uniformly into liver triglycerides but were incorporated more rapidly into phospholipids as their chain length increased. Incorporation of stearic acid into phosphatidyl choline compared to that of palmitic acid and was higher than for lauric or myristic acid.

Incorporation of stearic acid into phosphatidyl inositol or phosphatidyl ethanolamine was higher than for the other fatty acids. Wang and Koo obtained lymph chylomicrons containing carbon-labeled stearic, myristic or linoleic acids (18) or stearic, palmitic or oleic acids (19) from rats and injected them intravenously into recipient rats. Stearic acid was removed from the plasma more slowly than the other fatty acids and was incorporated more slowly into hepatic triglycerides. The percentage of [^{14}C]-stearate appearing in liver phospholipids was higher than for the other fatty acids.

Several investigators have studied the effects of stearic acid-rich fats *vis-à-vis* cholesterol and lipoprotein metabolism. Feldman *et al.* (20) studied cholesterol absorption and turnover in rats fed tristearin, triolein or safflower oil. Diets containing 0.025% cholesterol and 10% of the test fat were fed for 6 weeks. Tristearin-fed rats gained significantly less weight than did the control or other test groups. Rats fed tristearin absorbed significantly less cholesterol and removal of cholesterol from the plasma was most rapid in this group. They also exhibited the lowest cholesterol levels, 58 ± 2 mg/dl compared to 77 ± 1 mg/dl in triolein-fed rats and 80 ± 1 mg/dl in those fed safflower oil. Sterol turnover and synthesis were higher in rats fed tristearin. There were no differences in neutral or acidic steroid excretion.

In a second study (21), trilaurin, trimyristin, tripalmitin and tristearin were compared. Cholesterol was administered by gavage rather than by diet, but again absorption was lowest and turnover highest in the tristearin-fed group. Waterman *et al.* (22) compared the effects of tallow and safflower oil on growth, plasma lipids and lipogenesis in rats, pigs and chicks. Growth was similar in pigs fed the two fats; plasma cholesterol levels were also similar (116 ± 6 mg/dl in the safflower oil group and 125 ± 4 mg/dl in the tallow group) and plasma triglyceride levels were significantly lower in the tallow-fed group (37 ± 6 mg/dl) than in the safflower oil group (56 ± 6 mg/dl). Weight gains in rats or chicks fed either fat were similar. There were no significant differences in plasma cholesterol levels but rats fed tallow had significantly higher triglyceride levels than those fed safflower oil. In all three species, adipose tissue fatty acid synthesis was considerably lower in tallow-fed animals. Kritchevsky *et al.* (23) studied cholesterol metabolism in rats fed 14% cocoa butter, palm kernel oil, coconut oil or corn oil. Cholesterol absorption was lowest in rats fed cocoa butter and cholesterogenesis from either acetate or mevalonate was highest.

Monsma and Ney (24) fed rats diets containing 0.2% cholesterol and 16% fats containing increasing amounts of stearic acids. The fats were lard (15% stearic acid), beef tallow (19% stearic acid) and cocoa butter (35% stearic acid). The control fat was corn oil (2% stearic acid). They found reduced absorption of stearic acid as its level in the diet increased. In a subsequent study (25), rats were fed 0.035% cholesterol and 16% corn oil (2% stearic acid), butterfat (18% stearic acid), beef tallow (16% stearic acid), palm oil (4% stearic acid) and coconut oil (3% stearic acid) for 6 weeks. Data are presented in Table 5.5

Rats fed butterfat exhibited significantly lower cholesterol levels than those fed tallow or palm oil. Rats fed beef tallow had the highest triglyceride levels. Of interest is the observation that the highest amounts of plasma cholesteryl ester were found in rats fed the fats containing the highest levels of stearic acid. It would be interesting to know if the fatty acid composition of the plasma cholesteryl esters varied with dietary fat. High-density lipoproteins (HDL) from

Table 5.5 Plasma lipids of rats (eight to ten per group) fed fats of different stearic acid content for 6 weeks.

Fat (% 18:0)	Plasma lipid (mg/dl)		
	Cholesterol[1]	E/F[1,2]	Triglyceride[1]
Corn oil (1.8)	95 ± 4 abc[3]	4.2 ± 0.1 b	67 ± 4 c
Butter fat (17.6)	84 ± 4 c	5.6 ± 0.2 bc	81 ± 7 bc
Beef tallow (15.8)	97 ± 4 a	5.1 ± 0.5 a	101 ± 9 a
Palm oil (3.7)	101 ± 4 a	3.2 ± 0.3 c	90 ± 5 ab
Coconut oil (2.8)	89 ± 3 bc	3.5 ± 2 bc	90 ± 3 ab

Source: after Ney *et al.* (25).
[1] Means ± SE.
[2] E/F = esterified/free cholesterol.
[3] Means in vertical column bearing different letters are significantly ($P < 0.05$) different.

rats fed the saturated fats contained more free cholesterol, triglycerides and apolipoprotein E (apoE) than that from the corn oil-fed rats.

Imaizumi *et al.* (26) fed hamsters diets containing 8% fat with or without 2% cholesterol. The fats were prepared by co-randomization of soybean oil, high-oleic acid safflower oil, and trilaurin, trimyristin, tripalmitin and tristearin. The fats contained 28.4 ± 0.6% oleic acid and 19.5 ± 0.6% linoleic acid. The specific fats contained 51.5% lauric acid, 48.6% myristic acid, 48.9% palmitic acid and 40.1% stearic acid. On the cholesterol-free diets, the rats fed the stearic acid-rich fat had significantly lower plasma cholesterol levels. When the diet contained 0.2% cholesterol, cholesterol levels were similar in all groups. There were no differences among plasma triglyceride levels regardless of dietary cholesterol content. Hamsters fed the stearic acid-rich, cholesterol-free diet exhibited the lowest apparent fat digestibility. Salter *et al.* (27) fed hamsters diets containing 0.005% cholesterol and 10, 15 or 20% fat. The fats were triolein or triolein plus equal portions of trimyristin, tripalmitin or tristearin. Diets containing tristearin did not increase the cholesterol content of any major lipoprotein fraction.

Atherogenicity of stearic acid-rich fats

Kritchevsky and Tepper (28) studied the effects of various saturated fats (6%) on atherogenesis in rabbits fed 2% corn oil. The fats compared were corn oil, palm oil, coconut oil and cocoa butter. Cocoa butter was 17% less atherogenic than either palm or coconut oil (Table 5.6).

Connor *et al.* (29) fed rabbits diets in which 18.5% cocoa butter, coconut oil or hydrogenated vegetable oil was added to chow and fed for 4 (hydrogenated vegetable oil), 10 (coconut oil) or 12 (cocoa butter) months and found no gross aortic atherosclerosis. Addition of saturated fat to a commercial diet will not lead to atherosclerosis in rabbits even when added for a year (30). This is probably due

Table 5.6 Influence of cocoa butter on atherosclerosis in rabbits fed 2% cholesterol.

Dietary fat (6%)[1]	Survival	Serum cholesterol (mg/dl)	Average atherosclerosis[2]	
			Aortic arch	Thoracic arch
Coconut oil	21/22	1652 ± 158	2.24 ± 0.24	1.45 ± 0.21
Cocoa butter	21/22	1556 ± 130	1.83 ± 0.23	1.17 ± 0.18
Corn oil	21/22	1583 ± 178	1.36 ± 0.21	1.05 ± 0.18

Source: after Kritchevsky and Tepper (28).
[1] Rabbits fed 2% cholesterol and 6% fat for 2 months.
[2] Aortas graded visually on a 0–4 scale.

to the high fiber content of the diet. However, addition of saturated fat to a semi-purified diet containing casein, sucrose and cellulose provides an atherogenic diet (31, 32). When such a diet containing corn oil, palm kernel oil, cocoa butter or coconut oil was fed to rabbits for 9 months, the diet containing cocoa butter was significantly less atherogenic than diets containing palm kernel oil or coconut oil (33) (Table 5.7).

Table 5.7 Influence of cholesterol-free, semi-purified diet containing 14% fat on atherosclerosis in rabbits.

Dietary fat (survival)[1]	Serum cholesterol (mg/dl)	Average atherosclerosis[2]	
		Aortic arch	Thoracic arch
Coconut oil (5/12)	475 ± 104 ab[2]	2.10 ± 0.51 ab	1.10 ± 0.24 ab
Cocoa butter (5/12)	220 ± 32 ac	0.18 ± 0.21 ac	0.25 ± 0.09 ac
Corn oil (5/12)	62 ± 6 bc	0.21 ± 0.07 ab	0.08 ± 0.06 bc

Source: after Kritchevsky *et al.* (33).
[1] Diets contained 40% sucrose, 25% casein, 14% fat and 15% cellulose. Fed for 9 months.
[2] Aortas graded visually on a 0–4 scale.
[3] Values in vertical column bearing same letter are significantly different ($P < 0.05$).

In an earlier study of specific fatty acid effects on experimental atherosclerosis, rabbits were fed 2% cholesterol and 6% of six different fats. The fats were corn oil, randomized corn oil, or corn oil co-randomized with lauric, myristic, palmitic or stearic acid (34). The four co-randomized fats were designed to provide an excess of one specific fatty acid. As Table 5.8 shows, there were no significant differences in severity of atherosclerosis but the atherogenicity of the stearic acid-rich fat was within 5% of that of corn oil, whereas the other saturated fats were 10–20% more atherogenic. The most atherogenic fat was the one rich in palmitic acid.

Conclusion

The data show consistently that fats high in stearic acid are consistently less cholesterolemic and atherogenic than are fats containing the other common long-

Table 5.8 Influence of fats enriched in one specific fatty acid on atherosclerosis in rabbits.

Dietary fat[1]	Survival[2]	Serum cholesterol (mg/dl)[2]	Average atherosclerosis[2,3]	
			Aortic arch	Thoracic arch
Corn oil	43/46	2633 ± 363	1.65 ± 0.13	1.10 ± 0.11
Randomized corn oil	42/46	2022 ± 310	1.59 ± 0.13	1.08 ± 0.10
19% lauric[4]	41/46	2003 ± 284	1.98 ± 0.15	1.15 ± 0.13
18.2% myristic[4]	34/46	1883 ± 264	1.82 ± 0.12	1.24 ± 0.11
30% palmitic[4]	42/46	2080 ± 146	2.07 ± 0.14	1.30 ± 0.12
23.4% stearic[4]	40/46	1977 ± 204	1.74 ± 0.14	1.08 ± 0.09

Source: after Kritchevsky *et al.* (34).
[1] Diets contained 2% cholesterol, 6% fat, fed for 2 months.
[2] Values represent average of five experiments.
[3] Aortas graded visually on a 0–4 scale.
[4] Appropriate triglyceride randomized with corn oil.

chain fatty acids, namely lauric, myristic and palmitic. The data also suggest that absorbability of cocoa butter is reduced, probably due to its high stearic acid content.

Acknowledgment

Supported, in part, by a Research Career Award (HL00734) from the National Institutes of Health.

References

1. Keys, A., Anderson, J.T. and Grande, F. (1965) Serum cholesterol response to changes in the diet. IV. Particular saturated fatty acids in the diet. *Metabolism* **14**, 776–787.
2. Hegsted, D.M., McGandy, R.B., Myers, M.L. and Stare, F.J. (1965) Quantitative effects of dietary fat on serum cholesterol in man. *Am. J. Clin. Nutr.* **17**, 281–295.
3. Kritchevsky, D. (1988) Effects of triglyceride structure on lipid metabolism. *Nutr. Rev.* **46**, 177–181.
4. Small, D.M. (1991) The effects of glyceride structure on absorption and metabolism. *Annu. Rev. Nutr.* **11**, 413–434.
5. Arnschink, L. (1890) Versuche über die Resorption verschiedenen Fette aus dem Darmkanale. *Z. Biol.* **26**, 255–260.
6. Hoagland, R. and Snider, G.G. (1943) Digestibility of certain higher fatty acids and triglycerides. *J. Nutr.* **26**, 219–225.
7. Carroll, K.K. and Richards, J.F. (1958) Factors affecting digestibility of fatty acids in the rat. *J. Nutr.* **64**, 411–424.
8. Mattson, F.H., Nolen, G.A. and Webb, M.R. (1979) The absorbability by rats of various triglycerides of stearic and oleic acid and the effect of dietary calcium and magnesium. *J. Nutr.* **109**, 1682–1687.

9. Calloway, D.H., Kurtz, G.W., McMullen, J.J. and Thomas, L.V. (1956) The absorbability of natural and modified fats. *Food Res.* **21**, 621–629.
10. Mattson, F.H. (1959) The absorbability of stearic acid when fed as a simple or mixed triglyceride. *J. Nutr.* **69**, 338–342.
11. Bergstedt, S.E., Hayashi, H., Kritchevsky, D. and Tso, P. (1990) A comparison of absorption of glyceryl tristearate and glyceryl trioleate by rat small intestine. *Am. J. Physiol.* **259**, G386–G393.
12. Bergstedt, S.E., Bergstedt, J.L., Fujimoto, K., Mansbach, C., Kritchevsky, D. and Tso, P. (1991) Effects of glycerol tripalmitate and glycerol trioleate on intestinal absorption of glycerol tristearate. *Am. J. Physiol.* **261**, G239–G247.
13. Chen, I.S., Subramaniam, S., Vahouny, G.V., Cassidy, M.M., Ikeda, I. and Kritchevsky, D. (1989) A comparison of the digestion and absorption of cocoa butter and palm kernel oils and their effects on cholesterol absorption in rats. *J. Nutr.* **119**, 1569–1573.
14. Apgar, J.L., Shively, C.A. and Tarka, S.M., Jr (1987) Digestibility of cocoa butter and corn oil and their influence on fatty acid distribution in rats. *J. Nutr.* **117**, 660–665.
15. Elovson, J. (1965) Conversions of palmitic acid and stearic acid in the intact rat. *Biochim. Biophys. Acta* **106**, 291–303.
16. Emken, E.A. (1992) What is the metabolic fate of dietary long-chain fatty acids (especially stearic acid) in normal physiological states, and how might this relate to thrombosis? *Am. J. Clin. Nutr.* **56,** 798S.
17. Leyton, J., Drury, P.J. and Crawford, M.A. (1987) *In vivo* incorporation of labeled fatty acids in rat liver lipids after oral administration. *Lipids* **22**, 553–558.
18. Wang, S. and Koo, S.J. (1993) Plasma clearance and hepatic utilization of stearic, myristic, and linoleic acids introduced via chylomicrons in the rat. *Lipids* **28**, 697–703.
19. Wang, S. and Koo, S.J. (1993) Evidence for distinct metabolic utilization of stearic acid in comparison with palmitic and oleic acids in rats. *J. Nutr. Biochem* **4**, 594–601.
20. Feldman, E.B., Russell, B.S., Schnare, F.H., Miles, B.C., Doyle, E.A. and Moretti-Rojas, I. (1979) Effects of tristearin, triolein and safflower oil diets on cholesterol balance in rats. *J. Nutr.* **109**, 2226–2236.
21. Feldman, E.B., Russell, B.S., Schnare, F.H., Moretti-Rojas, I., Miles, B.C. and Doyle, E.A. (1979) Effects of diets of homogenous saturated triglyceride on cholesterol balance in rats. *J. Nutr.* **109**, 2237–2246.
22. Waterman, R.A., Romsos, D.R., Tsai, A.C., Miller, E.R. and Leveille, G.A. (1975) Influence of dietary safflower oil and tallow on growth, plasma lipids and lipogenesis in rats, pigs and chicks. *Proc. Soc. Exp. Biol. Med.* **150**, 347–351.
23. Kritchevsky, D., Tepper, S.A., Bises, G. and Klurfeld, D.M. (1983) Influence of cocoa butter on cholesterol metabolism in rats: comparison with corn oil, coconut oil and palm kernel oil. *Nutr. Res.* **3**, 229–236.
24. Monsma, C.C. and Ney, D.M. (1993) Interrelationship of stearic acid content and triacylglycerol composition of lard, beef tallow and cocoa butter in rats. *Lipids* **28**, 539–547.
25. Ney, D.M., Lai, H.-C., Lasekan, J.B. and Lefevre, M. (1991) Interrelationship of plasma triglycerides and HDL size and composition in rats fed different saturated dietary fats. *J. Nutr.* **121**, 1311–1322.
26. Imaizumi, K., Abe, K., Kuroiwa, C. and Sugano, M. (1993) Fats containing stearic acid

increase fecal neutral steroid excretion and catabolism of low density lipoproteins without affecting plasma cholesterol concentration in hamsters fed a cholesterol-containing diet. *J. Nutr.* **123**, 1693–1702.

27. Salter, A.M., Mangiapanc, E.H., Bennett, A.J., *et al.* (1998) The effect of different dietary fatty acids on lipoprotein metabolism: concentration-dependent effects of diets enriched in oleic, myristic, palmitic and stearic acids. *Br. J. Nutr.* **79**, 195–202.
28. Kritchevsky, D. and Tepper, S.A. (1965) Cholesterol vehicle in experimental atherosclerosis. VII. Influence of naturally occurring saturated fats. *Med. Pharmacol. Exp.* **12**, 315–320.
29. Connor, W.E., Rohwedder, J.J. and Armstrong, M.L. (1967) Relative failure of saturated fat in the diet to produce atherosclerosis in the rabbit. *Circ. Res.* **20**, 658–663.
30. Kritchevsky, D. and Tepper, S.A. (1964) Cholesterol vehicle in experimental atherosclerosis. VI. Long-term effects of fats and fatty acids in a cholesterol-free diet. *J. Atheroscler. Res.* **4**, 113–116.
31. Kritchevsky, D. and Tepper, S.A. (1965) Factors affecting atherosclerosis in rabbits fed cholesterol-free diet. *Life Sci.* **4**, 1467–1471.
32. Kritchevsky, D. and Tepper, S.A. (1968) Experimental atherosclerosis in rabbits fed cholesterol-free diets: influence of chow components. *J. Atheroscler. Res.* **8**, 357–369.
33. Kritchevsky, D., Tepper, S.A., Bises, G. and Klurfeld, D.M. (1982) Experimental atherosclerosis in rabbits fed cholesterol-free diets. 10. Cocoa butter and palm oil. *Atherosclerosis* **41**, 279–284.
34. Kritchevsky, D. and Tepper, S.A. (1967) Cholesterol vehicle in experimental atherosclerosis. X. Influence of specific saturated fatty acids. *Exp. Mol. Pathol.* **6**, 394–401.

Chapter 6

Cardiovascular Health: Role of Stearic Acid on Atherogenic and Thrombogenic Factors

Penny M. Kris-Etherton and Terry D. Etherton

Coronary heart disease (CHD) is the leading cause of morbidity and mortality in developed countries (1). CHD is a progressive disease characterized by atherosclerosis, which is a pathological condition resulting from the development of fibrous plaques or lesions in coronary arteries (1). The formation of these plaques results from complex biological processes initiated by chronic endothelial injury or damage (2). As atherosclerosis progresses, plaques increase in size, project into the arterial lumen, occlude the artery and block or reduce blood flow to the heart, brain and/or extremities, resulting in a heart attack, stroke or gangrene of the extremities. Although thought to begin in childhood, CHD usually is not manifested clinically until the fourth decade of life or later. Thus, CHD is a chronic disease that develops silently over decades.

CHD results from chronic exposure to both modifiable and non-modifiable risk factors (3). Modifiable risk factors include an elevated plasma total cholesterol level, specifically low-density lipoprotein (LDL) cholesterol, low levels of high-density lipoprotein (HDL) cholesterol as well as smoking, hypertension, obesity, physical inactivity and diabetes mellitus. In addition, elevated levels of some blood coagulation factors, fibrinogen and factor VII, and elevated plasma triglycerides and lipoprotein (a) (Lp(a)) levels also are considered to be modifiable risk factors for CHD. Furthermore, altered production of important bioactive compounds (e.g. prostaglandins, thromboxanes and leukotrienes) which affect platelet function, and in turn thrombosis tendency, also are thought to affect CHD risk. Non-modifiable risk factors include a strong family history and age (i.e. in men age 45 years and older, and in women due to menopausal status age 55 years and older, is a positive risk factor).

Diet plays a major role in the development and treatment of many modifiable CHD risk factors including an elevated total and LDL cholesterol level, hypertension, obesity and diabetes mellitus. In addition, diet also affects plasma triglycerides, HDL cholesterol and blood clotting tendency via effects on platelet function and hemostatic factors. Many of the effects attributable to diet reflect the type and amount of fat consumed. Much of the research conducted to date has focused on understanding the role of individual fatty acids on factors that affect atherosclerosis and thrombosis.

Specifically, numerous studies have evaluated the effects of fatty acids on plasma lipids, lipoproteins, platelet function and hemostatic functions. These studies have assessed the effects of both fatty acid classes as well as individual fatty acids, with much of the emphasis being on lipid and lipoprotein end points. Consequently, we have a much greater understanding of how dietary fat classes and fatty acids affect these parameters and, conversely, much less information about fat effects on thrombogenesis.

One of the major discoveries that has been made in the field is that fatty acids have remarkably diverse effects on the metabolic processes that modulate the development and progression of CHD. Whereas different metabolic effects of the fat classes have been identified, there can be marked variation in the biological effects of the individual fatty acids within a fat class. A notable example is stearic acid, which is a unique long-chain saturated fatty acid (SFA). Thus, the purpose of this paper is to discuss how stearic acid differs from the other long-chain SFAs and other fatty acids with respect to its effects on atherogenic and thrombogenic risk factors and, consequently, its role in the development of CHD.

Types of studies conducted to evaluate effects of stearic acid on CHD risk

Our knowledge of the effects of stearic acid on plasma lipids and lipoproteins, platelet function and hemostasis comes from epidemiological investigations, controlled clinical studies, animal studies and *in vitro* experimentation (e.g. cell culture, tissue perfusion and metabolic measurements made on cells isolated from tissues or biologic fluids). The research findings from these studies have been evaluated for the strength of an association between dietary stearic acid and CHD and/or an associated risk factor, consistency of the findings, specificity of results and biologic plausibility.

Epidemiological studies assess whether a relationship exists between parameters of interest (e.g. stearic acid and LDL cholesterol levels) but cannot establish causal relationships. Nonetheless, they are useful in developing hypotheses that can be tested experimentally. Controlled clinical studies are considered the gold standard for establishing cause and effect relationships; they are essential for providing information that is used to make dietary recommendations. Animal studies and *in vitro* experiments provide important information about the mechanisms of action of findings reported in controlled clinical studies.

Collectively, the different types of studies conducted to date have given us a good understanding of the effects of stearic acid on plasma lipids and lipoproteins. Our knowledge about the effect stearic acid has on platelet function and hemostasis is still growing. The results from a variety of experimental approaches provide a powerful means to evaluate the biological effects of stearic acid. This body of evidence is summarized in the subsequent sections.

Effects of stearic acid on plasma lipids and lipoproteins

Epidemiological studies

Historically, two types of epidemiological assessments have been conducted to evaluate diet effects on end points of interest, cross-population and within-population studies. In cross-population studies, because dietary habits and CHD risk factors can differ greatly between cultures, detection of differences in intake of key dietary variables is easier and the likelihood of finding significant associations increases (4). However, these groups also differ in many other ways that could possibly affect CHD risk and incidence. For example, some societies have a more relaxed lifestyle, less stress, more physical activity, a different quality of life, etc. Studies conducted within population groups control for important variables that could affect the end points of interest. A major limitation of within population studies is the relatively small variation in the end points of interest such as habitual diet and blood cholesterol levels. Nonetheless, the results from cross-population studies generally are consistent with the results of within-population studies.

The widely acclaimed Seven Countries Study (5) was the first epidemiological study to report highly significant relationships between diet and both the incidence of CHD and the level of serum cholesterol. Because this study was well controlled and conducted on a large, diverse sample of 12 000 men in 18 populations in seven countries (Finland, Greece, Italy, Japan, the Netherlands, the USA and Yugoslavia), it has provided very significant information about the relationship of diet to CHD and important risk factors. This study found that SFA intake (as a percentage of calories) was significantly correlated with the 5-year incidence of CHD, as well as serum cholesterol levels. Importantly, 80% of the variability in the level of serum cholesterol among the groups was due to the different amounts of SFA among the diets of the populations. Central to the objective of this paper is that the Seven Countries Study only dealt with fatty acid classes. Keys and his colleagues did not evaluate associations with individual fatty acids.

Interestingly, a recent study (6) has attempted to examine this association by determining the individual fatty acid composition of the diets of the study population (diets actually were prepared from food records collected at the time the study was conducted, and the fatty acid composition of the diets was determined by chemical analyses). As expected, saturated fat still was highly correlated with incidence of CHD and the level of serum cholesterol. The individual SFAs (lauric ($C_{12:0}$), myristic ($C_{14:0}$), palmitic ($C_{16:0}$), and stearic ($C_{18:0}$) acids) also were highly correlated with CHD risk (r = 0.81–0.86, $P < 0.001$); however, because the intake of individual SFA was highly correlated with total dietary SFA (r = 0.83–0.97), it was not possible to evaluate the independent effects of the individual SFA on CHD risk.

An earlier cross-population study corroborated the early Seven Countries findings that total dietary SFA (independent of total fat intake) increased the risk of CHD (7). Two more recent studies (8, 9) with data collected from 18 (8) or 40 countries (9) have also reported a significant relationship between SFA intake, cholesterol and monounsaturated fatty acid (MUFA) consumption with the incidence of CHD. However, the positive association of CHD with MUFA likely was confounded because intakes of SFA and MUFA were highly intercorrelated (r = 0.81) (8). After adjustment for SFA and cholesterol, MUFA and polyunsaturated fatty acid (PUFA) were significantly and negatively correlated (r = –0.33, $P < 0.05$) with CHD mortality (9).

Of interest are the recent studies that have reported an inverse association between PUFA intake (8, 9) and PUFA content of adipose tissue (10) and the incidence of CHD (8, 9). Thus, the epidemiological evidence from cross-population studies provides strong evidence that dietary SFAs are positively associated, whereas PUFA and possibly MUFA intakes are negatively associated with the incidence of CHD. Unfortunately these studies have provided little insight about associations between the intake of individual fatty acids (and specifically stearic acid) and the incidence of CHD.

Within-population studies showing a significant positive relationship of CHD risk with SFA (% of energy) include the Puerto Rico Heart Health Program (rural cohort only) (11), the Honolulu Heart Program, a study of men of Japanese ancestry (12), the Ireland–Boston Diet–Heart Study (13) and, more recently, the Belgium Study (14) and the Atherosclerosis Risk in Communities Study (15). In the latter study with over 13 000 black and white men, and women (ages 45 to 64 years) from four US communities, higher intakes of SFA (and cholesterol) were associated with a greater carotid artery wall thickness (a measure of coronary artery disease).

Studies not showing that SFAs are associated with CHD risk include the Israel Ischemic Heart Disease Study (16), the urban cohort of the Puerto Rico Heart Health Program (11), the Western Electric Study (17) and the Zutphen Study (18).

Interestingly, a strong inverse association between PUFA intake and CHD mortality was reported in three studies: the prospective Western Electric Study (17), the Honolulu Heart Program (12), and a longitudinal, within-population study of >21 000 men and women in Belgium (14).

In summary, the epidemiological evidence shows that dietary SFAs increase the risk of CHD. However, these studies shed little light about the role of individual SFAs. Because the epidemiological studies do not prove cause and effect relationships, it has been necessary to pursue other experimental paradigms to evaluate the effects of individual fatty acids.

Controlled clinical studies

The Seven Countries Study laid the groundwork for the remarkably productive era of clinical investigation that followed. Many well-controlled feeding studies have been conducted over the past 45 years that have advanced our understanding of the effects different fats, fat classes and individual fatty acids have on plasma lipids, lipoproteins and, in more recent years, measures of thrombogenesis. These studies have culminated in analyses that have summarized primarily the effects of fat classes and, to a lesser extent, individual fatty acids on plasma lipids and lipoproteins.

Most notable are the blood cholesterol predictive equations reported in 1965 by Keys *et al.* (19) and Hegsted *et al.* (20). More recently, blood cholesterol predictive equations have been developed to assess the effects of fat quality on plasma total *and* lipoprotein cholesterol levels (Table 6.1) (21, 22).

In addition, other modeling approaches have been used to summarize the literature and characterize the effects of individual fatty acids on plasma lipids and lipoproteins (Fig. 6.1).

Table 6.1 Predictive equations for estimating the changes in plasma cholesterol in response to changes in dietary fatty acids and cholesterol*.

Equation	Author
Original equations	
$\Delta TC = 1.2\,(2\Delta(C_{12:0} - C_{16:0})) - \Delta P) + 1.52\Delta Z$	Keys *et al.* (19)
$\Delta TC = 2.16\Delta S - 1.65\Delta P = 0.067\Delta C - 0.53$	Hegsted *et al.* (20)
Recent equations	
$\Delta TC = 1.51\Delta S - 0.12\Delta M - 0.60\Delta P$ $\Delta LDL\text{-}C = 1.28\Delta S - 0.12\Delta M - 0.60\Delta P$ $\Delta HDL\text{-}C = 0.47\Delta S + 0.34\Delta M + 0.28\Delta P$	Mensink and Katan (21)
$\Delta TC = 2.10\Delta S - 1.16\Delta P + 0.067\Delta C$ $\Delta LDL\text{-}C = 1.74\Delta S - 0.77\Delta P - 0.44\Delta C$ $\Delta HDL\text{-}C = 0.43\Delta S + 0.10\Delta M + 0.22\Delta P + 0.18\Delta C$	Hegsted *et al.* (23)
$\Delta TC = 2.02\Delta(C_{12:0} - C_{16:0}) - 0.03\Delta C_{18:0} - 0.48\Delta M - 0.96\Delta P$ $\Delta LDL\text{-}C = 1.46\Delta(C_{12:0} - C_{16:0}) + 0.07\Delta C_{18:0} - 0.69\Delta M - 0.96\Delta P$ $\Delta HDL\text{-}C = 0.62\Delta(C_{12:0} - C_{16:0}) - 0.06\Delta C_{18:0} + 0.39\Delta M + 0.24\Delta P$	Yu *et al.* (22)

*Where ΔTC = change in plasma total cholesterol in mg/dl; $\Delta LDL\text{-}C$ and $\Delta HDL\text{-}C$ = change in plasma LDL and HDL cholesterol in mg/dl; $\Delta C_{12:0}$, $\Delta C_{14:0}$, $\Delta C_{16:0}$, $\Delta C_{18:0}$ = change in percentage of daily calories from $C_{12:0}$, $C_{14:0}$, $C_{16:0}$; ΔS, ΔM, ΔP = change in percentage of daily calories from SFA, MUFA and PUFA; ΔZ = change in the square root of daily dietary cholesterol in mg/1000 calories; and ΔC = change in dietary cholesterol in mg/day.

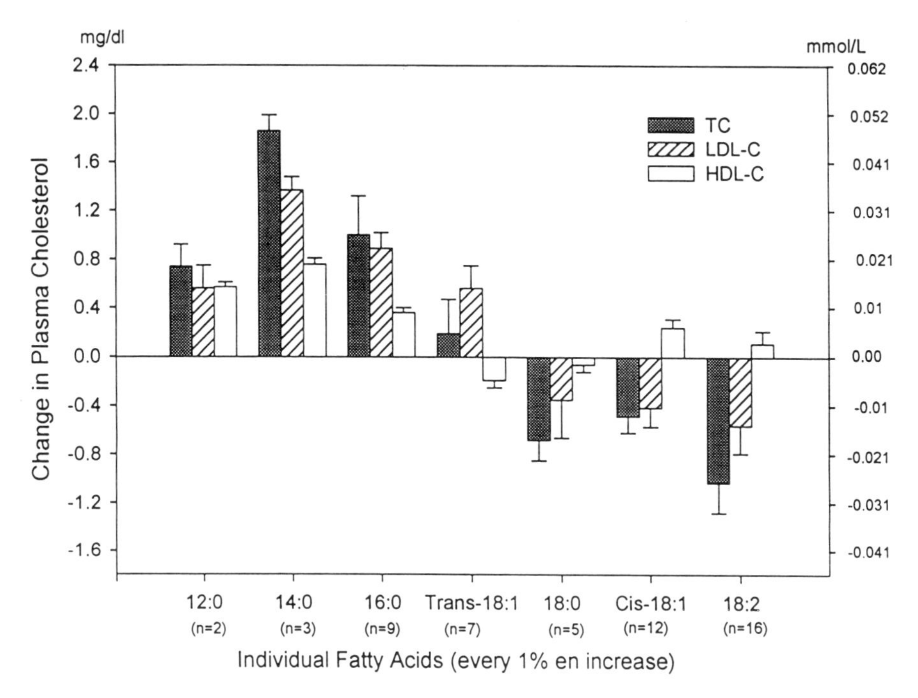

Fig. 6.1 Effects of individual fatty acids on serum total, LDL and HDL cholesterol when 1% of energy from carbohydrates in the diet is replaced by 1% of energy from the fatty acid in question.

The blood cholesterol predictive equations consistently show that SFAs raise blood cholesterol levels approximately twice as much as PUFAs lower them. MUFAs are either neutral or mildly hypocholesterolemic. In this paper, *neutral* is used to convey that blood cholesterol levels are neither increased nor decreased. As shown in the Yu *et al.* (22) equations, stearic acid is a unique SFA in that it elicits a neutral blood cholesterol response. In fact, both Keys *et al.* (19) and Hegsted *et al.* (20) did not include stearic acid in the grouping of SFA for the development of their predictive equations. Both investigators classified stearic acid as a neutral fatty acid (like oleic acid) with respect to its effect on plasma cholesterol level. (Of note is that Keys *et al.* (19) reported that the predictive equations 'failed' when they were used to estimate the plasma cholesterol response to diets high in stearic acid (e.g. cocoa butter)). Interestingly, the LDL cholesterol response parallels the total cholesterol response. Also of interest is that the regression equations demonstrate that all fatty acid classes raise HDL cholesterol levels. Stearic acid, however, has a neutral effect on HDL cholesterol. Thus, all other fatty acid classes raise HDL cholesterol compared with stearic acid.

Interest in the cholesterolemic effects of stearic acid was renewed in 1988 when Bonanome and Grundy (24) demonstrated that stearic acid was hypocholesterolemic compared to palmitic acid. This study raised important questions about the relative hypocholesterolemic effects of stearic acid compared to the other long-chain SFAs. Moreover, since MUFA and PUFA are hypocholesterolemic compared to SFA, questions were raised about stearic acid's relative effects. For example, does stearic acid elicit a neutral plasma cholesterol response similar to that of MUFA or does it have an independent cholesterol-lowering effect like PUFA?

Recent studies have compared stearic acid with other long-chain SFAs, oleic acid and linoleic acid (Table 6.2). These studies have shown that compared with the other long-chain SFAs, stearic acid (~15% of energy) significantly lowers plasma total (15–28%), LDL – (22–29%) and HDL (6–27%) cholesterol concentrations (24, 25, 26). One study showed that stearic acid and oleic acid elicit similar effects on total LDL cholesterol and HDL cholesterol (24). In contrast, when stearic acid is substituted for linoleic acid (18:2*n*-6), plasma total cholesterol and LDL cholesterol concentrations increase (3% and 6%, respectively) and HDL cholesterol concentrations decrease (4%).

Based on a limited data base, the small number of studies reported in Table 6.2 show that stearic acid's effects on total and LDL cholesterol levels are intermediate to those reported for SFA and PUFA; the effects are similar to those observed for MUFA, providing further evidence that stearic acid is a neutral fatty acid.

A summary of a rather limited number of studies that have examined the effects of individual fatty acids on plasma total and lipoprotein cholesterol levels is shown in Fig. 6.1. It is apparent that individual fatty acids elicit markedly different effects on plasma total and lipoprotein cholesterol levels. Among the SFAs, myristic acid is the most potent hypercholesterolemic long-chain SFA, followed by palmitic acid and then lauric acid. These SFAs also increase HDL cholesterol levels. In agreement with the regression equations, stearic acid is distinctly different, having essentially no effect on plasma total and lipoprotein cholesterol levels. Thus, in agreement with the regression equations, stearic acid appears to be a neutral long-chain SFA.

In summary, results from the recent studies conducted to examine individual fatty acid effects agree well with those derived from the blood cholesterol predictive equations. The studies conducted in just the last 10 years have advanced our understanding of the effects of individual fatty acids (in particular, the SFAs and specifically stearic acid) on both total and lipoprotein cholesterol levels. Further studies, however, are needed to confirm the results reported in Fig. 6.1 because of the very limited number of data points (i.e. small number of studies).

Mechanisms of action of stearic acid

The biological basis for the uniqueness of stearic acid is not entirely clear. It was once thought, based on animal data, that stearic acid was neutral because of incomplete absorption. However, human studies have shown either no difference

Table 6.2 Stearic acid substituted for other fatty acids: effects on plasma lipids[1].

Reference	18:0 substituted for another FA (% of energy)	High-18:0 diet								Comparison diet								Percentage change			
		12:0 (%)	14:0 (%)	16:0 (%)	18:0 (%)	18:1 (%)	18:2*n*-6 (%)	Fat (%)	Chol (mg/d)	12:0 (%)	14:0 (%)	16:0 (%)	18:0 (%)	18:1 (%)	18:2*n*-6 (%)	Fat (%)	Chol (mg/dl)	TC (%)	LDL-C (%)	HDL-C (%)	TG (%)
Tholstrup *et al.* (25)	12:0 + 14:0 (16%)	0.04	0.04	1.6	16.9	18.0	2.6	40	216	12.0	4.1	2.7	1.2	16.0	1.9	40	216	–28.1*	–29.0*	–27.4*	–5.4
Tholstrup *et al.* (26)	16:0 (15%)	0.04	0.04	1.6	16.9	18.0	2.6	40	216	0.3	0.5	17.3	1.9	15.3	4.2	40	212	–22.1*	–26.4*	–13.1*	–4.1
Bonanome and Grundy (24)	16:0 (15%)	NR	0.04	3.3	17.2	15.8	3.2	40	<100	NR	0.4	18.0	1.9	15.5	3.8	40	<100	–14.4*	–21.5*	–5.5	0.7
Bonanome and Grundy (24)	18:1 (16%)	NR	0.04	3.3	17.2	15.8	3.2	40	<100	NR	0.04	2.2	0.9	31.9	4.8	40	<100	–4.5	–7.5	–8.8	5.8
Zock and Katan (27)	18:2*n*-6 (9%)	0.5	1.0	5.7	11.8	15.4	3.9	40	388	0.7	0.9	5.8	2.8	14.7	12.0	40	402	3.2*	6.0*	–4.1*	9.5*

[1] FA fatty acid; Chol = cholesterol; TC = total cholesterol; LDL-C = low-density lipoprotein cholesterol; HDL-C = high-density lipoprotein cholesterol; TG = triglycerol; NR = nor reported.

* Significant difference between treatments ($P < 0.05$).

(28) or only slightly less absorption (29, 30) compared with other fatty acids. Even in the studies that show differences, greater than 90% of stearic acid is absorbed. Other evidence suggests that hypercholesterolemic SFAs suppress LDL receptor activity whereas stearic acid does not (31). It also has been postulated that stearic acid may be neutral because it is rapidly metabolized to oleic acid; but results obtained by stable-isotope tracer methods have shown that less than 20% of stearic acid is converted to oleic acid (32, 33).

Of interest are the results from recent studies which suggest that, compared to other SFAs (e.g. palmitic acid), stearic acid is preferentially incorporated into phospholipids instead of cholesteryl esters and triglycerides (32). This may account, in part, for its lack of a cholesterolemic effect. Other still unidentified mechanism(s) also may contribute to the uniqueness of stearic acid.

Effects of stearic acid on thrombosis

Epidemiological studies

There is epidemiological evidence that populations with a high intake of SFAs have increased platelet reactivity. This association, however, is based primarily on results from studies of *ex vivo* platelet aggregation in small groups of farmers living in Europe (34). Significant differences for *in vitro* platelet reactivity were observed among small groups of farmers from France, the UK, and Belgium. These results were attributed to differences in dietary SFA consumption; the stearic acid content of the diet, in particular, was shown to be most strongly associated with enhanced platelet reactivity. Of interest, however, is that in these studies the intake of SFA was high in all farmers, and the difference in stearic acid intakes between groups was small (<1% of energy).

A more recent and very comprehensive examination of the effects of diet and thrombosis/coagulation factors comes from the Atherosclerosis Risk in Communities (ARIC) study (35). Analysis of food intake data showed that, after controlling for common CHD risk factors, a high intake of total fat, cholesterol and SFA (from animal sources) was found to be associated with higher levels of fibrinogen and factor VII (which could increase thrombosis tendency). This study was, however, unable to identify a relationship of any individual SFA with coagulation factors.

Epidemiological studies have shown that a high intake of PUFA and omega-3 fatty acids, principally from fish, was associated with a reduced CHD risk and thrombosis tendency. In the ARIC study, a higher intake of fish was associated with lower levels of two coagulation factors (fibrinogen and factor VII).

Clinical and animal studies

Since the 1960s, scientists have been studying the potential relationship between dietary SFA and thrombosis tendency (36). These early studies, conducted primarily *in vitro* or with various animal models, demonstrated distinct effects of SFA on platelet aggregation and tendency for thrombosis. In these studies, fatty acids were added to whole blood or infused directly into the circulation of experimental animals. All SFAs shortened the thrombus formation time considerably compared with the unsaturated fatty acids; furthermore, the effects of individual SFAs ($C_{12:0}$–$C_{18:0}$) increased as chain length increased. Consequently, stearic acid has been considered to be the most thrombogenic SFA because it shortened thrombus time considerably when compared with other long-chain SFAs.

The techniques used in these studies have been criticized, however. For example, in some studies, fatty acids were added to whole blood *in vitro* or directly infused in animals. It is now clear that these effects are directly related to the insolubility of these fatty acids and thus the ensuing effects on blood clotting could not possibly be representative of what occurs *in vivo*. In support of this, most well-controlled human studies have failed to replicate the results of these early animal studies.

Few well-controlled studies have investigated the effects of total dietary SFA on platelet aggregation while keeping PUFA and total fat constant, and the results of these are equivocal; two studies observed no adverse effects of a diet rich in total SFA (20% of energy) on *in vitro* platelet aggregation compared with MUFA or PUFA, and one actually showed a beneficial effect (reduced sensitivity to aggregating compounds) when subjects consumed SFA-rich diets (37, 38).

In well-controlled studies that have examined the effects of PUFA and MUFA while maintaining the levels of SFA and total fat, the effects on platelet aggregation are variable. Some studies have demonstrated a reduction in *in vitro* platelet aggregation, whereas others have shown an increased tendency for blood clotting.

Literally hundreds of studies have been published on the antithrombotic effects of fish oil or long-chain omega-3 highly unsaturated PUFA. These have been reviewed in detail by Nordoy and Goodnight (36). In general, subjects given omega-3 PUFA consistently show a mild prolongation of the bleeding time and decreased reactivity to platelet aggregation. Other measures of platelet reactivity, such as platelet adhesion to surfaces, are also reduced.

Dietary long-chain omega-3 highly unsaturated PUFA have also been shown to affect thrombosis by altering the production of thromboxane, by replacing arachidonic acid in membranes and thereby reducing the precursor available for thromboxane synthesis. Thus, the overall result of long-chain omega-3 highly unsaturated PUFA consumption is a shift in eicosanoid balance with favorable effects on cardiovascular hemostasis.

Some studies have examined the effects of the plant source of omega-3 fatty acid, α-linolenic acid. No consistent or significant effects of diets rich in oils containing linolenic acid ($C_{18:3}$) such as canola or flaxseed oils have been found either on bleeding time or platelet function (36).

To gain a better understanding of the effects of stearic acid on platelet function *in vivo*, Mustad *et al.* examined the effects of diets rich in stearic acid on thrombosis tendency in healthy young men (39). They assessed thrombosis tendency by measuring the *in vivo* production of thromboxane A_2 (TXA_2; a proaggregatory eicosanoid). They measured the excretion of its stable metabolite, thromboxane B_2 (TXB_2), in urine and also the stable metabolite of prostaglandin I_2 (PGI_2; the antiaggregating eicosanoid), 6-keto-prostaglandin F_1 (PGF_1), to assess the effect of the diets on the ratio of TXA_2:PGI_2 production, a measure of hemostasis.

In this study (39), diets high in stearic acid (cocoa butter and milk chocolate, the richest food source of this fatty acid) were compared to one high in lauric acid and myristic acid (dairy butter). The source of stearic acid in the experimental diets was cocoa butter and milk chocolate. A mix of cocoa butter : butter (4:1) similar to that of milk chocolate was used to control for any putative thrombogenic factors, other than fatty acids, that may be present in chocolate.

This study reported that there was no effect of a diet (fed for 25 days) containing large quantities of stearic acid (~30 g/day) on the excretion of TXA_2 or PGI_2 metabolites. Thus, it was concluded that the diets rich in stearic acid do not increase the thrombosis tendency compared to either a typical American diet or a diet rich in lauric and myristic acids.

The most notable change in the fatty acid composition of the platelet phospholipids was in the TXA_2 precursor, arachidonic acid, which increased in the phosphatidylethanolamine subclass in subjects on the butter diet and decreased when subjects consumed the cocoa butter diet. Others also have reported an increase in membrane arachidonic acid content when diets rich in butter were consumed; these diets have also been associated with enhanced thrombosis tendency (40). Consequently, these changes in arachidonic acid suggest that the diets rich in lauric and myristic acids may increase and diets rich in stearic acid may decrease the potential for TXA_2 production in platelets stimulated *in vivo*.

A recent study conducted by Schoene *et al.* (41) showed that stearic acid may have a more favorable effect on platelet physiology than other long-chain SFAs. In this study, subjects who consumed diets high in either stearic acid or palmitic acid had similar bleeding times (an estimate of clotting time). However, the mean platelet volume (MPV) after the high-stearic acid diet was significantly smaller (MPV = 7.2 ± 0.23 fl) compared to the diets high in palmitic acid (MPV = 8.34 ± 0.2 fl), suggesting that platelets from subjects consuming stearic acid were actually less sensitive to *in vivo* activation. Thus, while the full significance of the effects on platelet physiology is not yet known, these data further suggest that dietary stearic acid is not a thrombogenic SFA.

Only a small number of studies have investigated the effects of stearic acid on measures of hemostasis. These studies have also shown mixed results. For example, one study (27) found diets rich in stearic acid (50 g/day) had lower factor VII coagulation activity (which would reduce thrombotic tendency) when compared with either their habitual diet or diets rich in other long-chain SFAs. In contrast, Mitropoulos *et al.* (42) reported that individuals self-selecting diets rich in total fat (60% of energy) and stearic acid (7%) had increased factor VII levels when compared with individuals consuming a diet rich in PUFA or one much lower in fat (15% of energy). However, in this study, the high-SFA diet also was higher in palmitic, myristic and lauric acids compared with the other diets. Consequently, it is difficult to attribute any effect on factor VII activity specifically to stearic acid.

More recently, the Dietary Effects on Lipoproteins and Thrombogenic Activity (DELTA) Study, a multicenter study funded by the National Institutes for Health/National Heart, Lung and Blood Institute (NIH/NHLBI) and designed to evaluate diet effects on lipids, lipoproteins and hemostatic factors reported that reductions in dietary SFA from 17% of energy on an average US diet to 10% of energy on a Step 1 diet and 6% of energy on a low-saturated fat diet decreased factor VII levels by 1.6% and 2.5%, respectively (43). Thus, a decrease in dietary SFA results in a modest decrease in factor VII levels. The contribution of individual long-chain SFAs including stearic acid is unclear. Surprisingly, fibrinogen levels increased modestly (2.5%) in subjects fed the low saturated fat diet versus the average US diet. (This increase would be expected to increase the risk of thrombosis.) The significance of these small changes in hemostatic factors in response to changes in diet remains to be established.

Food source and consumption of stearic acid

The major long-chain SFAs in the diet are lauric acid, myristic acid, palmitic acid and stearic acid. Based on the 1995 Continuing Survey of Food Intake by Individuals (CSFII) survey, intake of stearic acid was 2.9% of total energy. Meat, poultry, fish, grain products and full-fat dairy products are major food sources of stearic acid. Presently, approximately 0.2 g of stearic acid is provided by sugars and sweets in the US diet (44).

Common cooking oils contain relatively small amounts of stearic acid although hydrogenation of vegetable oils for the production of shortening can increase this amount. Cocoa butter contributes proportionally more stearic acid than any other naturally occurring fat. However, based on the CSFII data for sugars and sweets consumption, cocoa butter is not a major food source of stearic acid.

Changes in the consumption of stearic acid in the US diet will reflect changes both in the amount and type of fat consumed. The prevailing trend for total fat and saturated fat intake to decrease in recent years has been associated with a small decrease in the consumption of stearic acid. Increased consumption of fat sources high in stearic acid (principally synthesized fats and conceivably seed oils from transgenic plants high in stearic acid) could impact significantly on stearic acid intake. Therefore it is not unreasonable to presume that stearic acid intake could increase significantly in the future; however, this increase will probably be the result of an increase in the consumption of 'designer' fats and oils.

Summary

Stearic acid has come to assume a unique position in the SFA class. Its uniqueness relates to it having a neutral cholesterolemic effect unlike the other long-chain SFAs. It does not affect total or lipoprotein cholesterol levels even when fed at very high levels. A definitive explanation for these distinct biological effects remains to be determined.

With respect to thrombosis, stearic acid does not adversely affect platelet function and appears to play no role in thrombogenic events that contribute to the development of CHD. Moreover, there are no convincing data that show a specific effect of stearic acid on any of the hemostatic factors measured to date. However, because the data base is quite small, more studies are needed to gain a better understanding of the role of all individual fatty acids, including stearic acid, on thrombogenesis. Despite this, it is clear from a substantive data base that stearic acid is not an atherogenic fatty acid and, based on the limited data available, does not adversely affect thrombosis.

References

1. AHA (1998) *Heart and Stroke Facts: 1998 Statistical Supplement.* American Heart Association, Dallas.
2. Ross, R. (1993) The pathogenesis of atherosclerosis: a perspective for the 1990s. *Nature* **362**, 801–809.
3. NCEP (1993) *National Cholesterol Education Program. Second Report of the Expert Panel on Detection, Evaluation, and Treatment of High Blood Cholesterol in Adults.* National Institutes of Health, Bethesda.
4. Willett, W.C. (1990) Nutritional epidemiology. In *Monographs in Epidemiology and Biostatistics*, vol. 15. Oxford University Press, New York.
5. Keys, A. (1970) Coronary heart disease in seven countries. *Circulation* **41** (Suppl. 1), 1–221.
6. Kromhout, D., Menotti, A., Bloemberg, B., *et al.* (1995) Dietary saturated and *trans* fatty acid and cholesterol and 25-year mortality from coronary heart disease: the Seven Countries Study. *Prev. Med.* **24**, 308–315.
7. Kato, H., Tillotson, J., Nichaman, J.S., Rhoads, G.G. and Hamilton, H.B. (1973) Epidemiologic studies of coronary heart disease and stroke in Japanese men living in Japan, Hawaii and California. *Am. J. Epidemiol.* **97**, 372–385.
8. Hegsted, D.M. and Ausman, L.M. (1988) Diet, alcohol and coronary heart disease in men. *J. Nutr.* **118**, 1184–1189.
9. Artaud-Wild, S.M., Connor, S.L., Sexton, G. and Connor, W.E. (1993) Differences in coronary mortality can be explained by differences in cholesterol and saturated fat

intakes in 40 countries but not in France and Finland. A paradox. *Circulation* **88**, 2771–2779.
10. Riemersma, R.A., Wood, D.A., Butler, S., *et al.* (1986) Linoleic acid content in adipose tissue and coronary heart disease. *Br. Med. J. (Clin. Res. Ed.)* **292**, 1423–1427.
11. Garcia-Palmieri, M.R., Sorlie, P., Tillotson, J., Costas, R., Cordero, E., Rodrigues, M. (1980) Relationship of dietary intake to subsequent coronary heart disease incidence: the Puerto Rico Heart Health Program. *Am. J. Clin. Nutr.* **33**, 1818–1827.
12. McGee, D.L., Reed, D.M., Yano, K., Kagan, A., Tillotson, J. (1984) Ten-year incidence of coronary heart disease in the Honolulu Heart Program. Relationship to nutrient intake. *Am. J. Epidemiol.* **119**, 667–676.
13. Kushi, L.H., Lew, R.A., Stare, F.J., *et al.* (1985) Diet and 20-year mortality from coronary heart disease. The Ireland–Boston Diet–Heart Study. *N. Engl. J. Med.* **312**, 811–818.
14. Joossens, J.V., Geboers, J. and Kesteloot, J. (1989) Nutrition and cardiovascular mortality in Belgium. *Acta Cardiol.* **44**, 157–182.
15. Tell, G.S., Evans, G.W., Folsom, A.R., Shimakawa, T., Carpenter, M.A. and Heiss, G. (1994) Dietary fat intake and carotid artery wall thickness. The Atherosclerosis Risk in Communities (ARIC) Study. *Am. J. Epidemiol.* **139**, 979–989.
16. Medalie, J.H., Kahn, H.A., Neufeld, N.H., Riss, E. and Golbourt, U. (1973) Five-year myocardial infarction incidence. II. Association of single variables to age and birthplace. *J. Chronic Dis.* **26**, 325–349.
17. Schekelle, R.B., Shryock, A.M., Paul, O., *et al.* (1981) Diet, serum cholesterol, and death from coronary heart disease. The Western Electric Study. *N. Engl. J. Med.* **304,** 65–70.
18. Kromhout, D. and de Lezenne Coulander, C. (1984) Diet, prevalence and 10-year mortality from coronary heart disease in 871 middle-aged men. The Zutphen Study. *Am. J. Epidemiol.* **119**, 733–741.
19. Keys, A., Anderson, J.T. and Grande, F. (1965) Serum cholesterol response to changes in the diet: IV. Particular saturated fatty acids in the diet. *Metabolism* **14**, 776–787.
20. Hegsted, D.M., McGandy, R.B., Myers, M.L. and Stare, F.J. (1965) Quantitative effects of dietary fat on serum cholesterol in man. *Am. J. Clin. Nutr.* **17**, 281–295.
21. Mensink, R.P. and Katan, M.B. (1992) Effect of dietary fatty acids on serum lipids and lipoproteins. A meta-analysis of 27 trials. *Arterioscler. Thromb.* **12**, 911–919.
22. Yu, S., Derr, J., Etherton, T.D. and Kris-Etherton, P.M. (1995) Plasma cholesterol-predictive equations demonstrate that stearic acid is neutral and monounsaturated fatty acids are hypocholesterolemic. *Am. J. Clin. Nutr.* **61**, 1129–1139.
23. Hegsted, M.M., Ausman, L.M., Johnson, J.A. and Dallal, G.E. (1993) Dietary fat and serum lipids: an evaluation of the experimental data. *Am. J. Clin. Nutr.* **57**, 875–883.
24. Bonanome, A. and Grundy, S.M. (1988) Effect of dietary stearic acid on plasma cholesterol and lipoprotein levels. *N. Engl. J. Med.* **318**, 1244–1248.
25. Tholstrup, T., Marckmann, P., Jespersen, J. and Sandström, B. (1994) Fat high in stearic acid favorably affects blood lipids and factor VII coagulant activity in comparison with fats high in palmitic acid or high in myristic and lauric acids. *Am. J. Clin. Nutr.* **59**, 371–377.
26. Tholstrup, T., Marckmann, P., Jespersen, J., Vessby, B., Jart, A. and Sandström, B. (1994) Effect on blood lipids, coagulation, and fibrinolysis of a fat high in myristic acid and a fat high in palmitic acid. *Am. J. Clin. Nutr.* **60**, 919–925.

27. Zock, P. and Katan, M. (1992) Hydrogenation alternatives: effects of *trans* fatty acids and stearic acid versus linoleic acid on serum lipids and lipoproteins in humans. *J. Lipid Res.* **33**, 399–410.
28. Bonanome, A. and Grundy, S.M. (1989) Intestinal absorption of stearic acid after consumption of high-fat meals in humans. *J. Nutr.* **119**, 1556–1560.
29. Mitchell, D.C., McMahon, K.E., Shively, C.A., Apgar, J.L. and Kris-Etherton, P.M. (1989) Digestibility of cocoa butter and corn oil in human subjects: a preliminary study. *Am. J. Clin. Nutr.* **50**, 983–986.
30. Dougherty, R.M., Allman, M.A. and Iacono, J.M. (1995) Effects of diets containing high or low amounts of stearic acid on plasma lipoprotein fractions and fecal fatty acid excretion of men. *Am. J. Clin. Nutr.* **61**, 1120–1128.
31. Woollett, L.A., Spady, D.K. and Dietschy, J.M. (1992) Regulatory effects of the saturated fatty acids 6:0 through 18:0 on hepatic low-density lipoprotein receptor activity in the hamster. *J. Clin. Invest.* **89**, 1133–1141.
32. Emken, E.A. (1994) Metabolism of dietary stearic acid relative to other fatty acids in human subjects. *Am J. Clin. Nutr.* **60**, 1023S–1028S.
33. Emken, E.A., Adlof, R.O., Rohwedder, W.K. and Gulley, R.M. (1993) Influence of linoleic acid on desaturation and uptake of deuterium-labeled palmitic and stearic acids in humans. *Biochim. Biophys. Acta* **1170**, 173–181.
34. Renaud, S., Godsey, F., Dumont E., Thevenon, C., Ortchanian E. and Martin, J.L. (1986) Influence of long-term diet modifications on platelet function and composition in Moselle farmers. *Am. J. Clin. Nutr.* **43**, 136–150.
35. Shahar, E., Folsom, A.R., Ulu, K.K., *et al.* for the ARIC Investigators. (1993) Associations of fish intake and dietary omega-3 polyunsaturated fatty acids with a hypocoagulable profile; the Atherosclerosis Risk in Communities (ARIC) Study. *Arter. Thromb.* **13**, 1205–1212.
36. Nordoy, A. and Goodnight, S.H. (1990) Dietary lipids and thrombosis. Relationships to atherosclerosis. *Atherosclerosis* **10**, 149–163.
37. Foley, M., Ball, M., Chisholm, A., Duncan, A., Spears, G. and Mann, J. (1992) Should mono- or polyunsaturated fats replace saturated fat in the diet? *Eur. J. Clin. Nutr.* **46**, 429–436.
38. Kwon, J.-S., Snook, J.T., Wardlaw, G.M. and Hwang, D.H. (1991) Effects of diets high in saturated fatty acids, canola oil, or safflower oil on platelet function, thromboxane B_2 formation, and fatty acid composition of platelet phospholipids. *Am. J. Clin. Nutr.* **54**, 351–358.
39. Mustad, V.A., Kris-Etherton, P.M., Derr, J., Reddy, C.C. and Pearson, T.A. (1993) Comparison of diets rich in stearic acid versus myristic acid and lauric acid on platelet fatty acids and excretion of TXA_2 and PGI_2 metabolites in healthy young men. *Metabolism* **42**, 463–469.
40. Galli, C., Agradi, E., Petroni, A., *et al.* (1981) Differential effects of dietary fatty acids on the accumulation of arachidonic acid and its metabolic conversions through the cyclooxygenase and lipoxygenase in platelets and vascular tissue. *Lipids* **16**, 165–172.
41. Schoene, N., Allman, M.A., Dougherty, R.M., Denvir, E. and Iacono, J.M. (1993) Diverse effects of dietary stearic and palmitic acids on platelet morphology. In *Essential Fatty Acids and Eicosanoids* (Ed. by Gibson, R. and Sinclair, A.), pp. 290–291. AOCS, Champaign, IL.

42. Mitropoulos, K.A., Miller, G.J., Martin, J.C., Reeves, B.E.A. and Cooper, J. (1994) Dietary fat induces changes in factor VII coagulant activity through effects on plasma free stearic acid concentration. *Arter. Thromb.* **14**, 214–222.
43. Elmer, P.J., Stewart, P., Lefevre, M., *et al.* for the DELTA Investigators. (submitted) Effect of reducing dietary saturated fat on plasma hemostatic factors fibrinogen, factor VII and PAI-1: the DELTA study. *Circulation.*
44. Agricultural Research Service, Beltsville Human Nutrition Research Center, and Food Services Research Group (1995) *USDA Continuing Survey of Food Intakes by Individuals (CSFII).* US Department of Agriculture, Hyattsville, MD.

Chapter 7

Carbohydrate and Protein

Pierre Würsch and Paul-Andre Finot

The importance of carbohydrate and protein to the diet is fundamental to sustained good health. The energy supplied to the body in the form of carbohydrate is often considered more beneficial than that obtained from fat because carbohydrate, particularly complex carbohydrate, often carries with it greater amounts of other nutrients such as fibre. It is also absorbed more quickly by the body, although the presence of fat, such as in chocolate, tempers the absorption and the resultant glucose response, spreading it over a longer time interval.

Cocoa contains carbohydrate, more so as starch than sugars, but it is more common to consider the contribution from sugars mixed with cocoa during the manufacture of chocolate.

Proteins are present only in small amounts in cocoa. Accordingly, it is for the sensory experience that cocoa is consumed and not for the protein contribution. Nevertheless, since cocoa is often consumed as milk chocolate, the protein contribution will increase with milk protein, and the milk solids improve the taste of a nutritious food.

Carbohydrates

Sugars

Cotyledons of fresh cocoa beans contain only 2–4% of free sugars (1–3), beside traces of others sugars and sugar alcohols, such as galactose, raffinose, stachyose, melibiose, sorbose, mannitol, inositol, etc. (1,3). The final content of these sugars varies considerably in fermented beans of various origins, most likely owing to the type and extent of fermentation. Sucrose in well-fermented beans can decrease until near zero, whereas fructose and glucose increase correspondingly.

By contrast, traditionally poorly fermented cocoa contains about 1% sucrose (3, 4) as shown in Table 7.1. Invertase was found only in the testae of unfermented beans, but data show that sucrose is obviously hydrolysed during fermentation into glucose and fructose, but part of them might be metabolized.

Table 7.1 Changes of sugar content in shell-free cocoa bean during fermentation.

Days of fermentation	Sucrose	Fructose	Glucose	Total
Berbert (3)				
0	2.48	0.09	0.07	2.74*
2	1.22	0.16	0.04	1.49*
4	0.07	0.54	0.49	1.20*
6	0.01	0.54	0.12	0.80*
Bracco *et al.* (4)				
0	1.65	0.1	0.30	2.05
2	1.61	0.12	0.31	2.04
4	1.16	0.40	0.67	2.23
6	0.7	0.39	0.64	1.73

* Totals include other minor sugars not designated.

Reineccius *et al.* (2) suggest that the sugars are absorbed from the pulp during fermentation. During roasting, most of the reducing sugars disappear and non-reducing sugars decrease as well (5).

Starch

Schmieder and Keeney (6) reported a mean value of 5.30% starch (4.5–7.0%) for 12 lots of cocoa beans (including shell) representing major geographic regions of production (Table 7.2). The mean particle size of starch was of 4.6 μm with a range of 2.0–12.5 μm, with 36% amylose. Two variables, which might affect the final starch content of cocoa beans, are fruit ripeness at harvesting and fermentation process. Starch in cocoa beans increased progressively from 4.3% to 6.8% between 4.5 months and 5.0 months, but then decreased to 6.3% at 5.5 months when pods were harvested.

Reduction of starch content does not mean it was consumed or synthesis stopped. More likely is that other constituents were accumulated at a faster rate.

Table 7.2 Starch content of cocoa beans representing regions of production.

Cocoa bean source	% starch in whole dry bean including shells
Ghana	7.00; 5.80
Bahia	4.78; 4.86
Samoa	4.77
Ecuador	5.51; 5.08
Venezuela	4.66; 5.63
Jamaica	4.50
Trinidad	4.77; 6.02

Source: Schmieder and Keeney (6).

Depending upon growing conditions, synthesis of fat is especially pronounced during this period (7). Analysis of samples from a fermentation trial in Brazil revealed a gradual increase in starch content of cocoa bean throughout the fermentation. At the start, 5.5% of the bean solids was starch; after 6 days when fermentation was concluded, starch accounted for 6.5% of the solid content. The apparent increase in starch content involves the loss of soluble, non-starch solids through exudation, sweating and sugar fermentation, as shown above.

Geilinger *et al.* (8) found 8.4% of starch in cocoa liquor. The egg-shaped starch granules have a mean size of 4.4 μm, ranging from 1.5–8 μm, an amylose content of 30.4% and a gelatinization temperature range of 53–68°C. In another lot of cocoa mass from Ghana, they found 6.4% starch in fermented beans (9). No change occurred during roasting and diverse preparations of the cocoa mass.

Data for starch analysed in various commercial products are shown in Table 7.3. When starch content was expressed as starch per unit of cocoa mass, values were quite similar. This indicates that manufacturing processes employed in making these products did not alter the amount of cocoa starch present in cocoa beans. There is no information on the digestibility of starch in roasted beans.

Table 7.3 Starch content.

Product	% starch	% starch in cocoa mass (non-fat)
Regular cocoa	15.5	18.6
Dutched cocoa	15.9	18.9
Regular chocolate liquor	6.9	16.3
Dutched chocolate liquor	7.0	16.3
Dark, sweet chocolate	3.1	20.4
Milk chocolate (a)	1.1	18.3
Milk chocolate (b)	1.3	18.2

Source: Schmieder and Keeney (6).

Fibre

The dietary fibre (DF) fraction in a foodstuff is that part regarded as indigestible or resistant to the action of human digestive enzymes (10). Various methods have been developed and applied, and in the case of cocoa, very different data were obtained. Valiente *et al.* (11) found 17.8% and 16.1% DF in raw and roasted cocoa bean, respectively, using the method according to Prosky *et al.* (12). Approximately 20% of DF is soluble. Roasting, however, almost doubled the content of klason lignin from 4.4% in raw bean to 8.5% in roasted product due to formation of Maillard browning products, which are insoluble in the 72% sulphuric acid used for the lignin isolation (13). Bartolomé *et al.* (14), using the same procedure, found in defatted nib 7.8% soluble dietary fibre (SDF) and 42.2% insoluble dietary fibre (IDF), after deduction of the residual protein, which accounted for 14% of IDF. This represents roughly 22.5% DF in cocoa nib. Cocoa bean contains also significant amount of polyphenols (5.9%, expressed as tannic acid), but very little has been found in the fibre fractions (0.2% in SDF and 1.6% in IDF).

Geilinger *et al.* (9) determined the dietary fibre content using the neutral detergent method (15) and found 9.1% in fermented cocoa bean, 12.0% after roasting, and 12.1–15.4% in cocoa mass. This increase of apparent fibre during roasting and processing of the beans was probably due to condensation reactions between protein and polyphenols (16). These fibre contents are, however, lower than the value reported elsewhere, in part because it does not take into account the soluble fibre fraction.

Sugars in chocolate

The main carbohydrate present in chocolate is normally sucrose, and lactose in milk chocolate. Nutritive carbohydrate sweeteners (including sucrose) are permitted for use in chocolate products in the USA, under Food and Drug Authority (FDA) Standards of Identity, section 163 of CFR Title 21. In Europe, sucrose may also be present up to 55% in accordance with the EC directive on cocoa and chocolate products and consequential national legislation (17), although typically it is present at about 45%. Glucose syrups are not normally used in the manufacture of chocolate and indeed can create great problems in processing if present in more than small amounts, as they cause an increase in the viscosity of the molten chocolate. In plain chocolate, which does not contain milk solids, lactose is not normally included, the only sugar normally used being sucrose, which typically may be present at about 50%.

The use of sugar in chocolate manufacture is vital to provide bulking properties and to offset the bitterness of raw cocoa; it is also essential to chocolate's unique sensory experience. Nonetheless, sugar in our diet has been questioned in regard to the potential for health problems such as obesity, cardiovascular disease and cancer. The reader is directed to chapters in this book dealing specifically with these conditions in respect to chocolate consumption (Chapters 5, 6, 8 and 11). Additionally, the report of a expert consultative body appointed by the World Health Organization/Food and Agriculture Organization (WHO/FAO) to study carbohydrates in human nutrition has recently been released (18). Its findings further reinforce the already strongly held view within the scientific community that properly designed scientific studies have been unable to find any link between sucrose consumption and the prevalence of such chronic diseases.

It is permissible within the EU to use other sugars in chocolate as well as sucrose. Dextrose, fructose, lactose and maltose are specifically permitted up to 5% of the weight of the product without declaration. Dextrose may be incorporated at levels between 5 and 20%, in which case the name of the product has to be accompanied by a declaration of its presence. In Europe, polyols are permitted for use in cocoa-based products – energy reduced or with no added sugar – as a total replacement of sucrose (17). In the USA, regulation specifies that neither polyols nor polydextrose are allowed in chocolate of any type.

Polyols used in cocoa-based products are crystalline and obtained by hydrogenation of sugars. They are hydrogenated monosaccharides (sorbitol, mannitol, xylitol) or hydrogenated disaccharides (isomalt, maltitol and lactitol). These bulk sweeteners differ from the common sugars by the fact that they are poorly metabolized by the bacteria of the dental plaque and slowly digested or absorbed in the small intestine of mammals (19). One can thus produce non-cariogenic chocolates, if the other ingredients of the formulation do not provide cariogenic sugars. In the case of non-cariogenic milk and white chocolate, milk added needs to be lactose free.

The replacement of sucrose by polyols affects both the flavour and texture. The 'snap' characteristic of chocolate with sorbitol, isomalt and lactitol is generally judged higher than from chocolate with sucrose, maltitol and mannitol. Chocolate with mannitol is significantly more doughy in the mouth than the others. This is probably due to the low solubility of mannitol. Combining xylitol and mannitol can modify this texture (20). On the taste side, chocolate containing sorbitol, mannitol, lactitol or isomalt are significantly less sweet than those with sucrose, maltitol or xylitol. This is why adjustment of sweetness with intense sweeteners is recommended (21). Chocolate with sorbitol, and more with xylitol, exhibits an important refreshing effect (20). Although the choice of polyols is quite large, very little chocolate containing them is found on the market. Reasons might be the reduced quality both in texture and sweetness, the high price of these polyols as compared with sucrose, and the limited intestinal tolerance with some of them (19).

Protein

Cocoa and chocolate are mainly eaten because of their organoleptic properties and not as protein sources, which is certainly a very secondary aspect. However, a qualitative and quantitative evaluation of the contribution of cocoa-based products as sources of proteins can be made, taking into account not only the cocoa powder *per se* but also another protein source – milk – which is very often associated with cocoa, i.e. cocoa powder-based drinks and chocolate.

Nitrogenous compounds in cocoa beans

Cocoa contains two classes of nitrogenous compounds: proteins, which contribute up to about 80% of total nitrogen and methylxanthines. Their quantity and

Table 7.4 Distribution of nitrogen compounds in dry beans.

Bean type	Total nitrogen (g/100 g)	Protein nitrogen		Methylxanthine nitrogen (g/100 g)	
		(g/100 g)	(% of total nitrogen)	Theobromine	Caffeine
Criollo (Trinidad)	2.086	1.636	78.4	0.312	0.138
Forastero (Trinidad)	2.282	1.852	81.1	0.367	0.063
Forastero (West African)	2.280	1.720	75.4	0.530	0.030

relative proportion vary with the bean variety as presented in Table 7.4 (22).

The large quantity of non protein nitrogen makes difficult the *in vivo* nutritional evaluation of cocoa proteins, as a large proportion of methylxanthines are absorbed and excreted in the urine, which interferes with the nitrogen balance parameters digestibility and biological value.

As shown by Zak and Keeney (23), cocoa proteins of fresh beans are distributed in different protein classes: albumin, globulin, prolamine and glutenin. Their proportion changes with the bean variety; in the non-pigmented variety Criollo, the percentage distribution of these protein classes is 32, 25, 12 and 31% respectively, while in the pigmented variety Nacional the percentage distribution is 51, 25, 12 and 12% respectively. These differences in the protein classes influence the protein solubility, which is about 30% in the non-pigmented varieties and more than twice as much in the pigmented varieties. Tanning reactions of polyphenols most likely contribute to these differences.

A certain amount of protein nitrogen is present as free amino acids. According to Marvalhas (24) only 3% of the amino acids of cocoa beans are free. After fermentation, this level increases to higher levels (Table 7.5).

Amino acid composition

The amino acid composition of cocoa beans appears to be relatively constant from one variety to another. Zak and Keeney (23) did not find discernible differences among three non-pigmented and three pigmented varieties and Offem (25) did not find any significant differences in amino acid composition of three varieties of cocoa beans from south-eastern Nigeria.

The fermentation step does not significantly change the total amino acid composition as measured after acid hydrolysis. However, during fermentation, some enzymatic hydrolysis occurs, liberating peptides and free amino acids. The total amino acid content remains the same, but the free amino acids are multiplied by a factor of three (Table 7.5). During roasting, these free amino acids are partly destroyed through thermolysis and Maillard-type reactions.

Table 7.5 Total and free amino acids in unfermented and fermented Bahia defatted cocoa beans.

Amino acid	Unfermented (g/100 g)		Fermented (g/100 g)	
	Total	Free	Total	Free
Lysine	2.34	0.027	2.36	0.204
Histidine	0.49	0.034	0.47	0.026
Arginine	1.62	0.024	1.48	0.134
Aspartic acid	2.60	0.189	2.71	0.325
Threonine	0.66	0.009	0.63	0.074
Serine	0.60	0.035	0.62	0.101
Glutamic acid	3.60	0.122	3.59	0.190
Proline	1.13	0.045	1.14	0.092
Glycine	1.00	0.032	1.00	0.074
Alanine	0.90	0.042	0.88	0.154
Valine	1.28	0.008	1.32	0.083
Methionine	0.21	0.012	0.21	0.026
Isoleucine	0.71	0.018	0.74	0.067
Leucine	1.50	0.018	1.48	0.307
Tyrosine	0.61	0.025	0.65	0.116
Phenylalanine	1.18	0.024	1.17	0.228
Total	20.43	0.664	20.45	2.201

Protein modifications during cocoa processing

From the fresh beans to cocoa powder, cocoa proteins are chemically modified mainly by polyphenols, which affect their functional and nutritional properties.

The fermentation process modifies proteins and polyphenols. Proteins are partially hydrolysed into peptides and free amino acids, which are much more sensitive to chemical modifications than protein-bound amino acids. In the anaerobic step of fermentation, polyphenols are hydrolysed by β-galactosidases, which liberate the aglycones (mainly anthocyanes) and sugars (galactose, arabinose). In the aerobic phase of fermentation, anthocyanes polymerize into tannins, which can bind proteins through hydrogen bridges. In addition, ortho-diphenols, the most sensitive polyphenols to oxidative reactions (like epicatechin), oxidize through the action of polyphenol oxidases into quinones. Quinones are able to covalently bind the free amino groups of every free amino acid and the epsilon amino group of lysine and the sulphydryl group (–SH) of cysteine, bound to proteins (26). As a conclusion, during fermentation, three types of protein–polyphenol combinations occur:

- Reversible hydrogen bridges between the proton of the phenolic –OH and the oxygen of the peptide bonds.
- Irreversible hydrogen bonds between the peptide bonds and leucoanthocyans, procyanidins or polymers of flavanols (tanning effect).
- Irreversible covalent bonds between amino acids and quinones.

All these reactions contribute to reduce the solubility of the cocoa proteins, which decreases from 170 to 65 mg/g fat-free beans during fermentation (27). The protein quality is therefore expected to decrease similarly.

During roasting, the Maillard reaction takes place between the amino groups of the small peptides and of the free amino acids liberated during fermentation and the free reducing sugars, some of them also being liberated during fermentation. This Maillard reaction is responsible for the development of the cocoa aroma at the expense of the amino acids involved, and to a minor extent, of the nutritional quality.

Nutritional value of cocoa powder proteins

The composition of the essential amino acids of the Criollo variety is presented in Table 7.6 and compared with the essential amino acid profile proposed as a reference for a good protein quality according to FAO/WHO (28).

Table 7.6 Amino acid composition of fresh, fermented and roasted cocoa beans. Chemical score calculated from the proposed FAO/WHO amino acid pattern (28).

Amino acid	Criollo variety (g/16 g N)	FAO/WHO pattern (g/16 g N)	Chemical score (%)
Histidine	4.3	(1.9)	>100
Isoleucine	4.1	2.8	>100
Leucine	6.8	6.6	>100
Lycine	4.3	5.8	74
Methionine + cysteine	2.0	2.5	80
Phenylalanine + tyrosine	6.2	6.3	98
Threonine	5.2	3.4	>100
Tryptophan	?	1.1	—
Valine	7.7	3.5	>100

Cocoa proteins are limiting in lysine and methionine with a chemical score of 74, which can be considered as being good for a protein of plant origin. The value for tryptophan is never mentioned and might also be limiting. However, due to multiple modifications during processing, the protein quality is reduced and *in vivo* tests are necessary for its absolute evaluation. This evaluation has been made on cocoa powder under very specific conditions due to the nature of this food ingredient.

The protein quality of a food is generally evaluated in young rats fed a standard diet containing 10% protein either in a growth test to measure the *protein efficiency ratio* (PER) or in a nitrogen balance test to measure the *nitrogen digestibility* and the *biological value*. In these tests casein is taken as the reference protein. However, because of its palatability, no more than 25% cocoa powder can be given in a rat diet, which limits the protein level to around 5% in the diet.

An additional problem is that about 20% of cocoa powder nitrogen is composed of non-protein nitrogen coming from methylxanthines.

To overcome these limitations, Shahkhalili *et al.* (29) compared the nitrogen digestibility of a diet containing 25% cocoa powder with the digestibility of a diet containing the same levels of casein and of methylxanthines. They found that the true digestibility of total cocoa nitrogen was 28–30%. When this value was corrected by the digestibility of methylxanthines, the digestibility of cocoa proteins was reduced to 16–17%. This value is very low as compared to the digestibility of animal proteins (95–100%) and that of plant proteins (72–85%) (28). We can therefore consider that the proteins of cocoa powder have little practical value.

Cocoa powder as a milk modifier

Cocoa powder is currently used to give a pleasant taste to raw milk. This contributes to increase milk consumption in populations, like adolescents and the elderly, who need an extra supply of good quality proteins. A glass of 200 ml milk contributes significantly to the daily requirement of proteins (about 13%) and also of calcium (about 30%).

Milk chocolates

Chocolate bars are very often associated with milk to change or to improve their palatability. The proportion of milk can vary significantly according to the recipe. Table 7.7 gives an indication of the (available) proteins and calcium present in some chocolate bars expressed in quantity (g) and percentage of their respective daily requirement (RDA).

In conclusion, cocoa powder cannot be considered as a source of protein because its proteins are not digested due to their reactions with cocoa polyphenols. However, cocoa is often associated with milk, as a milk modifier or as

Table 7.7 Protein and calcium contents in 100 g milk chocolate compared to black chocolate bars.

	White chocolate	Milk chocolate	Dark chocolate	Cocoa powder
Calories (kcal)	550	550	490	357
Protein (g)	8.5	9.2	5.3	19.8
Available protein				
(g)	8.5	8.5–8.8	0.8	3.3
(% RDA)	16	16–17	1.5	6.2
Calcium				
(mg)	301	214	63	114
(% RDA)	37	27	8	14

milk chocolate. It is in these forms, having improved taste over cocoa powder alone, that cocoa provides much pleasure to people, and also results in increased consumption of milk proteins.

Carbohydrates are naturally present in cocoa as sugars, starches and fibres. Fermentation and processing usually causes their concentrations to change as chemical reactions occur. Cocoa powder contains appreciable concentrations of dietary fibre, although in chocolate, these levels are substantially reduced after blending with cocoa butter and adding sugar. Consequently, the greatest contribution of carbohydrates found in chocolate is as added sugars, usually sucrose, which is present to offset the bitterness of cocoa, thereby enhancing its palatability; and as lactose from milk solids used to make milk chocolate. It is technically possible to replace sucrose by polyols, but it tends to be at the expense of various desirable characteristics valued in quality chocolate. This practice is permissible in Europe, but not in North America. In any event, there are few examples to be found on the European market, suggesting little real success with the concept.

References

1. Forsyth, W.G.C. and Quesnel, V.C. (1963) The mechanism of cocoa curing. *Adv. Enzymol.* **25**, 457–492.
2. Reineccius, G.A., Andersen, D.A., Kavanagh, T.E. and Keeney, P.G. (1972) Identification and quantification of free sugars in cocoa beans. *J. Agric. Food Chem.* **20**, 199–202.
3. Berbert, P.R.F. (1979) Contribuiçao para o conhecimento dos açucares componentes da amendoa e do mel de cacau. *Rev. Theobroma (Brazil)* **9**, 55–61.
4. Bracco, U., Grailhe, N., Rostagno, W. and Egli, R.H. (1969) Analytical evaluation of cocoa curing in the Ivory Coast. *J. Sci. Fd Agric.* **20**, 713–717.
5. Rohan, T.A. and Stewart, T. (1966) The precursors of chocolate aroma: changes in the sugars during the roasting of cocoa beans. *J. Food Sci.* **31**, 206–209.
6. Schmieder, R.L. and Keeney, P.G. (1980) Characterization and quantification of starch in cocoa beans and chocolate products. *J. Food Sci.* **45**, 555–558.
7. Lehrian, D.W. (1978) Chemical composition and melting characteristics of cocoa butter related to the stage of maturity of cocoa seeds and microclimate temperature. PhD thesis. Pennsylvania State University, Philadelphia.
8. Geilinger, I., Amado, R. and Neukom, H. (1981) Isolation and characterization of native starch from cocoa beans. *Starch* **33**, 76–79.
9. Geilinger, I., Amado, R., Neukom, H., Kleinert, J. and Mikle, H. (1984) Einfluss verschiedener Verfahren zur Schocoladeherstellung auf Kakao-Inhalstoffe, speziell Kakaostärke. II. Chemische und physikalische Analysen der *Cacao* massen. *Lebensm.-Wiss. u.-Technol.* **17**, 201–204.
10. Trowell, H.C. (1974) Definition of fibre. *Lancet* **i**, 503.
11. Valiente, C., Esteban, R.M., Molla, E. and Lopez-Andreu, F.J. (1994) Roasting effects on dietary fiber composition of cocoa beans. *J. Food Sci.* **59**, 123–124.

12. Prosky, L., Asp, N.-G., Schweizer, T.F., DeVries, J.W. and Furda, I. (1988) Determination of insoluble, soluble and total dietary fiber in foods and food products: interlaboratory study. *J. Assoc. Off. Anal. Chem.* **71**, 1017–1023.
13. Van Soest, P.J. and Wine, R.H. (1967) Use of detergents in the analysis of fibrous feeds. IV. Determination of plant cell wall constituents. **50**, 50–55.
14. Bartolomé, B., Jiménez-Ramsey, L.M. and Butler L.G. (1995) Nature of the condensed tannins present in the dietary fractions in foods. *Food Chem.* **53**, 357–362.
15. Robertson, J.B. and Van Soest, P.J. (1977) *J. Anim. Sci.* **45**, 254–262.
16. Saura-Calixto, F. (1988) Effect of condensed tannins in the analysis of dietary fibre in carob pods. *J. Food Sci.* **53**, 176–180.
17. Dransfield, J. and Flowerdew, D.W. (1993) *EC Food Legislation*. British Food Manufacturers and Industries Research Association 1–10.
18. FAO/WHO (1998) Dietary carbohydrate and disease. In *Carbohydrates in Human Nutrition, FAO Food and Nutrition Paper 66*, pp 19–23. Food and Agriculture Organization, Rome.
19. Würsch, P. (1991) Metabolism and tolerance of sugarless sweeteners. In *Sugarless, the Way Forward* (Ed. by Rugg-Gunn, A.J.), pp. 32–51. Elsevier Applied Science, London.
20. Olinger, P.M. (1990) Sweetening the sugar-free challenge. *The Manufacturing Confectioner* **70**, 127–131.
21. Irwin, W.E. (1990) Sugar substitute in chocolate. *The Manufacturing Confectioner* **70**, 150–154.
22. Timble, D.J. (1977) Studies on the protein and purine alkaloids of cocoa beans. PhD thesis. Pennsylvania State University, Philadelphia.
23. Zak, D.L. and Keeney, P.G. (1976) Extraction and fractionation of cocoa proteins as applied to several varieties of cocoa beans. *J. Agric. Food Chem.* **24**, 479–482.
24. Marvalhas, N. (1972) Amino acids in fermented and unfermented cocoa beans. *Rev. Int. Choc.* **27**, 22–23.
25. Offem, J.O. (1990) Individual variation in the amino acid and chemical composition of defatted cocoa bean meal of three varieties of cocoa from south-eastern Nigeria. *J. Sci. Food Agric.* **52**, 129–135.
26. Hurrell, R.F., Finot, P.A. and Cuq, J.L. (1982) Protein polyphenols reactions. 1. Nutritional and metabolic consequences of the reaction between oxidized caffeic acid and the lysine residues of casein. *Br. J. Nutr.* **47**, 191–211.
27. Zak, D. L. (1973) Studies on the proteins of cocoa beans. PhD thesis. Pennsylvania State University, Philadelphia.
28. FAO/WHO (1991) *Protein Quality Evaluation. FAO Food and Nutrition Paper No. 51*. Food and Agriculture Organization/World Health Organization, Rome.
29. Shahkhalili, Y., Finot, P.A., Hurrell, R. and Fern, E. (1990) Effects of food rich in polyphenols on nitrogen excretion in rats. *J. Nutr.* **120**, 346–352.

Micro-Nutrients

Chapter 8

Phytochemicals and Phenolics

Nicholas J. Jardine

Cocoa is unusually rich in 'phytochemicals'. In unfermented cocoa beans, pigment cells make up about 11–13% of the tissue. These cells are rich in polyphenolic compounds, in particular catechins although the characteristic purple colour of unfermented cocoa beans is due to anthocyanins (1). When the beans are undergoing the process of fermentation, the walls of the pigment cells break down and their contents are exposed to other constituents within the bean. The colour of the bean changes from purple to brown in a well-fermented bean.

The various chemical changes that take place during fermentation have attracted the attention of those interested in the flavour development of cocoa. The colour change from purple to brown is merely the most visible sign of significant changes which take place in the chemistry of the bean. The polyphenols themselves undergo a variety of reactions, including self-condensation, and reaction with proteins and peptides. In addition to these various changes, there are those which take place during roasting and other steps of chocolate making. It is therefore easy to see why chocolate contains such a diversity of chemical compounds.

It has long been known that chocolate has excellent keeping qualities, derived from constituents in the cocoa which have antioxidant properties. It was presumably this that was the driver for the development of the milk chocolate crumb process in the 1920s. In this process, cocoa, sugar and milk are dried together to form a starting material for chocolate making. The antioxidants in cocoa beans provide powerful protection against the milk fats becoming rancid.

Much of the work on phenolic compounds involved in cocoa flavour was done in the period 1950–70, and the focus of work done at that time was to elucidate the involvement of these compounds in the development of chocolate flavour. Since that time, interest has grown dramatically in the antioxidant properties of plant phenolics, as the realization has grown that such compounds may have significant positive effects on human health. There is now considerable evidence that plant foods can protect against diseases like cancer and heart disease (see, for example, (2–4)). Various foods and beverages rich in such compounds have been investigated for health-related effects, as have individual isolated phenolics. Likewise, medicinal plants and plant-derived foods with presumed protective properties have been analysed in order to identify compounds with possible biological activity.

Despite having quite potent antioxidant activity, very little attention has been paid to the potential for cocoa to have protective properties. Only a handful of papers describe such effects where cocoa-containing material has been the food under observation. This short review summarizes the meagre information that is available. It also compiles those phenolic compounds that have been identified in cocoa (mostly in the course of flavour work) and points to the extensive information on potential health benefits of these compounds.

Studies of potential protective effects of cocoa

LDL oxidation

A major risk factor for heart disease is an elevated level in the blood of the cholesterol-containing fraction, low-density lipoprotein (LDL). High levels predispose to formation of atherosclerotic plaques in coronary arteries and elsewhere. It is now recognized however, that an important step in this process is the prior oxidation of LDL, which leads to its being scavenged by macrophages in artery walls (5, 6). In turn, the macrophages develop into cholesterol-filled *foam* cells, which are an essential component of plaque formation. It is believed that one way in which a diet high in vegetables and fruits protects from heart disease and other circulatory disease is that the antioxidants from foods in some way inhibit the process of LDL oxidation.

There is reasonable support for this notion. It has been repeatedly shown that flavonoids can prevent the oxidation of LDL *in vitro* (reviewed by Leake (7)). The *French paradox*, where the level of heart disease is much lower than in other northern European countries despite similar levels of fat intake, has often been ascribed to the protective effect of components in the diet, such as wine, which is rich in flavonoids. It has been shown that *in vitro* wine is capable of inhibiting LDL oxidation (7).

Extracts of cocoa have also been shown to have significant effect *in vitro* in inhibiting LDL oxidation (8). This was also investigated by Waterhouse *et al.* (9), who presented evidence that cocoa polyphenols had more antioxidant activity than that of red wines at similar concentrations. The question then arising is whether cocoa could have a similar effect *in vivo*. Kondo *et al.* (8) fed 35 g of defatted cocoa to 12 male volunteers, and showed a small but significant increase in the resistance to oxidation of LDL sampled 2 hours after cocoa ingestion. Natsume (10) has also presented evidence that LDL oxidation could be retarded in cholesterol-fed rabbits (a model for hypercholesterolaemia) when 1% cocoa polyphenols were added to the feed. Such experiments provide evidence that cocoa polyphenols can be absorbed and could therefore have a beneficial effect in retarding atheroma formation in humans. Further work will be needed to confirm the magnitude of the effect at normal consumption levels.

Other antioxidative effects of cocoa

There are a number of preliminary reports emanating from Japan which examine the effects of cocoa antioxidants.

Osawa (11) described the identification of a number of polyphenols in cocoa including catechin, epicatechin, quercetin and clovamide, and confirmed that these compounds had antioxidation potential in a number of test systems based on linoleic acid, red blood cell ghosts or liver microsomes. Crude polyphenol fractions had a pronounced effect in protecting against alcohol-induced ulcerative damage to the gastric mucosa in experimental rats. Evidence was presented for a strong effect of the polyphenols against lipid peroxidation in the mucosa. Osawa also examined the antioxidative activity of cocoa polyphenols in rats on a vitamin E-deficient diet designed to induce a state of oxidative stress in the body (that is, a lowered potential to resist the effects of oxidizing agents). Lipid peroxidation in blood plasma and various body organs was lower in the cocoa-fed rats, being particularly pronounced in the liver. The effect was dose dependent. Osawa concluded that antioxidants were absorbed from the diet and that they were dispersed widely through the body of the animals.

Antimutagenic effects

Osawa (11) has also presented preliminary evidence for antimutagenic activity of cocoa extracts in Ames test systems and an *in vivo* mouse test system. In the absence of details, evaluation of the significance of these experiments will have to wait for full publication.

Immune regulation

Sanbongi *et al.* (12) presented evidence from a series of experiments on isolated human T and B lymphocytes and granulocytes that cocoa polyphenols have the capacity to down-regulate excessive immune reactions. Likewise a suppressant effect was shown on neutrophil leucocytes, which had a much reduced capacity to generate the 'reactive oxygen' species hydrogen peroxide (H_2O_2) and superoxide free radical ($O_2\bullet$-) which are employed in cell-killing activities. More work will be needed to elucidate the biological significance of such immunoregulatory effects by cocoa polyphenols.

The same group (13, 14) has also claimed in preliminary publications that cocoa extracts were active against rheumatism-causing factors produced by B lymphocytes from rheumatoid arthritis patients. They also limited the levels of the inflammatory cytokines in patients' synovial cells. A restriction of immunoglobulin E (IgE) production was also found in B lymphocytes of patients suffering from atopic dermatitis and bronchial asthma patients.

Stress alleviation

In a series of experiments involving rats, Takeda (15) found evidence that feeding cocoa polyphenols could effectively reduce the effects of stress due to confinement, as measured by locomotor activity and other activity signs. It was also claimed that plasma corticosterone levels were maintained in the cocoa-fed rats but that they were elevated in the others. Likewise, these rats did not show the increase in lipid peroxide in liver and kidney seen in the stressed rats on the normal diet.

Physiological effects of individual cocoa phytochemicals

It will be clear from the foregoing that although some research on the physiological effects of cocoa constituents is currently in progress, very little information has yet reached the open scientific literature. These preliminary results appear promising, but there is a lack of information on how the extracts were made, their constituents or the concentration at which they were active. Consequently it is difficult to properly appraise the significance of the results.

This situation contrasts with other foodstuffs containing similar constituents. This is particularly true of tea, which has been under increasingly intense scrutiny for a couple of decades. In recent years, the phenolic compounds in plants generally have attracted interest for their antioxidative and other properties, in the hope that the secrets of their protective effects on human health might be unlocked. Thanks to this, it is possible to say something about the potential that cocoa-containing materials might have on human health. The remainder of this chapter therefore provides an overview of this rapidly expanding area of research.

Phytochemicals in cocoa

There is no 'official' definition of the word *phytochemical*. In its broadest sense, phytochemistry is the study of the chemistry of plants. Nowadays, the word phytochemical has taken on a more restricted meaning in some quarters, being defined as *a non-nutritive plant constituent able to confer a health benefit*. It is this meaning which is employed in this chapter.

The chemistry of cocoa has been extensively investigated in the past, but usually the objective was to further understanding of the chemical basis of cocoa and chocolate behaviour. There appear to have been no surveys for phytochemicals in cocoa. While there are a number of phytochemicals in cocoa of potentially great interest, there may be others as yet undiscovered or unreported. The main phytochemicals are listed in Tables 8.1, 8.2 and 8.3, together with estimates of concentration where available. A problem with the published information is that starting materials may vary between unfermented beans, fermented beans, beans or cocoa mass after roasting, and, in rare instances, cocoa powder or chocolate. Unfortunately, the materials of greatest interest to health, namely cocoa powder and chocolate, have been subject to the least analysis in respect of phytochemicals.

Table 8.1 Phytochemicals and polyphenolics.

Compound	Dry seeds (g/100 g)	After roasting and conching[1] (g/100 g)	In milk chocolate[2] (mg/100 g)
Flavanols			
Catechins	3.0		
(+)-Catechin	1.6–2.75	0.03–0.08	0.02
(−)-Epicatechin, (+)-Gallocatechin, (+)-Epigallocatechin	0.25–0.45	0.3–0.5	
Leucocyanidins			
L_1–L_4	2.7	L_1: 0.08–0.17	
Polymeric leucocyanidins	2.1–5.4		
Anthocyanins		0.01	
3-α-L-arabinosidyl cyanidin	0.3		
3-β-D-galactosidyl cyanidin	0.1		
Flavonols			
Quercetin			
Quercetin-3-arabinoside[3]			
Quercetin-3-glucoside[3]			
Total phenolics	13.5		

Source: Ziegleder and Biehl (16).
[1] Mohr (17).
[2] S. Tarka (personal communication).
[3] Sanbongi *et al.* (18).

The phytochemicals in cocoa, which are examined below, fall into a number of types. The polyphenolics are quantitatively the largest group (Table 8.1), comprising around 13.5% of the dried unfermented cocoa beans. Most of these are flavanols and flavonols (Fig. 8.1). The former mostly comprise catechins, and of these, by far the greatest single component is (−)-epicatechin. The catechins are colourless, while the anthocyanins, present in somewhat lower amounts, give rise to the colour (*cocoa purple*) in unfermented cocoa beans. As a result of fermentation and drying, much of the anthocyanin is converted to quinonic compounds, giving the beans their characteristic brown colour.

Table 8.2 Phenolic compounds identified in cocoa and milk chocolate.

Compound	mg/100 g dry weight[1]	mg/100 g milk chocolate[2]
Aesculetin		
Phloroglucinol	0.4	0.05
p-Hydroxybenzoic acid		
Protocatechuic acid	0.34	0.3
Vanillic acid	0.07	0.05
Syringic acid		
Phenylacetic acid		
o-Hydroxyphenylacetic acid	0.06	
p-Hydroxyphenylacetic acid		
Phloretic acid		
p-Coumaric acid	0.05	
p-Coumaryl quinic acid		
Caffeic acid	1.1	0.002
Caffeoyl quinic esters		
Ferulic acid	0.32	0.03
Chlorogenic acid		
Neo-chlorogenic acid		
Clovamide[3]		
Deoxyclovamide[3]		

[1] Ziegleder and Biehl (16).
[2] S. Tarka (personal communication).
[3] Sanbongi *et al.* (12).

Table 8.3 Phytosterols in cocoa butter.

Phytosterol	(g/100 g)
Sterols	
β-Sitosterol	0.10[1]–0.14[2]
Stigmasterol	0.04[1]-0.06[2]
Campesterol	0.01[1]–0.02[2]
Di- and triterpenes	
Phytol	0.03[1]
Cycloartenol	0.01[1]–18.9[3,4]
Aliphatic alcohols	
Lignoceryl alcohol	0.006[1]
Behenyl alcohol	0.005[1]
Ceryl alcohol	0.002[1]

[1] BIBRA (19).
[2] Chan *et al.* (20).
[3,4] Staphylakis and Gegiou (21, 22).

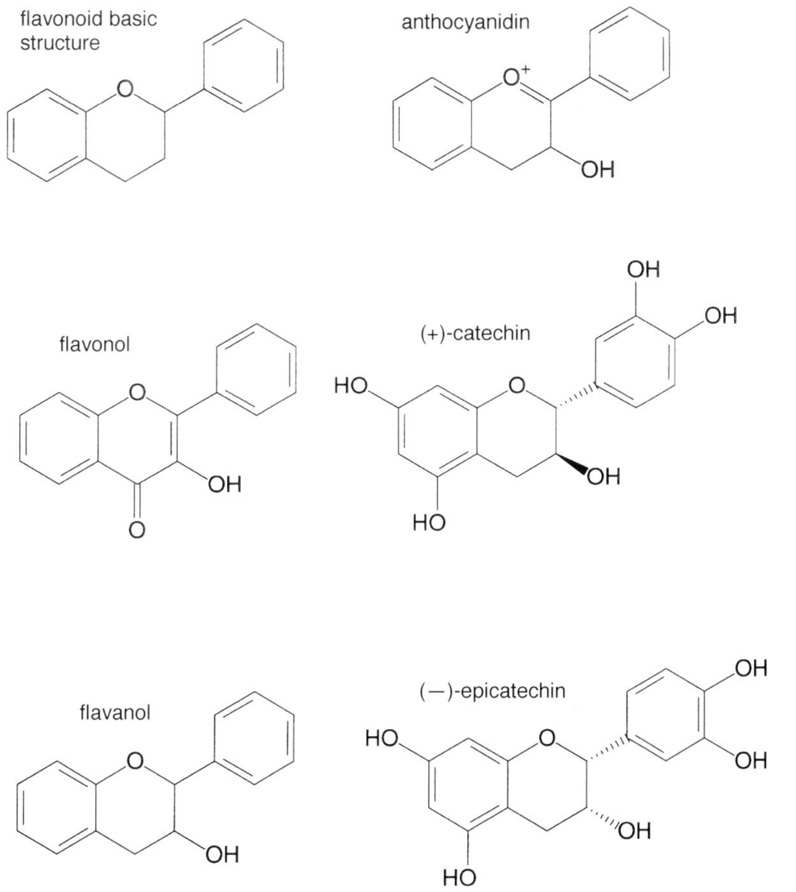

flavonoid basic structure

anthocyanidin

flavonol

(+)-catechin

flavanol

(−)-epicatechin

Fig. 8.1 Structures of common flavonoid types found in cocoa.

Quercetin (a flavonol) and two derivatives (a glucoside and an arabinoside) have also been described in cocoa (17) (Table 8.1, Fig. 8.2). The compounds are structurally similar to the simple flavanols.

The leucocyanidins (also termed *procyanidins*) are complex polymeric forms

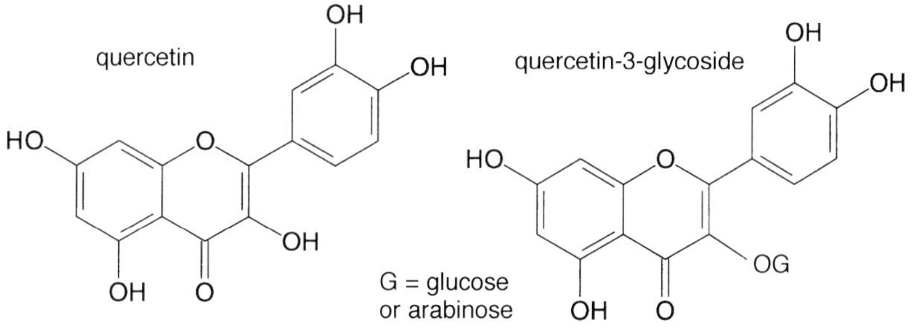

quercetin

quercetin-3-glycoside

G = glucose or arabinose

Fig. 8.2 Structure of quercetin and its glycoside derivatives.

of flavanols (Fig. 8.3). They are produced during fermentation, resulting from simple flavonols undergoing condensation reactions (23, 24). The most common building block is epicatechin. Such oligomeric molecules are often termed *tannins*, being originally extracted from galls and used in the conversion of hides to leather (25). Leather making relies on the ability of tannins to react with proteins to stabilize their structures. Leucocyanidins are the source of astringency in cocoa products (26).

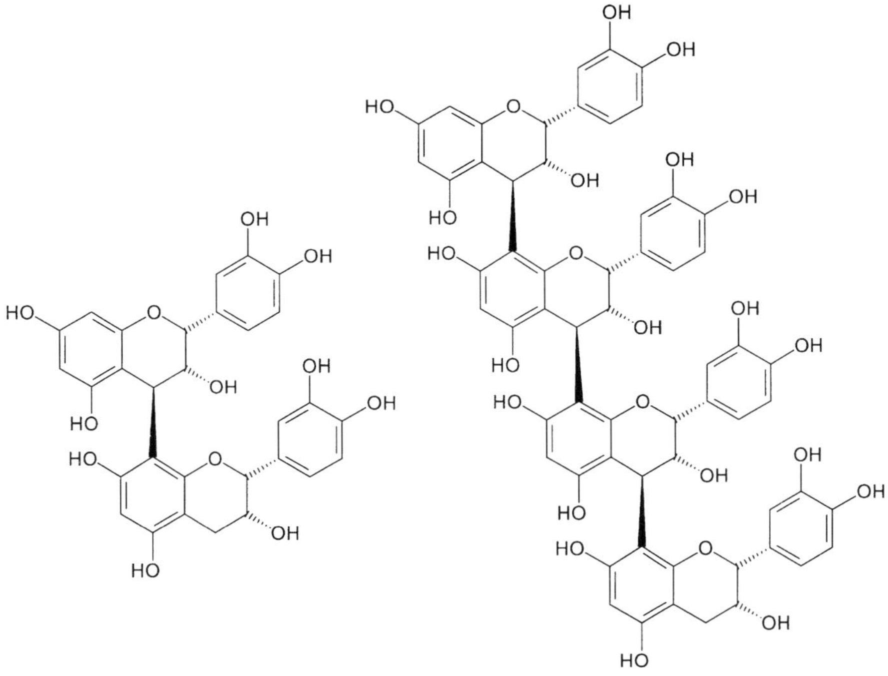

Fig. 8.3 Examples of leucocyanidins (procyanidins) found in cocoa; most are based on epicatechin. *Source:* Porter *et al.* (23).

Somewhat different to the above flavonoids but still containing phenolic structures are clovamide and dideoxyclovamide (Fig. 8.4). Their presence in cocoa has been recently described by Sanbongi *et al.* (18).

Various simple phenolic compounds have been described in cocoa at one time or another (Table 8.2). Where quantified, these have been found at somewhat lower concentrations than the flavonoids.

A number of minor components have been found in cocoa butter (Table 8.3, Fig. 8.5). These include sterols, terpenes and aliphatic alcohols.

Beneficial effects of flavonoids

There is now an extensive literature on a variety of biological effects of flavonoids. We have seen in an earlier section that little work has been done directly

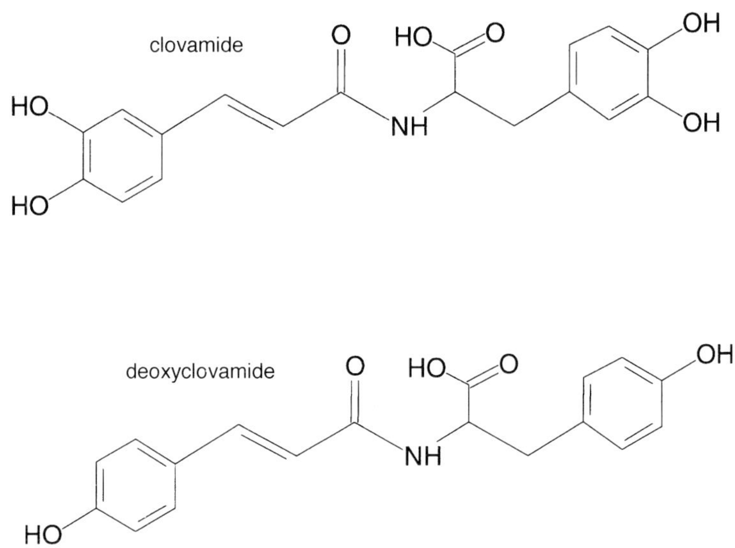

Fig. 8.4 Structures of clovamide and deoxyclovamide.

with cocoa materials themselves. However, the flavonoids identified in cocoa can be found in a wide variety of plant materials, and some of these, like tea, have been the subject of many experiments. Purified compounds have also been extensively studied. The literature on the biological effects of flavonoids has recently been reviewed by the British Industrial Biological Research Association (BIBRA) (27) on behalf of the UK's Biscuit, Cake, Chocolate and Confectionery Alliance. The beneficial effects were reviewed under three main headings: antigenotoxicity, anticarcinogenic activity and antioxidant activity. Other protective effects were also noted as appropriate, but many of these probably relate to antioxidant properties. In fact, antioxidant effects are now believed to be fundamental to most of the protective effects of flavonoids as well as a number of other phytoprotectant molecules.

(–)-Epicatechin – anticancer activity

Epicatechin is the flavonoid at greatest concentration in fermented cocoa beans (leaving aside the polymeric forms for which there is no quantitative data), and this discussion therefore concentrates on this constituent.

The development of cancer is a multi-stage process (28). In the initial stage, the genome (that is, the DNA) is damaged as a result of the action of chemical carcinogens, viruses, radiation, replication errors, etc. Much of this damage is oxidative in nature. For example, a potential chemical carcinogen may be converted by the body's detoxification enzymes to a chemical, which reacts with cellular DNA to cause a mutation. While repair mechanisms within the cell remove much of the damage, some can remain in the genome, resulting in a

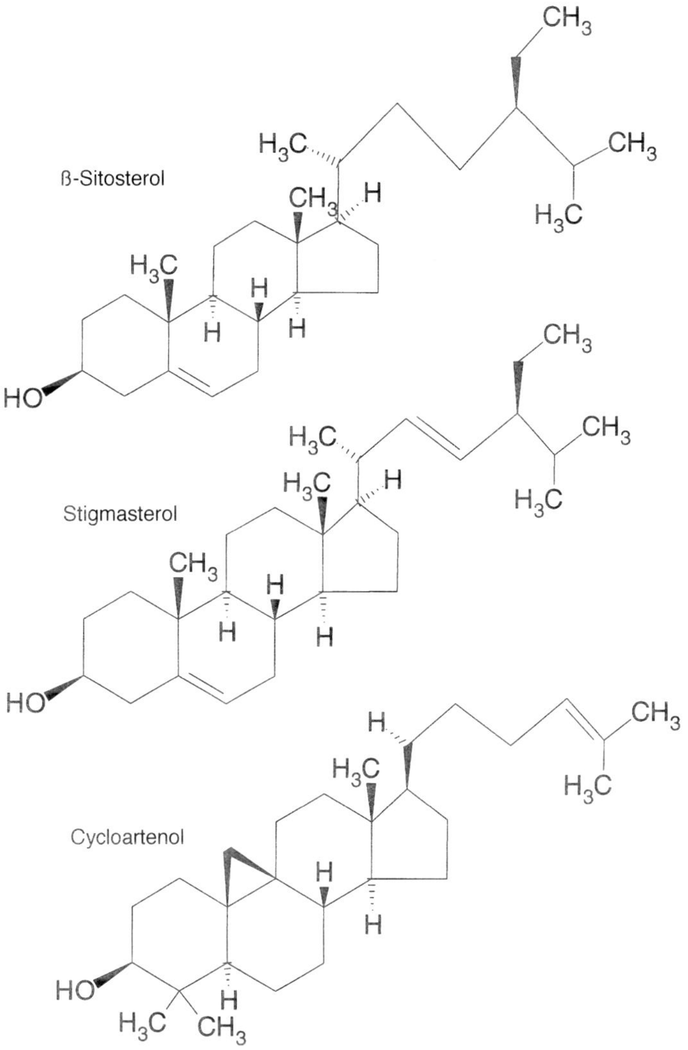

ß-Sitosterol

Stigmasterol

Cycloartenol

Fig. 8.5 Major sterols found in cocoa butter.

mutation in the DNA. It may seem perverse that the body's own detoxification system creates the ultimate carcinogen, but this needs to be seen in the light of the body's strategy for removing foreign chemicals, particularly insoluble ones. The mechanism involves oxidizing such compounds in order to make them easier to remove in the urine, and although the result is usually benign, sometimes it can result in the formation of a carcinogen. In turn, other detoxification (*phase II*) enzymes may increase the chemical's solubility further, for example by stimulating reaction with glutathione, thereby removing the carcinogen.

The next stage in the development of cancer is promotion. This stage is thought to be reversible, and involves division of the initiated cells to increase numbers.

Promotion may be mediated by chemicals which are not themselves genotoxic, but there are also endogenous mechanisms which can act. In the final stage – progression – the cells undergo further changes, losing the capacity for normal growth control so being able to grow into tumours, and they may also acquire the ability to invade distant sites (that is, to metastasize).

(–)-Epicatechin has been shown to have both antigenotoxic and anti-carcinogenic activity, pointing to a possible role in cancer prevention. It offered significant protection against aflatoxin B_1 (a potent mutagen and carcinogen) in model systems employing bacteria (the Ames test). *In vitro* it was shown to prevent the binding of aflatoxin B_1 and benzo[α]pyrene to DNA. It also has the ability to prevent the oxidation of one of the components of tobacco smoke to active forms that interact with DNA.

In vivo, single doses of (–)-epicatechin fed to gamma-irradiated rats prevented the formation of micronuclei in blood. Micronuclei would normally be seen in blood cells after gamma irradiation, and indicate damage done to chromosomes by ionizing radiation.

There is evidence that (–)-epicatechin can influence the later stages of the cancer process, having anticarcinogenic activity in experimental animal systems. It has been shown to act at the tumour promotion stage, for example inhibiting the formation of papillomas (benign skin tumours) promoted by application of croton oil, a potent tumour promoter, in carcinogen-treated mice. More generally, there are various reports of anticancer effects in experimental animals.

(–)-Epicatechin – other biological activity

Various experiments have shown that (–)-epicatechin is able to suppress lipid peroxidation in a variety of circumstances, for example in mammalian cells and in model cellular systems (29, 30). Vinson *et al.* (31) showed that (–)-epicatechin is able to suppress copper-catalysed oxidation of human LDL. It was also able to protect nerve cells from newborn mice from death induced by glucose oxidase (32).

Other demonstrated effects of (–)-epicatechin include an ability to decrease the solubility of cholesterol, effecting a reduction of cholesterol absorption from the intestine in a model rat system (33). Intravenous injection of (–)-epicatechin directly into the brains of mice inhibited the memory impairment induced by low brain oxygen levels.

Protective effects of other cocoa flavonoids

The previous section on (–)-epicatechin serves to illustrate the diverse nature of the biological effects of the flavonoids. (–)-Epicatechin was spotlighted as it is the most common of the flavonoids in cocoa, but the others have also been the subject of many studies and much information has been accumulated on them. Each has a focus of interest as components of other foods or beverages believed to have health benefits. For example, (+)-catechin, (–)-epicatechin and (–)-epigallocatechin are found in green tea. Onions are high in quercetin. Wine contains a variety of flavonoids including anthocyanins.

Table 8.4 summarizes information about the protective effects shown for the flavonoids found in cocoa (27). Although many effects have been shown for these compounds in test and animal systems, one important question that remains to be resolved is the extent to which humans absorb these compounds. To have much effect on human health, flavonoids must be taken up by the body. However, the information that is available on this point is fragmentary, but it has been well reviewed by Hollman and Katan (34). The degree of absorption of catechins from a major source, namely tea, is hardly known. Various experiments, mostly with animals, have shown a variable absorption of flavonoid aglycones when administered at very high doses. The figure varies between 4–58% depending on the flavonoid. Although it might be expected that glycosidically linked flavonoids would be less well absorbed than the aglycone (that is, the 'straight' compound not linked to a sugar molecule), recent studies comparing absorption of quercetin with quercetin glucoside surprisingly showed that the latter was much better absorbed. It would therefore seem that conventional wisdom needs to be re-evaluated.

Table 8.4 Summary of protective effects of cocoa polyphenols.

Chemical	Effect			
	Antigenotoxic	Anticarcinogenic	Antioxidant	Other
(+)-Catechin	+	+	+	ISC
(–)-Epicatechin	+	+	+	CHO, ISC
(+)-Gallocatechin	+	+	+	THR
(–)-Epigallocatechin	+	+	+	CHO
Leucocyanidin	a	a	a	LAT, BL
Polymeric leucocyanidin	a	a	a	
3-α-L-Arabinosidyl cyanidin	a	a	a	
3-β-D-Galactosidyl cyanidin	a	a	a	
Quercetin	a	+	+	IMM
Quercetin-3-galactoside	a	a	+	THR
Quercetin-3-glucoside	a	a	+	CHD

Source: BIBRA (27).
+ = protective activity reported in the literature.
a = no relevant data identified.
ISC = improvement in memory impairment caused by reduced blood oxygen levels.
CHO = reduction in cholesterol absorption.
THR = antithrombotic activity.
LAT = protection against toxicity from intake of lathymus plant seeds.
BL = increased vascular tissue strength.
IMM = anti-immunogenic effects.
CHD = epidemiological evidence of protection against coronary heart disease incidence and mortality.

Since so little is known about the absorption of flavonoids, it is not surprising that the factors affecting absorption are hardly understood. It is well known that polyphenols can bind strongly to protein. There is evidence that antioxidant polyphenols are not absorbed from tea drunk with milk (35). This could also be a factor affecting cocoa polyphenols. It is clear that much of the polyphenolic material is complexed with protein during cocoa fermentation. Furthermore, much of the chocolate that is eaten is milk chocolate where additional proteins are introduced from the milk. However, there is very little information on the fate of the complexed polyphenols after ingestion.

Another important issue for chocolate is the amount of flavonoids present. Again, there is not very much information available, but what there is would suggest a large loss of monomeric flavonoids at all stages from fermentation to chocolate production (Table 8.1). This makes it all the more crucial to learn more about the nature of the cocoa polyphenol adducts, and to determine their metabolic fate. Until we know this, it will be hard to judge their effects *in vivo*. Even the simplest adducts are quite complex, and recent patents (36–38) have claimed the presence of polymeric flavonoids up to 12 units or more. The patents have claimed a variety of biological effects for these complex molecules which would lead to health benefits, but the nature of patent disclosure and the information available from them makes it impossible to fully judge the efficacy of the claims.

To give some idea of the amounts of flavonoids needed to have beneficial effects in experimental animals, Table 8.5 is reproduced from the BIBRA (27) review.

It should also be pointed out that the relationship between disease and

Table 8.5 Oral doses of cocoa flavonoids needed to demonstrate a protective effect.

Chemical	Effect	Effective dose (mg/kg body weight/day)	Species
(+)-Catechin	Antigenotoxic	2.4–435	Rodents
	Anticarcinogenic	0.058–1500	Mice
	Inhibition of lipic peroxidation	2.4	Rats
	Protection from kidney injury	0.4	Rats
(–)-Epicatechin	Antigenotoxic	1.45	Mice
	Anticarcinogenic	0.058–100	Mice
	Reduced cholesterol absorption	70	Rats
	Protection from kidney injury	0.8	Rats
(–)-Epigallocatechin	Reduced cholesterol absorption	160	Rats
Leucocyanidin	Antidiabetic	100	Rats
Quercetin	Anticarcinogenic	1000–2500	Rodents

Source: BIBRA (27).

intake of flavonoids remains unclear. In the case of cancer, there is strong epidemiological evidence for a protective effect of vegetables and fruit. Identification of the active molecules is, however, another matter. In a review of the potential health effects of quercetin – the most common dietary flavonoid – Hertog and Hollman (39) and the recent Committee on the Medical Aspects of Food and Nutrition Policy (COMA) report (28) point out that there is a lack of convincing evidence for an anticancer effect of quercetin and other flavonoids in human populations, even though such effects can be seen in *in vitro* studies. Populations with a high intake of quercetin are not characterized by low mortality from cancer in a number of organs. In contrast, the epidemiological evidence that flavonoids like quercetin have a protective effect against heart disease is rather stronger, if not completely consistent. The mechanism for such a protective effect remains to be clarified. Janssen *et al.* (40) present evidence that the effect is not mediated in the main by haemostatic variables. Thus there remains much to be learned about how phytoprotective effects operate.

Biological effects of non-saponifiable components of cocoa butter

Cocoa beans contain around 55% lipid, called cocoa butter, most of which is triacylglycerol. A small fraction of cocoa butter (0.3%) comprises non-saponifiable material (Table 8.3). Thirty-nine sterols or triterpenes were identified or tentatively assigned by Staphylakis and Gegiou (21, 22). They estimated the content of non-esterified sterol in cocoa butter at 216 mg/100 g, much of it β-sitosterol (123 mg/100 g) and stigmasterol (60 mg/100 g). They also found 15 mg/100 g of methyl sterols and 35 mg/100 g of dimethylsterols, predominantly cycloartenol.

Sterols are very similar in structure to cholesterol (Fig. 8.5), and are capable of interfering with the absorption of dietary cholesterol. Sterols may therefore act to lower blood cholesterol, thereby reducing one of the important risk factors for heart disease.

The significance of the non-saponifiables in cocoa butter in lowering cholesterol has recently been reviewed by BIBRA (19). The following summarizes the assessment of this report. There have been a number of studies over the years demonstrating the ability of ß-sitosterol to lower cholesterol levels. However, it should be noted that most of the studies were carried out on subjects suffering from primary hypercholesterolaemia or hyperlipoproteinaemia. Such patients are genetically predisposed to synthesise cholesterol excessively, resulting in high blood cholesterol levels. Few studies have been done on subjects with normal or only slightly raised cholesterol levels. Two studies involving normal subjects demonstrated a reduction in cholesterol absorption of 40–50% when high levels of β-sitosterol were consumed. However it should be noted that in normal subjects dietary cholesterol is less of a determinant of cholesterol levels than the level of cholesterol-raising fatty acids in the diet.

There appears to be no information on the diterpene phytol, but there is quite a lot of information on the triterpenes including cycloartenol which is found in cocoa butter at levels around 0.01%. It is also found in rice bran oil, an edible oil used extensively in India and the Far East. Rice bran oil was found to reduce cholesterol levels, an effect which was later found to be associated with the non-saponifiable fraction which has cycloartenol as a major component. Cycloartenol is also found in olive oil, an oil also known to reduce the levels of LDL in humans. It may be concluded that the cholesterol-reducing properties of rice bran oil can be partly ascribed to the triterpene content.

The BIBRA review referred to above made an assessment of effects the above non-saponifiables were likely to have on cholesterol levels from their consumption via chocolate. The conclusions are summarized in Table 8.6.

Table 8.6 Cocoa butter non-saponifiables in chocolate.

Compound	Dose (mg/100 g chocolate)	Lowest effective dose
ß-Sitosterol	35.0	360 mg (healthy humans)
Stigmasterol	12.6	3500 mg (from rat study)
Camposterol	4.9	105 g (from mouse study)
Phytol	10.5	No data
Cycloartenol	4.9	1750 mg (from rat study)

Rather disappointingly, it would seem that the quantities of the compounds that may be ingested from eating even quite high amounts of chocolate are considerably less than the amounts that are needed before an appreciable effect on levels of cholesterol in the blood might be seen. Note that in the case of β-sitosterol, the observed biological effect is on cholesterol absorption. At the low levels found in chocolate, this is unlikely to translate into an observed effect on blood cholesterol in subjects with normal levels of blood cholesterol (as distinct from those suffering from hypercholesterolaemia).

One could speculate as to whether these compounds might act synergistically. The low levels of these compounds make this rather unlikely, and results from human studies of cocoa butter and milk chocolate on blood cholesterol (41) would seem to confirm this.

Phenolic acids in cocoa

Over the years a variety of phenolic acids have been identified in cocoa (16), Table 8.2, Fig. 8.6). There seems to be relatively little recent information on these constituents of cocoa, although recent additions to the list have been clovamide

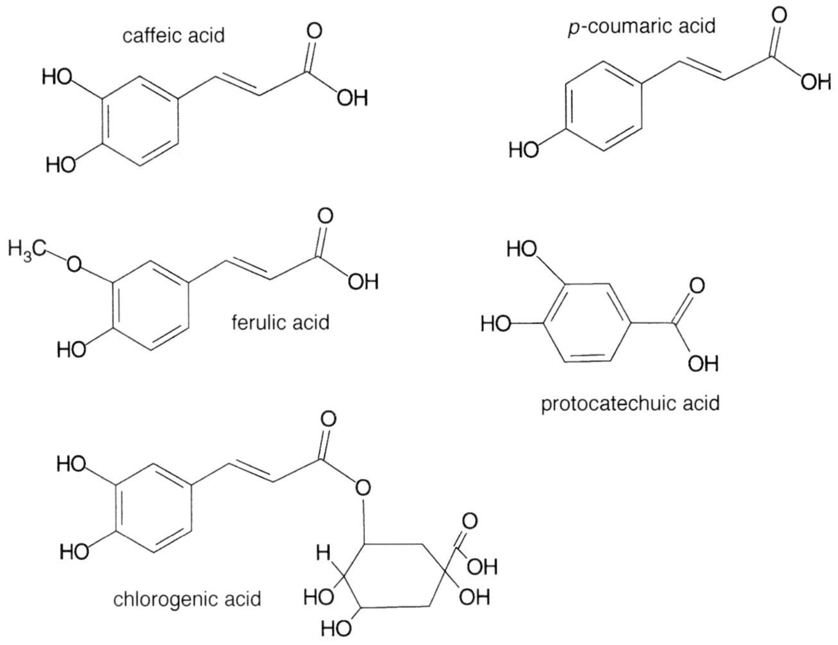

caffeic acid

p-coumaric acid

ferulic acid

protocatechuic acid

chlorogenic acid

Fig. 8.6 Structures of some phenolic acids found in cocoa.

and deoxyclovamide ((18), Fig. 8.4). There is a dearth of information on the quantities present in cocoa beans, fermented or unfermented, and even less on what is present in chocolate. This may be because these compounds were identified in the course of work on flavour aspects of cocoa and chocolate. As many of these compounds hold little interest from a flavour perspective, they have not been systematically studied or quantified in cocoa.

However, many of the compounds found in cocoa are widespread in the plant kingdom, where they are involved in general metabolic processes. Some of them are now beginning to attract interest elsewhere as potentially beneficial phytochemicals. For example, many have antioxidant properties, and this may be a major factor in providing protective effects.

Antioxidant properties of plant phenolics

A good example of the focus of recent work is ferulic acid, a phenolic which has been identified in cocoa. A great deal is known about ferulic acid in terms of its metabolism in plants and its chemistry, but only recently has its antioxidant activity been investigated. Graf reviewed this in 1992 (42) and identified two main areas of interest, namely radical scavenging and ultraviolet (UV) absorption. The ability of ferulic acid to scavenge free radicals derives from its structure permitting abstraction of a hydrogen atom to form a phenoxy radical. The resulting ferulate radical is resonance stabilized, making it unlikely to propagate further radical formation; its likely fate is condensation with another radical, including another ferulate radical, to yield the dimer curcumin. Inhibition of lipid peroxidation by ferulic acid has been shown in rat brain homogenates *in vitro*, although under the conditions employed, caffeic acid (also found in cocoa) was a lot more effective. It has also been reported that ferulic acid, unlike caffeic acid, scavenged superoxide anion radical, and also inhibited lipid peroxidation induced by superoxide, an effect similar in magnitude to that seen with superoxide dismutase (42).

Chlorogenic acid, caffeic acid and protocatechuic acid were shown by Ohnishi *et al.* (43) to have significant radical-scavenging activities in an *in vitro* test system. Chlorogenic, caffeic, ferulic and protocatechuic acids displayed activity in preventing the early stages of peroxidation of linoleic acid. These compounds were also tested to see whether they could prevent lipid peroxidation and haemolytic damage to isolated mouse erythrocytes during exposure to hydrogen peroxide. Chlorogenic and caffeic acids were both effective in this respect, showing more activity than α-tocopherol.

Ohnishi *et al.* (43) propose that these phenolics react with lipid peroxides (LOO•) as follows:

$$\text{PheOH} + \text{LOO}\bullet \rightarrow \text{PheO}\bullet + \text{LOOH}$$

In addition, there is evidence that some compounds, like the chlorogenic acid radical, may react with lipid peroxides:

$$\text{PheO}\bullet + \text{LOO}\bullet \rightarrow \text{LOO-Phe}$$

This activity is potentially important. When lipid peroxides are formed, they can react with other molecules in the vicinity – in membranes, this will often be with other lipids – thereby propagating a chain reaction. The above reaction demonstrates how phenolic antioxidants can act as chain terminators, breaking the cycle of peroxidation.

The activity of caffeic acid against reactive oxygen species was confirmed by Chimi *et al.* (44), who demonstrated that it could scavenge both hydroxyl radicals (•OH) and superoxide (O_2•-). It was also tested in a hepatocyte cell system supplemented with iron, inducing an oxidative stress which mimics certain pathological conditions. It was shown that caffeic acid was able to inhibit iron-induced lipid peroxidation. The authors comment that although free-radical scavenging might be the explanation for this protective effect, it is possible that the mechanism might also include iron chelation, effectively preventing the iron from reacting in this iron-loaded system. Such a mechanism has been demonstrated for polyphenols, for example flavonoids.

Such *in vitro* systems must, however, be interpreted carefully. In some circumstances, phenolics (including flavonoids) can act to promote oxidative damage, giving rise to cytotoxicity and genotoxicity in *in vitro* assay systems. For example, in the presence of transition metal ions like iron (Fe^{3+}) and copper (Cu^{2+}), caffeic acid can induce cellular and DNA damage (45). This was ascribed to autoxidation, which can occur to phenolics in the presence of such metal ions, resulting in the formation of H_2O_2 which mediates the damage. It would appear that where such ions are protein-bound, the capacity for damage is much less. There are a number of other factors which also need to be taken into account when evaluating the potential bioactivity of dietary phenolics (46).

Phenolics as protectants against LDL oxidation

The antioxidant potential of ferulic acid was investigated in lipid peroxidizing systems by Castelluccio *et al.* (47), alongside, among others, chlorogenic acid, caffeic acid and *p*-coumaric acid, each of which has been identified in cocoa. These workers investigated the relative effectiveness of these phenolic compounds in preventing peroxidation of LDL, protecting LDL cholesterol from oxidation, and preventing the oxidative modification of apoprotein B_{100}. The most effective were caffeic acid and chlorogenic acid, followed by ferulic acid and then *p*-coumaric acid. Vinson *et al.* (31) also noted that chlorogenic acid had the capacity to protect LDL from oxidation *in vitro*.

In another *in vitro* study of LDL oxidation, Laranjinha *et al.* (48) confirmed that chlorogenic acid and caffeic acid rapidly scavenge peroxyl (ROO•) radicals in the aqueous phase, thereby preventing the chain initiation of LDL peroxidation. In contrast, the behaviour of protocatechuic acid in their system was different, suggesting that this compound was not acting as an antioxidant but as a retardant; that is, it reacts with ROO• to slow down the initiation and propagation steps of lipid peroxidation to some extent. The same group (49) have further elucidated some of the reactions occurring within LDL and illustrated possible mechanisms by which phenolic antioxidants could protect against lipid peroxidation.

The investigators employed a test system using ferrylmyoglobin as an oxidant, and examined α-tocopherol consumption in LDL. α-Tocopherol is considered to be the major antioxidant present in LDL. In the test system, α-tocopherol disappeared as a function of ferrylmyoglobin concentration. After depletion of α-tocopherol, caffeic and *p*-coumaric acids were able to prevent lipid peroxidation, by a mechanism that converts ferrylmyoglobin back to metmyoglobin. Caffeic acid also delayed α-tocopherol consumption when present before the oxidation challenge or added during it, whereas *p*-coumaric acid accelerated the rate of α-tocopherol consumption. Thus these two phenolics may play different roles, the results suggesting that caffeic acid acts synergistically with α-tocopherol, by recycling it after oxidation, thereby extending the antioxidant capacity of LDL. *p*-Coumaric acid, despite increasing α-tocopherol consumption, nevertheless would appear to have a protectant effect, since LDL particles enriched with α-tocopherol still had enhanced protection against lipid peroxidation. This added protection may be on the basis of chain-breaking activity.

One interpretation of these findings is that α-tocopherol recycles *p*-coumaric acid from its radical form, making it available again to prevent propagation of peroxidation. The authors comment that it remains to be seen whether such reactions occur *in vivo*. However, these compounds are found in many foods including cocoa, with caffeic acid at quite high levels in the diet. Caffeic acid has furthermore been identified in human blood plasma, so a biological role is possible.

Anticarcinogenic effects of phenolics

Phenolic compounds like caffeic acid and ferulic acid can both block the production of carcinogens and prevent their action on DNA (50). Phenolics can induce the formation of 'phase II' enzymes, increasing detoxification potential (51).

Caffeic, chlorogenic and ferulic acids were found to inhibit carcinogenesis induced by the chemical 4-nitroquinoline-1-oxide (NQO) (52). This chemical when administered to rats in drinking water leads to the formation of carcinomas of the tongue. When the phenolics were fed at levels of 250–500 parts per million in the diet just prior to and during the NQO administration, then the number of carcinomas and precancerous changes were significantly reduced. At the levels fed, the phenolics were judged to have no toxic effects. This experiment is consistent with the hypothesis that the phenolics block the early stages of carcinogenesis.

The efficacy of protocatechuic acid in reducing carcinogenic activity of a variety of chemical carcinogens including NQO at various target organs in experimental rat studies has also been demonstrated (53). The results suggest that protocatechuic acid, like the phenolics in the study above, acts to block carcinogenesis. There is evidence that it can also suppress the later cell proliferation stages of tumour formation since the number of tumours was reduced if protocatechuic acid was fed after the initiation stage, that is, after administration of the chemical carcinogen.

Aflatoxin B_1 is a potent liver carcinogen in animals and possibly in humans. Aboobaker *et al.* (54) investigated the effect of both simple phenolics (caffeic and chlorogenic acids) and flavonoids (e.g. catechin) on some of the significant steps involved in aflatoxin carcinogenesis. Aflatoxin is first activated by oxidative enzymes to produce an epoxide, which in turn can react with DNA to cause the genotoxic damage, which leads ultimately to liver cancer. Diets containing the phenolics were fed to rats for 3 weeks, after which cell fractions containing the activating enzymes were extracted from the livers. The ability of the enzymes to form the epoxide from aflatoxin B_1 was then tested *in vitro*. It was found that the enzymes derived from rats fed the two phenolic acids (and catechin) had a reduced capacity to form the aflatoxin epoxide and also reduced DNA adduct formation. The ability to increase detoxification, in this case as shown by increasing capacity to detoxify the aflatoxin epoxide by reaction with glutathione, was also tested. Caffeic and chlorogenic acids were found to be inactive, but catechin stimulated this 'phase II' detoxification.

These results encourage the notion that simple phenolic compounds found in plant foods including cocoa could provide a degree of protection against cancer, but it is clear that there remains much to be learned about the mechanisms involved. The steps leading to cancer are complex and phytoprotectants could act at various steps in the process.

Other effects of phenolics

Due to its conjugated unsaturation, ferulic acid strongly absorbs UV radiation, and at high concentrations may protect other light-sensitive compounds against oxidative damage by preferentially absorbing UV. As pointed out by Graf (42) this may have biological significance, as the highest concentrations of ferulic acid occur in leaves, seeds and other organs with a high surface to volume ratio. This effect has already been harnessed for human health, by its topical use in skin lotions and sunscreens. It has been investigated as an agent to lessen the side-effects of chemo- and radiotherapies of carcinoma by increasing the natural immune defence. There are also reports (cited by Graf (42)) of anti-inflammatory activity, inhibition of chemically induced carcinogenesis as well as an anti-diarrhoeal activity.

A beneficial effect of ferulic acid was also seen (55) in a study of nutritional muscular deficiency, a disease induced in chicks under conditions of vitamin E deficiency. When ferulic acid (and also quercetin, a flavonoid found in cocoa) were fed to chicks reared on a diet deficient in vitamin E and essential fatty acid, there appeared to be a marked conservation or production of essential fatty acids in tissues. Ferulic acid and quercetin were found to provide partial protection against the disorder.

Conclusion

This review presents an intriguing scenario. Underlying the pathologies of many human diseases are mechanisms which depend on oxidative reactions. On the other hand there is also considerable epidemiological evidence that plant materials have protective effects against many of these diseases. At the same time, it has become increasingly appreciated that certain constituents in plants have antioxidant properties. It is therefore not surprising that there is a widespread presumption that at least some of the protective effects of plants depend upon these antioxidant compounds. It may be speculated that plant materials that are rich in antioxidants are also beneficial to human health.

However, various pieces in this intriguing jigsaw will have to be put into place before we can be sure that foods of plant origin like cocoa have a measurable beneficial effect on human health. Despite all the optimism in the field of phytochemicals, there remain large and tantalizing gaps in our knowledge. Although effects can be seen *in vitro*, we need to know whether and how much of these active compounds are absorbed. In most cases our knowledge about the absorption of phytochemicals is fragmentary, and thus we know little about their target tissues or their effects *in vivo*. Where these compounds do show protective activity, we will need to understand the mechanism of action. Studies of cocoa in this field lag some way behind other plant materials, so that less is known about the effects of cocoa than some other plant materials. Even our knowledge of the constituents of the most commonly consumed form of cocoa, namely chocolate, is very limited. Yet we know already from early work on cocoa – raw and fermented – that it has a high content of antioxidants. Cocoa is conceivably a treasure chest of compounds with potentially beneficial effects on human health. It will be fascinating to watch the development of the field of phytochemicals, and to learn its implications for cocoa products.

References

1. Forsyth, W.G.C. and Quesnel, V. (1957) Cacao polyphenolic substances. IV. The anthocyanin pigments. *Biochem. J.* **65**, 177–179.
2. Nutrition Society Symposium (1996) Physiologically active substances in plant foods (non-nutrient nutrients). *Proc. Nutr. Soc.* **55**, 370–446.
3. Reddy, S., Thane, C.W. and Jones, S.A. (1996) BFMIRA scientific and technical survey no. 182: *Fruit and vegetables: nature's functional foods*. BFMIRA, London.
4. Steinmetz, K.A. and Potter, J.D. (1996) Vegetables, fruit, and cancer prevention: a review. *J. Am. Diet. Assoc.* **96**, 1027–1039.
5. Ross, R. (1993) The pathogenesis of atherosclerosis: a perspective for the 1990s. *Nature* **362**, 801–809.
6. Witztum, J.L. (1994) The oxidation hypothesis of atherosclerosis. *Lancet* **344**, 793–795.
7. Leake, D.S. (1998) The effects of flavonoids on the oxidation of low-density lipoproteins. In *Flavonoids in Health and Disease* (Ed. by Rice-Evans, C.A. and Packer, L.), pp. 253–276. Marcel Decker, New York.
8. Kondo, K., Hirano, R., Matsumoto, A., Igarashi, O. and Itakura, H. (1996) Inhibition of LDL oxidation by cocoa. *Lancet* **348**, 1514.
9. Waterhouse, A.L., Shirley, J.R. and Donovan, J.L. (1996) Antioxidants in chocolate. *Lancet* **348**, 834.

10. Natsume, M. (1997) *Effects of cacao-mass polyphenol on the LDL oxidation capacity.* Annual Meeting of the Japan Society for Bioscience, Biotechnology and Agrochemistry.
11. Osawa, T. (1995) *Antioxidation and antimutagenic reactions of polyphenols contained in chocolate cocoa.* International Symposium on Nutrition of Chocolate and Cocoa, Japan.
12. Sanbongi, C., Suzuki, N. and Sakane, T. (1997) Polyphenols in chocolate, which have antioxidant activity, modulate immune functions in humans *in vitro. Cell. Immunol.* **177**, 129–136.
13. Sakane, T. (1995) *Immunity-regulating reaction by cacao-mass antioxidant.* International Symposium on Nutrition of Chocolate and Cocoa, Japan.
14. Sakane, T. (1996) *Control effects of cacao-mass polyphenols over chronic and allergic irritation.* International Symposium on Nutrition of Chocolate and Cocoa, Japan.
15. Takeda, H. (1996) *Stress-alleviating effect of constituents of the cacao bean.* International Symposium on Nutrition of Chocolate and Cocoa, Japan.
16. Ziegleder, G. and Biehl, B. (1988) Analysis of cocoa flavour components and flavour precursors. In *Analysis of Non-Alcoholic Beverages* (Ed. by Linskens, H.F. and Jackson, J.F.), pp. 321–393. Springer-Verlag, Berlin.
17. Mohr, W. (1962) Über das Verkommen von Polyhydroxyphenolen in Schokoladenmassen und ihr Verhalten beim Conchieren. *Fette Seifen Anstrichmittel Ernährungs* **69**, 831–844.
18. Sanbongi, C., Osakabe, N., Natsume, M., Tazikawa, T., Gomi, S. and Osawa, T. (1998) Antioxidative polyphenols isolated from *Theobroma cacao. J. Agric. Food Chem.* **46**, 454–457.
19. BIBRA (1996) Cholesterol-lowering effects of the non-saponifiable components of cocoa butter. Report commissioned by the Biscuit, Cake, Chocolate and Confectionery Alliance, London.
20. Chan, W., Brown, J. and Buss, D.H. (1994) In *Miscellaneous Foods: Supplement to McCance and Widdowson's The Composition of Foods*, p. 166. Royal Society of Chemistry and Ministry of Agriculture, Fisheries and Food, London.
21. Staphylakis, K. and Gegiou, D. (1985) Sterols in cocoa butter. *Fette Seifen Anstrichmittel* **87**, 150–153.
22. Staphylakis, K. and Gegiou, D. (1985) Free, esterified and glucosidic sterols in cocoa butter. *Lipids* **20**, 723–728.
23. Porter, L.J., Ma, Z. and Chan, B.G. (1991) Cacao procyanidins: major flavanoids and identification of some minor metabolites. *Phytochemistry* **30**, 1657–1663.
24. Forsyth, W.G.C. and Quesnel, V. (1963) The mechanism of cacao curing. *Adv. Enzymol.* **25**, 457–492.
25. Kim, H. and Keeney, P. (1983) *Polyphenols – tannins in cocoa beans.* 37th PMCA Production Conference, pp. 60–63.
26. Bonvehi, J.S. and Coll, F.V. Evaluation of bitterness and astringency of polyphenolic compounds in cocoa powder 1997. *Food Chem.* **60**, 365–370.
27. BIBRA (1996a) *An evaluation of the beneficial effects of cocoa nib constituents.* Report commissioned by the Biscuit, Cake, Chocolate and Confectionery Alliance, London.
28. COMA (1998) Nutritional aspects of the development of cancer. Report on Health and Social Subjects, 48. UK Department of Health, London.

29. Katiyar, S.K., Agarwal, R. and Mukhtar, H. (1994) Inhibition of spontaneous and photoenhanced lipid peroxidation in mouse epidermal microsomes by epicatechin derivatives from green tea. *Cancer Lett.* **79**, 61–66.
30. Shimoi, K. (1996) Radioprotective effects of antioxidant plant flavonoids in mice. *Mutat. Res.* **350**, 153–161.
31. Vinson, J., Jang, J. Dabbagh, Y.A., Serry, M.M., and Songhuai, C. (1995) Plant polyphenols exhibit lipoprotein-bound antioxidant activity using an *in vitro* oxidation model for heart disease. *J. Agric. Food Chem.* **43**, 2798–2799.
32. Matsuoka, Y. (1995) Ameliorative effects of tea catechins on active oxygen-related nerve cell injuries. *J. Pharmacol. Exp. Ther.* **274**, 602–608.
33. Ikeda, I. (1992) Tea catechins decrease micellar solubility and intestinal absorption of cholesterol in rats. *Biochem. Biophys. Acta* **1127**, 141–146.
34. Hollman, P.C.H. and Katan, M.B. (1998) Absorption, metabolism, and bioavailability of flavonoids. In *Flavonoids in Health and Disease* (Ed. by Rice-Evans, C.A. and Packer, L.), pp. 483–522. Marcel Decker, New York.
35. Serafini, M., Ghiselli, A. and Ferro-Luzzi, A. (1996) *In vitro* antioxidant effect of green and black tea in man. *Eur. J. Clin. Nutr.* **50**, 28–32.
36. M&M Mars (1996) *Antineoplastic cocoa extracts, methods for making, using.* Patent: WO 96/10404.
37. M&M Mars (1997) *Cocoa extract compounds and methods for making and using the same.* Patent: WO 97/36497.
38. M&M Mars (1997) *Cocoa extract compounds and methods for making and using the same.* Patent: WO 97/36597.
39. Hertog, M.G.L. and Hollman, P.C.H. (1996) Potential health effects of the dietary flavonol quercetin. *Eur. J. Clin. Nutr.* **50**, 63–71.
40. Janssen, P.L.T.M.K., Mensink, R.P., Cox, F.J.J., *et al.* (1998) Effects of the flavonoids quercetin and apigenin on hemostasis in healthy volunteers: results from an *in vitro* and a dietary supplement study. *Am. J. Clin. Nutr.* **67**, 255–262.
41. Kris-Etherton, P.M., Derr, J., Mitchell, D.C., *et al.* (1993) The role of fatty acid saturation on plasma lipids, lipoproteins, and apolipoproteins: 1. Effects of whole food diets high in cocoa butter, olive oil, soybean oil, dairy butter, and milk chocolate on the plasma lipids of young men. *Metabolism* **42**, 121–129.
42. Graf, E. (1992) Antioxidant potential of ferulic acid. *Free Radical Biol. Med.* **13**, 435–448.
43. Ohnishi, M., Morishita, H., Iwahashi, H., *et al.* (1994) Inhibitory effects of chlorogenic acids on linoleic acid peroxidation and haemolysis. *Phytochemistry* **36**, 579–583.
44. Chimi, H., Morel, I., Lescoat, G., Pasdeloup, N., Cillard, P. and Cillard, J. (1995) Inhibition of iron toxicity in rat hepatocyte culture by natural phenolic compounds. *Toxic. In Vitro* **9**, 695–702.
45. Nakayama, T. (1994) Suppression of hydroperoxide-induced cytotoxicity by polyphenols. *Cancer Research* (Suppl.) **54**, 1991s–1993s.
46. Decker, E.A. (1997) Phenolics: pro-oxidants or antioxidants? *Nutr. Rev.* **55**, 396–398.
47. Castelluccio, C., Paranga, G., Melikian, N., *et al.* (1995) Antioxidant potential of intermediates in phenylpropanoid metabolism in higher plants. *FEBS Lett.* **368**, 188–192.
48. Laranjinha, J.A.N., Almeida, L.M. and Madeira, V.M.C. (1994) Reactivity of dietary

phenolic acids with peroxyl radicals: antioxidant activity upon low-density lipoprotein peroxidation. *Biochem. Pharmacol.* **48**, 487–494.

49. Laranjinha, J., Vieira, O., Madeira, V. and Almeida, L. (1995) Two related phenolic antioxidants with opposite effects on vitamin E content in low-density lipoproteins oxidized by ferrylmyoglobin: consumption vs regeneration. *Arch. Biochem. Biophys.* **323**, 373–381.
50. Wattenberg, L.W. (1985) Chemoprevention of cancer. *Cancer Res.* **45**, 1–8.
51. Stich, H.F. and Rosin, M.P. (1984) Naturally occurring phenolics as antimutagenic and anticarcinogenic agents. *Adv. Exp. Med.* **177**, 1–29.
52. Tanaka, T., Kojima, T., Kawamori, T., *et al.* (1993) Inhibition of 4-nitroquinoline-1-oxide-induced rat tongue carcinogenesis by the naturally occurring plant phenolics caffeic, ellagic, chlorogenic and ferulic acids. *Carcinogenesis* **14**, 1321–1325.
53. Tanaka, T., Kojima, T., Kawamori, T. and Mori, H. (1995) Chemoprevention of 4-nitroquinoline-1-oxide-induced oral carcinogenesis by dietary protocatechuic acid during initiation and postinitiation phases. *Cancer* (Suppl. 6) **75**, 1433–1439.
54. Aboobaker, V.S., Balgi, A.D. and Bhattacharya, R.K. (1994) *In vivo* effect of dietary factors on the molecular action of aflatoxin B1: role of non-nutrient phenolic compounds on the catalytic activity of liver fractions. *In Vivo* **8**, 1095–1068.
55. Jenkins, K.J. and Atwal, A.S. (1995) Flavonoids increase tissue essential fatty acids in vitamin E-deficient chicks. *Nutr. Biochem.* **6**, 97–103.

Chapter 9

Minerals in Cocoa and Chocolate

Ian Knight

The importance of minerals to the body and their form in the diet have been the subject of great discussion over the years. The role of minerals is still being uncovered as research progresses, although it is clear that macro-minerals such as calcium, phosphorus and magnesium play an important part in maintaining the integrity of human physiology. Hand in hand with the beneficial effect of maintaining the correct level of macro-nutrients themselves, of course, are the confounding effects of oxalate and phytate, which interfere with the bio-availability of certain minerals.

Historical studies have revealed that 14 elements are required by the body in quantities of less than a few milligrams per day; they are chromium, cobalt, copper, fluorine, iodine, iron, manganese, molybdenum, nickel, selenium, silicon, tin, vanadium and zinc. Their actions on the physiology of the human body can be so diverse as to have defied rational efforts to classify them. Nonetheless, Table 9.1. shows the broad physiological areas in which macro- and micro-minerals function as far as they are presently understood.

Generally, the UK Ministry of Agriculture, Fisheries and Food (MAFF) (1) broadly describes three main functions:

(1) As constituents of the bones and teeth. These include calcium, phosphorus and magnesium.
(2) As soluble salts which help control the composition of body fluids and cells. These include *sodium* and *chloride* in the fluids outside the cells (e.g. blood), and *potassium*, *magnesium* and *phosphorus* inside the cells.
(3) As essential adjuncts to many enzymes and other proteins such as hemoglobin, which are necessary for the release and utilization of energy. *Iron*, *phosphorus*, *zinc* and most of the remaining elements act in this way.

As enormous advances have been made in analytical techniques, so progress has been made to demystify some of these mechanisms, although in many cases short-term deprivation of specific minerals is not noticeable, only becoming

Table 9.1 Physiological contribution of minerals.

Function	Mineral*														
	Fe	Ca	P	Mg	Zn	Na	K	Cl	Cr	Cu	F	I	Mn	Se	Mb
Building bones	X														
Maintaining bone strength		X	X	X											
Bone maintenance		X									X				
Blood clotting		X													
Blood synthesis	X									X					X
Blood pressure regulation						X	X								
Body cell function				X											
Energy utilization			X												
DNA/RNA function			X												
Regulation of body fluids						X	X	X							
Digestive function								X							
Insulin/glucose function									X						
Macro-nutrient utilization					X										
Enzymes				X						X					X
CNS function		X				X	X	X					X		
Cell energy										X					
Cell growth					X									X	
Thyroid function												X			
Antioxidant function														X	
Tissue repair					X										
Tooth maintenance											X				
Muscle function		X				X	X								

* Fe = iron; Ca = calcium; P = phosphorus; Mg = magnesium; Zn = zinc; Na = sodium; K = potassium; Cl = chlorine; Cr = chromium; Cu = copper; F = fluorine; I = iodine; Mn = manganese; Se = selenium; Mb = molybdenum.

obvious after long periods. As the biochemical roles for mineral micro-nutrients are being unraveled, it is becoming clear that cofactor actions with enzymes play a central part and that obvious deficiency effects may appear as secondary or even tertiary consequences of the primary cause.

Because many trace elements can exert antagonistic effects on other elements, the effect of a deficiency in one specific element can be confusing, as the observed effect may in fact be the result of the relaxing of an antagonism. This could take the form of competitive absorption or it may be competition for limited amounts of plasma transport vehicles.

The physiological role of minerals

All minerals are absorbed in the intestines and then transported and stored in different ways. Because they are stored, consuming excess minerals can be harmful. On the other hand, deficiencies of dietary minerals are associated with many apparently unrelated problems. The extent to which they are absorbed differs greatly. Iron, for example, is absorbed to differing degrees depending upon the source of food supplying it. Heme and non-heme iron are generally obtained from animal and vegetable sources, respectively. Heme iron is well absorbed (20–30%), whereas non-heme iron is less well absorbed unless the body's stores are depleted; however, this is positively influenced by vitamin C. Similarly, only 30–40% of calcium in the diet is normally absorbed. This level is also positively influenced by vitamin D and negatively influenced by phytate and oxalate, which are present in certain foods.

The role of cocoa and chocolate

Cocoa is almost unique in being a veritable storehouse of minerals. According to MAFF, it contains more iron than almost any other vegetable (10.5 mg/100 g); only curry powder contains more (58.3 mg/100 g). Table 9.2 shows the contribution of minerals made by cocoa powder.

Table 9.2 Mineral composition of cocoa powder.

Mineral	(mg/100 g)
Ash	6330
Potassium	2058
Sodium	9
Calcium	170
Magnesium	594
Iron	14
Phosphorus	795
Zinc	8
Copper	5
Manganese	5

Source: CMA (2).

This can be contrasted with Table 9.3, showing UK intakes of minerals according to MAFF (1); Table 9.4, showing US recommended daily allowances (RDAs); and Table 9.5, expressing US estimated safe and adequate daily dietary intakes (ESADDIs) of selected minerals (3).

It is clear that cocoa is a rich source of minerals (4). Copper was not included in either of Tables 9.1 or 9.2; however, it has become the subject of more intense study recently as its deficiency has been suggested in the prevalence of biochemical parameters associated with cardiovascular disease (5). Dark chocolate candies contribute high levels of copper (22.1% of ESADDI), whereas milk chocolate candies contribute 11.3% of ESADDI. It was shown, in a study using 3-day dietary records, that chocolate is a major source of dietary copper in the North American diet, making the highest contribution to mean daily copper

Table 9.3 Daily intake and total body content of minerals for an adult man.

Mineral	Daily intake	Total body content
Major minerals	**(mg)**	**(g)**
Calcium	900	1000
Phosphorus	1500	780
Potassium	3200	140
Sodium	3400	140
Chloride	5200	95
Magnesium	300	19
Iron	14.0	4.2
Zinc	11.4	2.3
Trace elements	**(mg)**	**(mg)**
Fluoride	1.82	2600
Copper	1.63	72
Selenium	0.06	>15
Iodine	0.24	13
Manganese	5.0	12
Chromium	0.09	<2
Cobalt	0.3	1.5

Source: Crown copyright is reproduced with the permission of the Controller of Her Majesty's Stationery Office (1).

Table 9.4 US recommended dietary allowances.

	Age	Calcium (mg)	Phosphorus (mg)	Magnesium (mg)	Iron (mg)	Zinc (mg)	Iodine (μg)	Selenium (μg)
Infants	0–6 months	400	300	40	6	5	40	10
	6–12 months	600	500	60	10	5	50	15
Children	1–3	800	800	80	10	10	70	20
	4–6	800	800	120	10	10	90	20
	7–10	80	800	170	10	10	120	30
Males	11–14	1200	1200	270	12	15	150	40
	15–18	1200	1200	400	12	15	150	50
	19–24	1200	1200	350	10	15	150	70
	25–50	800	800	350	10	15	150	70
	51+	800	800	350	10	15	150	70
Females	11–14	1200	1200	280	15	12	150	45
	15–18	1200	1200	300	15	12	150	50
	19–24	1200	1200	280	15	12	150	55
	25–50	800	800	280	15	12	150	55
	51+	800	800	280	10	12	150	55
Pregnant		1200	1200	320	30	15	175	65
& lactating								
	1st 6 months	1200	1200	355	15	19	200	75
	2nd 6 months	1200	1200	340	15	16	200	75

Source: adapted with permission from NAS (3).

Table 9.5 Estimated safe and adequate daily dietary intakes (ESADDI) for minerals.

	Age	Copper (mg)	Manganese (mg)	Fluoride (mg)	Chromium (μg)	Molybdenum (μg)
Infants	0–6 months	0.4–0.6	0.3–0.6	0.1–0.5	10–40	15–30
	6–12 months	0.6–0.7	0.6–1.0	0.2–1.0	20–60	20–40
Children & Adolescents	1–3	0.7–1.0	1.0–1.5	0.5–1.5	20–80	25–50
	4–6	1.0–1.5	1.5–2.0	1.0–2.5	30–120	30–75
	7–10	1.0–2.0	2.0–3.0	1.5–2.5	50–200	50–150
	11+	1.5–2.5	2.0–5.0	1.5–2.5	50–200	75–250
Adults		1.5–3.0	2.0–5.0	1.5–4.0	50–200	75–250

Source: adapted with permission from NAS (3).

intake (5). Other major contributing sources were various chocolate-containing foods and drinks.

Sodium, potassium and chloride are virtually ubiquitous in the diet and so are always present in excess. Calcium is perhaps the only essential element for which cocoa powder is not a rich source; however, since cocoa is often consumed with milk in either drinks or milk chocolate, this is of little concern. Equally, however, consideration of chocolate as the dietary form of cocoa consumed may reduce some of the other mineral contributions, as the cocoa component is only a part of the formulation, albeit a significant part. Even so, both forms are capable of providing not insignificant amounts of many essential elements, although clearly chocolate should not be thought of as the sole dietary component for the attainment of mineral levels.

Requirements for magnesium are between 120 and 350 mg for most demographics. Cocoa contains almost 600 mg of magnesium per 100 g compared with 140, 165, 121 and 119 mg per 100 g for peas, white wheat, corn and rice, respectively (6). In fact, it was shown that the regular use of a cocoa product effectively prevented negative nutritional effects of a diet moderately deficient in magnesium (7). Iron present in cocoa powder is 93% usable and 85% of the phosphorus is reported to be bioavailable, although 15% of this is present as phytate (6). This tends to form complexes with both calcium and magnesium, thereby reducing their bioavailability. Cocoa or chocolate can aid in providing significant proportions of the RDA for iron, magnesium and zinc, as well as significant proportions of the ESADDI for copper and manganese.

Cocoa appears to have an affinity for minerals and its composition is reflective of its environment, being a natural product. It was feared that cocoa might also contain high amounts of heavy metals, since they are widespread in the environment (e.g. lead and cadmium in tea, yeast, crustaceans and molluscs). This was studied by Prugarova and Kovac (8), who found levels of lead and cadmium to be variable, but within acceptable levels.

There appears to be no argument as to the high mineral content of cocoa and chocolate; however, the extent to which they are available to the body has provoked discussion and further research. In reality, there is no reason to doubt that the body efficiently utilizes most minerals present in cocoa.

Interactions and bioavailability

The study of bioavailability of minerals is complex because of the wide range of variables that seem to affect the extent to which many of them are absorbed. This can range from macro-chemical interactions, such as oxalates with calcium and phosphorus as phytate with zinc, to micro-chemical reactions, such as ascorbic acid with iron and mineral/mineral interactions. Additionally, there are lifestyle factors such as fasting, which can influence mineral absorption.

Oxalates/calcium

Cocoa and consequently chocolate are rich in oxalic acid. In the presence of oxalic acid or its anion oxalate, calcium easily forms water-insoluble calcium oxalate, which can form into renal calculi or kidney stones in predisposed individuals (9). Cocoa was found to contain 0.72–1.18% oxalic acid on a fat-free basis (6). Kasidas and Rose (10) determined the oxalate content of common foods and found high levels in rhubarb, spinach, tea, beetroot (beets), peanuts, chocolate and parsley in approximate descending concentration.

Since calcium is essential to health and has come under greater scrutiny as more is known about osteoporosis, the possible effect of oxalate interfering with the availability of calcium takes on a greater significance. Indeed, there are examples in calcium deficiency during which, if insufficient calcium is present to provide for its various non-bone functions, the body can withdraw more calcium from the bones than is being deposited, leading to potentially serious reductions in bone strength (11).

A study by Nguyen *et al.* (12) measured a number of physiological parameters, notably calciuria, oxaluria, phosphaturia, as well as diabetes parameters such as plasma glucose insulin, C peptide, etc., following ingestion of a single 100 g dark chocolate bar against a 55 g sucrose control. They found a striking increase in oxaluria after chocolate ingestion with no change in the sucrose control group. This was consistent with previous findings by Balcke *et al.* (13), who also found increased oxaluria, but found no significant differences in oxalate excretion between groups consuming 50 g or 100 g chocolate bars, suggesting a plateau level beyond which no greater oxaluria is caused. Nguyen *et al.* also found calciuria increased after both chocolate and sucrose ingestion (12).

The authors suggest that these urinary changes may favor the formation of calcium oxalate calculi. These findings, and the work of Lagemann *et al.* (9), concluded that cocoa and chocolate consumption should be of concern only to those with a predisposition to hypercalciuria. Furthermore, counseling these stone formers led to a marked drop in daily oxalate excretion, presumably due to avoidance of foods rich in oxalates, although this is not specifically stated. Moreover, there were no recurrences of calculus formation in some 6 years in these subjects. There are foods containing higher amounts of oxalic acid than cocoa and chocolate, notably spinach, rhubarb and tea. The high consumption of the latter in the UK diet renders most other sources apparently unimportant (4).

Zinc/phytate

Interesting work by Aremu and Abara (4) in Nigeria on phytate – the calcium magnesium salt of inositolhexaphosphoric acid – examined both past and present findings on the relation between phytate:zinc molar ratios and the bioavailability of zinc in cocoa. The authors reported phytate:zinc ratios of 89 to 132 in five brands of cocoa beverage. They cited Turnland *et al.* (14), who suggested that phytate:zinc ratios of 15:1 lead to reduced bioavailability of zinc in humans and observed that were this strictly the case, high ratios in their own study would render all the zinc unavailable. Wise (15) found the relationship also extended to dietary calcium levels. In this model, phytate precipitation is incomplete until calcium:phytate molar ratios reach approximately 6:1. At ratios lower than this, phytate precipitation is incomplete, leaving some dietary zinc in solution and thereby available. The level of zinc remaining in solution then increases with decreasing calcium:phytate molar ratios. In any event, the literature agrees that high dietary phytate levels can lead to reduced bioavailability of zinc.

Iron/polyphenols

Non-heme iron absorption is dependent to a large degree on its binding to competing ligands in the digestive tract. Certain ligands, such as ascorbic acid, can enhance iron absorption while others may inhibit it. Consumption after fasting significantly increased absorption of iron in rats and it was found that some polyphenols, typically found in herb teas, cocoa and coffee, can inhibit iron absorption (16). There are references in the literature to work done to fortify cocoa drinks with iron (17, 18).

Miscellaneous

There are some examples of excess of one mineral causing deficiency of another. High doses of zinc can interfere with copper absorption, but such high levels could only be accomplished through supplementation and not through dietary means (19). An interesting study by Valiente *et al.* (20) investigated the binding capacity of cocoa and total dietary fiber (TDF) with cadmium *in vitro*. Cadmium is considered a toxic metal, competing with iron, copper, zinc, manganese and selenium for ligands in biological systems, and is one of the most effective elements capable of decreasing intestinal absorption of iron in the human body. Once absorbed, it accumulates in the kidneys, having a mean life of 18–30 years. Concentrations of cadmium ion greater than 200 μg/g damage the kidney irreversibly. Cadmium is found in many foods, especially in cereals and seafood, but cocoa appears to be high in cadmium (1.8 μg/g). However, it was found that cocoa possesses a great affinity for cadmium and will bind it to itself, thereby rendering it insoluble and unavailable to the kidneys. In an experiment using buffered solutions of cocoa and isolated TDF, designed to simulate physiological pH conditions, cocoa did not become saturated even at cocoa:cadmium ratios of 500:1, a level as high as 100 μg/ml of cadmium. As cocoa binds excess cadmium, it releases other beneficial minerals – iron, zinc, calcium and magnesium – particularly at digestive pH levels. This led the authors to propose cocoa as an agent for eliminating a proportion of toxic metals from the human body.

Summary

Cocoa and chocolate are both very rich sources of minerals, more so than almost any other food. The extent to which these minerals are actually used by the body seems to be the only question, and the experimental evidence persuades against generalizing since there are so many factors that seem capable of affecting the absorption of minerals.

While it is established that the presence of oxalic acid will limit the absorption of calcium, this does appear to be the only effect of any magnitude worthy of cautionary note. However, since cocoa is not a good source of calcium anyway, the point may seem academic, although milk chocolate does contain a significant amount of calcium. Nonetheless, only those people having a predisposition for hypercalciuria (i.e. stone formers) are advised to exercise more than normal dietary caution, as they should avoid all foods high in oxalic acid.

All of the literature reiterates the need to follow a balanced diet in order to obtain all of the necessary minerals in a sensible proportion. It would not be reasonable to offer other than this advice; however, in subjects with renal stone disorders, it would be prudent to avoid excessive consumption of foods rich in oxalates, such as cocoa, chocolate, beets, spinach and teas.

Generally, cocoa can provide a significant proportion of those minerals the human body needs to function properly, particularly copper, for which it is the major source in the North American diet. Moreover, in the case of minerals harmful to the body such as cadmium – a renal toxin – research suggests cocoa's propensity to help scour it from the body, by binding it harmlessly to itself and thereby facilitating its elimination.

References

1. MAFF (1995) Minerals. In *Manual of Nutrition – Reference Book 342*, 10th edn. HMSO, London.
2. CMA (1997) *Nutrient Database for Three Selected Major Ingredients Used in the NCA/CMA Recipe Modeling Database: Chocolate Liquor, Cocoa Powder and Cocoa Butter.* National Confectioners Association and Chocolate Manufacturers Association, McLean, VA.
3. National Academy of Sciences (1989) *Recommended Dietary Intakes*, 10th edn. National Academy Press, Washington, DC.
4. Aremu, C.Y. and Abara, A.E. (1992) Hydrocyanate, oxalate, phytate, calcium and zinc in selected brands of Nigerian cocoa beverage. *Plant Foods Hum. Nutr.* **42**, 231–237.
5. Joo, S.-J. and Betts, N.M. (1996) Copper intakes and consumption patterns of chocolate foods as sources of copper for individuals in the 1987–88 nationwide food consumption survey. *Nutr. Res.* **16** (1), 41–52.
6. Cook, L.R. and Muersing, E.H. (1982) Nutrition and physiology. In *Chocolate Production and Use*. Harcourt Brace Jovanovich, New York.
7. Planells, E., Monserrat, R., Carbonell, J., Mataix, J. and Llopis, J. (1997) Ability of a cocoa product to prevent chronic Mg deficiency in rats. *J. Agric. Food Chem.* **45**, 4017–4022.
8. Prugarova, A. and Kovac, M. (1987) Lead and cadmium content in cocoa beans. *Nahrung* **31** (5–6), 635–636.
9. Lagemann, M., Anders, D., Graef, V. and Bodeker, R.H. (1985) Effect of cocoa on excretion of oxalate, citrate, magnesium and calcium in the urine of children. *Monatsschr. Kinderheilkd.* **133** (10), 754–759.
10. Kasidas, G.P. and Rose, G.A. (1980) Oxalate content of some common foods: determination by an enzymatic method. *J. Hum. Nutr.* **34**, 255–266.
11. ADA (1996) Vitamins and minerals. In *The American Dietetic Association's Complete Food and Nutrition Guide* (Ed. by Duyff, R.L.). Chronimed Publishing, Minneapolis.
12. Nguyen, N.U., Henriet, M.T., Dumoulin, G., Widmer, A. and Regnard, J. (1994) Increase in calciuria and oxaluria after a single chocolate bar load. *Horm. Metab. Res.* **26**, 383–386.
13. Balcke, P., Zazgornik, J., Sunder-Plassmann, G., *et al.* (1989) Transient hyperoxaluria after ingestion of chocolate as a high risk factor for calcium oxalate calculi. *Nephron* **51** (1), 32–34.
14. Turnland, J.R., King, J.C., Keyes, W.R., Gong, B. and Michel, M.C. (1984) A stable isotope study of zinc absorption in young men: effects of phytate and α-cellulose. *Am. J. Clin. Nutr.* **40**, 1071–1077.
15. Wise, A. (1983) Dietary factors determining the biological activities of phytate. *Rev. Clin. Nutr.* **53**, 791–806.
16. Brown, R.C., Klein, A., Simmons, W.K. and Hurrell, R.F. (1990) The influence of Jamaican herb teas and other polyphenol-containing beverages on iron absorption in the rat. *Nutr. Res.* **10**, 343–353.
17. Fairweather-Tait, S.J., Minski, M.J. and Richardson, D.P. (1983) Iron absorption from a malted cocoa drink fortified with ferric orthophosphate using the stable isotope ^{58}Fe as an extrinsic label. *Br. J. Nutr.* **50**, 51–60.

18. Hurrell, R.F., Reddy, M.B., Dassenko, S.A. and Cook, J.D. (1991) Ferrous fumarate fortification of a chocolate drink powder. *Br. J. Nutr.* **65** (2), 271–283.
19. IFT Expert Panel on Food Safety and Nutrition (1984) Food nutrient interactions. *Food Technol.* October, 59–63.
20. Valiente, C., Molla, E., Martin-Cabrejas, M.M., Lopez-Andreu, F.J. and Esteban, R.M. (1996) Cadmium binding capacity of cocoa and isolated total dietary fiber under physiological pH conditions. *J. Sci. Food Agric.* **72**, 476–482.

Chapter 10

Methylxanthines

Joan L. Apgar and Stanley M. Tarka Jr

Cocoa beans contain a variety of biologically active components, including the methylxanthines theobromine, caffeine and theophylline. Theobromine (3,7-dimethylxanthine) is the predominant methylxanthine in cocoa beans. Caffeine (1,3,7-trimethylxanthine), the most common methylxanthine in the food supply, is found in relatively low concentrations in cocoa beans. Theophylline (1,3-dimethylxanthine) has been detected in cocoa beans, but at such low concentrations that its presence generally is ignored.

This paper is intended to provide a general overview of the physiological effects of these naturally occurring methylxanthines in cocoa and chocolate. The majority of available studies have focused on caffeine due to its widespread consumption in beverages, its therapeutic uses, and continuing public concern over its physiological and health-related effects. In comparison, there is relatively little published work on theobromine and cocoa, presumably due to the long-term consumption of cocoa-related products with no major health effects. A recently published comprehensive text on caffeine and methylxanthine foods and beverages is the basis for much of this review (1). Finally, theophylline has been extensively studied for its therapeutic applications, but is mentioned only briefly in this review due to the extremely small amount present in cocoa beans.

Content in cocoa/chocolate products

After fermenting and roasting, cocoa beans are broken and cleaned to separate shell from the kernel or nibs. The nibs are then finely ground, producing a thick fluid known as chocolate liquor or chocolate, and sold commercially as unsweetened baking chocolate. Cocoa powder, or simply cocoa, is prepared by removing some of the fat (cocoa butter) from chocolate liquor and pulverizing the remaining material. Bean variety, maturity and fermentation conditions can influence methylxanthine content, although some of these differences are minimized by blending of beans during processing.

Literature values for the methylxanthine content of cocoa beans and chocolate products were summarized recently (2). Chocolate liquor contains about 1.22% theobromine and 0.21% caffeine (3, 4) while cocoa contains 1.89–2.69% theobromine and 0.16–0.31% caffeine (4–7). Since methylxanthines are only slightly soluble in fat, their concentration in cocoa, with its lower fat content, generally is higher than in chocolate liquor. Only small amounts of theobromine (0.008%) and caffeine (0.038%) have been measured in cocoa butter (7).

Chocolate liquor and cocoa typically are not consumed alone, but mixed with other ingredients to make sweetened confectionery, baked products or other chocolate-containing foods. The percentage of chocolate liquor and/or cocoa decreases in the finished food as the percentage of other ingredients increases. As a result, the methylxanthine content of the finished product also is lowered.

Most chocolate liquor is used in the production of chocolate and chocolate-coated confections. Milk chocolate, the most popular form of chocolate confectionery in the US, contains a minimum of 10% chocolate liquor. Sweet chocolate contains at least 15% chocolate liquor; semi-sweet or bitter-sweet chocolate must contain at least 35% chocolate liquor. The theobromine and caffeine contents of several commercial chocolate confectionery products are presented in Table 10.1. Milk chocolate contains approximately 65 mg theobromine and 10 mg of caffeine per serving, whereas sweet chocolate provides about 185 mg of theobromine and 30 mg of caffeine per serving.

Table 10.1 Theobromine and caffeine content of commercial chocolate confectionery.

Product	Theobromine		Caffeine		Reference
	(%)	(mg per serving)*	(%)	(mg per serving)*	
Milk chocolate	0.160	64	0.025	10	4
	0.188	75	0.019	8	6
Milk chocolate w/peanuts	0.122	49	0.010	4	7
Milk chocolate w/rice	0.150	60	0.018	7	7
Sweet/dark chocolate	0.463	185	0.069	28	4
	0.441	176	0.054	22	6
Chocolate-covered confections					
Wafers w/chocolate creme	0.106	42	0.013	5	7
Caramel cookie bar	0.054	22	0.007	3	7
Peanut butter cup	0.072	29	0.009	4	7

* Serving based on 40 g reference amount defined by the US Food and Drug Administration (FDA).

Methylxanthine contents of some chocolate-containing foods and beverages have been reported in several studies (2, 4, 6, 8, 9), some of which are presented in Table 10.2. Theobromine contents of 228–284 mg and caffeine of 6–42 mg per serving have been reported for some beverages made with unsweetened cocoa

Table 10.2 Theobromine and caffeine content of chocolate-containing foods and beverages.

Food/beverage	Serving[1]	Theobromine		Caffeine		Reference
		(%)	(mg per serving)	(%)	(mg per serving)	
Brownies	40 g	0.142	57	0.014	6	6
Chocolate cake	80 g	0.162	130	0.016	13	6
Chocolate cake w/chocolate frosting	80 g	0.114	91	0.008	7	6
Chocolate chip cookies	30 g	0.070	21	0.014	4	8
Hot cocoa[2]	5 fl oz	0.042	62	0.003	4	9
Chocolate milk[3]	8 fl oz	0.024	58	0.002	5	4
Chocolate pudding	½ cup	0.062	88	0.005	7	6
Chocolate ice cream	½ cup	0.062	41	0.003	2	6
Chocolate syrup	2 tbsp	0.212	85	0.027	11	6
Chocolate fudge topping	2 tbsp	0.158	63	0.009	4	6
Cocoa rice/corn cereals	30 g	0.070	21	0.007	2	6

[1] Serving based on reference amounts defined by the US FDA.
[2] Hot cocoa made from a single-serve package of instant hot cocoa mix.
[3] Chocolate milk from instant mix as directed.

(10). However, most chocolate beverages are prepared from sweetened cocoa mixes that contain only about 65 mg of theobromine and 4 mg per serving of caffeine (4, 9). Comparison of results between studies has been complicated by a lack of documentation for preparation and serving sizes. Due to the limited amount of data, additional analyses of chocolate-containing foods and beverages are needed to determine dietary methylxanthine intakes from all food sources more accurately.

Consumption from chocolate products

Data on the methylxanthine intake from chocolate-containing foods are very scarce. As expected, most studies have focused on caffeine, the majority of which is consumed through coffee, tea, and soft drinks. Average daily caffeine intakes of about 200 mg per day (or 2.4 mg/kg) for adults have been estimated (11, 12). Two 14-day surveys conducted in 1972–73 and 1987–89 showed that the average caffeine intake from all sources was 2.4 mg/kg for adults and 0.67–0.75 mg/kg for children 12–17 years of age (Table 10.3) (13). Chocolate products, however, provided only about 0.03 mg/kg of caffeine for adults and 0.09 mg/kg for children 12–17 years of age.

These results agreed with those reported by Morgan *et al.*(14) in which average caffeine consumption from all food sources was 37 mg (0.9 mg/kg) among children 5–18 years of age, although some individuals consumed as much as 250 mg/day. Among all the subjects, an average of 6.4 mg (17%) of the total

Table 10.3 Average daily intake of caffeine from all foods and chocolate-containing foods.

Year and source of data	Age	All food		Chocolate foods		Reference
	(years)	(mg)	(mg/kg)	(mg)	(mg/kg)	
1972–73, 14-day household menu census	6–11	—	0.86	—	0.15	13
	12–17	—	0.75	—	0.09	
1977, 7-day food diaries	9–10	25	0.8	—	—	14
	5–18	37	0.9	6.4	—	
1981–82, 24-hour dietary recall	10	61	1.7	—	—	15
1987–89, 14-day household menu census	6–11	—	0.67	—	0.12	13
	12–17	—	0.65	—	0.08	
	25+	—	2.40	—	0.03	
	All ages	—	1.90	—	0.06	
1991–93, 3-day food diaries	6–10	16	0.51	6.2	—	16

caffeine intake came from chocolate foods. When caffeine consumption was calculated for days when consumed, caffeine intake from chocolate foods was only 11.7 mg.

Based on 24-hour dietary recalls, caffeine intake of children 10 years old participating in the Bogalusa Heart Study (15) was 1.7 mg/kg/day, almost twice that reported by Morgan *et al.* (14). When caffeine-containing foods were separated into five categories – coffee, tea, soft drinks, chocolate-containing food and chocolate milk beverages/pudding/ice cream – the most commonly consumed sources were soft drinks and chocolate-containing foods. However, due to the relatively low amount of caffeine in chocolate-containing foods compared to other food sources, the major source of caffeine was soft drinks.

More recently, Ellison *et al.* (16) analyzed 3-day food diaries from participants in the Framingham Children's Study. Average caffeine intake was 16 mg/day (0.51 mg/kg). Chocolate foods and beverages contributed 6.2 mg/day of caffeine (38%) – 17% from baked products, 11% from dairy beverages, 5% from candy and 5% from dairy desserts.

The methodology used to assess dietary intakes as well as regional or socioeconomic differences may contribute to the variation in caffeine intakes among these studies. Despite the differences and the scarcity of data, the contribution of chocolate-containing products to the average daily caffeine consumption appears to be insignificant in light of the amount of caffeine consumed from other sources. Additional studies are needed to determine more current information, especially concerning the contribution of chocolate to both caffeine and total methylxanthine intake.

Metabolism/disposition

The metabolism of both caffeine and theobromine has been extensively characterized in both animals and humans (17–24). These compounds are readily absorbed from the gastrointestinal tract, and quickly dispersed throughout all tissues and organs. Equilibrium is quickly established between mother and fetus, and between blood and all tissues including the brain and testes. Once absorbed, the methylxanthines are partially demethylated and oxidized, and excreted primarily as methyluric acids or as methylxanthines. Approximately 1–5% of a caffeine dose and 11–17% of a theobromine dose are excreted unchanged in the urine (17, 21, 22).

The half-life of caffeine in humans is about 2.5–4.5 hours (21) while that of theobromine is about 10 hours (23). Although repeated doses of a theobromine solution increased theobromine's half-life in humans (24), consumption of dark chocolate for 7 days did not alter the metabolic disposition of theobromine (22). This difference was partially attributed to a reduced rate of absorption of theobromine from cocoa and chocolate matrices compared to an oral solution (25). However, Mumford *et al.* (26) recently reported that chocolate appeared to delay caffeine absorption, but increase theobromine absorption relative to capsules.

The elimination and half-life of methylxanthines are prolonged in neonates and infants, and in the later stages of pregnancy. The half-life of transplacentally acquired caffeine is approximately 80 hours, and caffeine is largely eliminated unchanged for the first 3 months of life, apparently due to a slower rate of *N*-demethylation in the neonate and infant (27). Recommendations for women to consume moderate amounts of caffeine during pregnancy are partially based on these observed differences in metabolism.

Physiological effects

Theobromine, caffeine and theophylline share several physiological actions, including stimulation of the central nervous system (CNS), diuretic effects, stimulation of cardiac muscle, and relaxation of smooth muscle, notably bronchial muscle. However, they differ greatly in the intensity of their actions on the various systems (Table 10.4). For example, caffeine appears to exert its strongest effect on brain and skeletal muscle; theophylline's most potent effects are on the heart, bronchia and kidneys. In general, theobromine tends to exert the least physiological effect of these three methylxanthines.

CNS and endocrine

Caffeine is well recognized as a powerful CNS stimulant, acting on the cortex, medulla and, in large enough amounts, even the spinal cord. About 150–250 mg

Table 10.4 Relative strength of physiological effects exerted by caffeine, theobromine and theophylline.

System	Caffeine	Theobromine	Theophylline
Brain	Strong	Weak	Moderate
Heart	Weak	Moderate	Strong
Bronchia	Weak	Moderate	Strong
Skeletal muscle	Strong	Weak	Moderate
Kidneys	Weak	Moderate	Strong

Source: Czok (25).

of caffeine can stimulate the cortex, resulting in reduced fatigue, shortened reaction time, sensory stimulation and increased motor activity. High blood concentrations produced by the same amount of caffeine administered parenterally can also stimulate the medullary respiratory, vasomotor and vagal centers. Amounts of 200–500 mg can cause headaches, tremulousness, nervousness and irritability (28).

Delay in sleep has been noted after caffeine doses of 100 mg, although there appears to be considerable variation between and even within individuals (29). Estimated time to go to sleep after 300 mg of caffeine varied from 15–240 min for one individual, but never exceeded 45 min after consuming a placebo (30). In other studies, regular coffee consumers were relatively insensitive to caffeine's effect on their sleep habits while non-coffee drinkers experienced a delayed onset of sleep (31, 32). Sleep patterns can be influenced by temperature, noise and fatigue as well as habitual caffeine consumption, complicating the evaluation of study results.

In contrast to caffeine, theobromine is virtually inactive as a CNS stimulant, although headaches have been reported after ingestion of 50 g of cocoa per day, and sweating, trembling and severe headaches after long-term ingestion of 100 g of cocoa per day (25). Even in large doses, theobromine appears to be ineffective in relieving drowsiness and fatigue, and does not prolong the time required to fall asleep (29).

Interest in the behavioral and autonomic effects of caffeine led to investigations of the hormonal effects of caffeine. Increases in brain levels of serotonin and 5-hydroxyindoleacetic acid were observed after both acute and chronic administration of caffeine (33). Plasma epinephrine and norepinephrine were increased after 220–250 mg caffeine consumption (34, 35). Higher doses on the order of 500 mg can cause endocrine stress symptoms characterized by increased serum adrenocorticotropic hormone (ACTH) and cortisol (36). After 7 days of caffeine consumption, no increases in hormone levels were observed, suggesting that habitual consumers are the least sensitive to caffeine's effects on the endocrine system. Based on the high doses of caffeine required to produce these effects, it is unlikely that normal consumption of cocoa and chocolate-containing foods would produce any observable endocrine effects.

Cardiovascular/circulatory

The methylxanthines can affect cardiovascular function directly or indirectly influence it through their effects on the neurocrine and endocrine systems. Although caffeine is the weaker of the two, both theobromine and caffeine stimulate cardiac muscle, increasing the force of contraction, heart rate and cardiac output (28, 34). At the same time, these compounds can also stimulate the medullary vagal nuclei, producing a decrease in heart rate. The result of these opposing actions may be no change in heart rate, bradycardia or tachycardia. Occasionally, arrhythmias have been observed in persons who use excessive amounts of caffeine.

Caffeine and theobromine briefly dilate coronary, pulmonary and general systemic blood vessels, and thereby increase blood flow. The methylxanthines have been therapeutically used to dilate coronary arteries and increase coronary blood flow, although there is controversy over the effectiveness of this practice. They also stimulate the medullary vasomotor center to constrict blood vessels and decrease cerebral blood flow and oxygen tension. The vasoconstrictive action is believed to be responsible for the efficiency of xanthines in relieving hypertensive headaches. Conversely, sudden withdrawal of caffeine results in a dilatation of cerebral blood vessels and reports of headaches (37).

Due to the multiple physiological effects of the methylxanthines, their overall impact on systemic blood pressure is usually minimal or, at worst, unpredictable (28, 38). Stimulation of the central vasomotor and myocardium tend to increase blood pressure; central vagal stimulation and peripheral vasodilation tend to decrease blood pressure. Pulse pressure is increased by the vasodilation and enhanced cardiac output while pulmonary arterial pressure is usually reduced.

As with the neurocrine and endocrine effects, cardiovascular effects appear to be minimized by habitual methylxanthine intake. Blood pressure, heart rate, plasma renin activity and plasma and urinary catecholamines increase after a single dose of caffeine is ingested by caffeine-naïve individuals (35). However, levels return to baseline within 4–7 days of repeated caffeine administration (38).

Caffeine consumption has been considered as a possible risk factor for coronary heart disease, although results from both animal and human studies have not been consistent. Interpretation of study results is confounded by a host of factors, including age, sex, adiposity, alcohol consumption, exercise, smoking habits, occupational stress and ethnicity. Two studies by the Boston Collaborative Drug Surveillance Program suggested that coffee consumption is associated with an increased risk of myocardial infarction (39, 40). In another study, coffee raised cholesterol, phospholipids and triglycerides, but tea caused a fall in all three fractions (41). Retrospective and prospective studies since that time have failed to consistently find a correlation between coffee consumption and ischemic heart disease (42-44). A number of the reported correlations actually may have been due to other confounding factors such as brewing techniques, non-caffeine components of coffee, or behaviors associated with coffee drinkers such as smoking (45-47).

Results of both animal and human studies indicate that, under normal conditions, the methylxanthines do not adversely affect heart disease, high blood pressure or irregular heart beat. The effects of the methylxanthines on blood pressure, heart rate, blood cholesterol and blood glucose levels, and their possible role in heart disease, are reviewed in more detail in other references (47, 48).

Respiratory and muscle

The methylxanthines are recognized for their effectiveness in stimulating respiration and relaxing the smooth muscles of the bronchi; thus, they have been used therapeutically to treat asthma and neonatal apnea (28, 34, 49). Theobromine has a more pronounced effect than caffeine, but is much less effective than theophylline.

The methylxanthines also strengthen voluntary muscle contraction, increasing the capacity for muscular work and reducing muscle fatigue (50). This effect may be due to increased release of acetylcholine. Since the level of caffeine required to increase muscular work also stimulates the CNS, there is debate as to whether the effect is a result of caffeine's action on the CNS or directly on voluntary muscle. Caffeine is the most effective while theobromine has little if any effect on the strength of voluntary muscle contraction.

Renal

In the past, the methylxanthines have been administered therapeutically for their diuretic effect, but generally have been replaced with more efficacious compounds. The stimulatory effect of the methylxanthines on kidney function is attributed to increased renal blood flow, increased glomerular filtration, and reduced reabsorption of sodium and chloride in the proximal tubules (25, 34, 37, 51).

The methylxanthines' effects on renal dynamics have prompted investigations to determine their effect on mineral excretion. Consumption of 150 or 300 mg of caffeine by women has been linked with increased urinary excretion of calcium, magnesium, sodium and potassium (52–56). In one study, 300 mg of theobromine in a fudge did not increase urinary calcium or sodium (54). The actions of the methylxanthines as adenosine antagonists and possible involvement of prostaglandins have been proposed as mechanisms for the increased urinary calcium excretion (55, 56).

With the current public awareness of osteoporosis, methylxanthine consumption has been suggested as a risk factor for osteoporosis due to the increased calcium loss. The data fail to show any consistent effect of caffeine consumption on the incidence of bone fracture (57). The majority of studies have found no association between caffeine intake, and bone density changes. A few studies have reported small but significant increases in either fractures or bone loss associated with caffeine intake although confounding variables were not controlled in most of these studies. A recent study in postmenopausal women found no association between caffeine intake and any bone measurements (58). The literature suggests that a high caffeine intake over one's lifetime contributes negatively to calcium balance. However, no adverse physiological bone effects are expected from moderate caffeine consumption and adequate calcium intake (54, 57).

Other

Caffeine quantities of 100 mg and higher can increase the basal metabolic rate for at least 3 hours after administration (21, 59), whereas theobromine appears to have no significant effect on the rate of metabolism.

Moderate amounts of caffeine have been shown to increase the secretion of stomach acids and relax muscles and blood vessels in the digestive tract (34). As such, caffeine has been implicated in the aggravation of peptic ulcers. However, since decaffeinated coffee creates the same effects, some other component may be a factor (60). Theobromine has no reported effect on the secretion of gastric acids (28).

Behavior, mood and performance

Mood

Caffeine consumption can result in either positive or negative mood changes, depending on the dose and possible other factors such as genetic predisposition (61). Subjects reported increased energy and a feeling of well-being after 75–300 mg of caffeine, but increased anxiety and tension after 500 mg. Evans and Griffiths (62) showed that moods were related to the expectation of the subjects. Those who knew they were consuming caffeine reported positive mood effects. If the same amount was consumed unknowingly, negative moods were reported. In a comparison of theobromine and caffeine, Mumford *et al.* (63) found that caffeine, but not theobromine, produced changes in the group ratings of subjective effects, including increased well-being, energy and alertness.

Goldstein *et al.* (32) reported that the response to caffeine corresponded to the habitual caffeine consumption. Heavy consumers of caffeine were less nervous and reported fewer headaches with increasing consumption of caffeine. Conversely, nonusers reported more complaints of nervousness and gastrointestinal complaints with increased doses of caffeine. While methylxanthines may slightly disrupt the behavior in the nonuser, they may produce desirable effects for the habitual user (62).

Numerous studies have attempted to link methylxanthine intake with behavioral and attention problems in children. One study compared the behavioral effects of caffeine in children who were either 'high' (500 mg/day or more) or 'low' caffeine consumers (64). Children received 5 mg/kg of caffeine or a placebo twice daily for 2 weeks. When not receiving caffeine, 'high' consumers scored higher on an anxiety questionnaire and showed lower autonomic arousal than 'low' consumers. Parents perceived the 'low' consumers as more emotional, inattentive and restless when consuming caffeine, while they rated 'high' consumers as not changed. It was suggested that these results indicate a possible physiological basis in children for dietary caffeine preference.

Results from a recent meta-analysis indicated that caffeine is not associated with any significant adverse effects on behavior or cognition (65). Additionally, the results support previous conclusions that caffeine may have a small, beneficial effect in some children, decreasing behavior that is characterized as aggressive.

Sudden withdrawals of caffeine among habitual consumers can result in headaches, fatigue, dysphoric mood changes and sometimes nausea and vomiting (61, 62, 66–68). Among individuals whose typical daily caffeine intake averaged 235 mg per day, 52% reported moderate or severe headaches and 8–11% reported symptoms of anxiety and depression when caffeine intake was discontinued. Moderate to heavy caffeine consumers also reported irritability, inability to work effectively, nervousness, restlessness and lethargy. Medical experts agree that these 'withdrawal' symptoms can be avoided by progressively decreasing caffeine intake over a few days.

Mental and physical performance

In that caffeine stimulates the CNS, alleviates fatigue and strengthens voluntary muscle contraction, it has been associated with enhanced mental and physical performance. Performance of rapid information processing tasks such as problem solving, logical reasoning and mental arithmetic tasks was improved after consumption of 150 mg of caffeine, but impaired after 600 mg (61, 69). Caffeine appears to negatively affect the performance of complex tasks. The literature indicates consistent benefits of moderate doses (32–256 mg) of caffeine on auditory and visual vigilance, although high doses may initially impair vigilance (70, 71). Caffeine can counteract the effects of fatigue on reaction time. Subjects who consumed caffeine doses of 75–300 mg demonstrated shortened auditory reaction times and reported feeling more alert and active. However, objective measures of alertness and psychomotor co-ordination were not improved (72, 73).

There are conflicting results and controversy as to caffeine's ergogenic benefits (74). No significant effects on short-term, high-intensity performance have been reported (75). The effects of caffeine on strength are variable, and may be related to sensitivity differences based on muscle type. Other studies suggested that caffeine enhances endurance in submaximal, long-term exercises including cross-country skiing, cycling and running (75–80). Subjects consuming caffeine before exercise generally increased their total work output and time to exhaustion (76, 77, 80). Numerous other studies failed to demonstrate any significant effects. These inconsistent results suggest that caffeine's effects may be mediated by other variables including training, dietary patterns, type of exercise and prior use of caffeine.

Other health-related effects

Public concern with the safety and health-related effects of the methylxanthines has grown over the last 20–30 years. Much of the public concern originated in the early 1970s as a result of the Food and Drug Administration's (FDA's) directive to re-examine the safety of food additives that were generally recognized as safe (GRAS), including caffeine. In 1978, the Select Committee on GRAS Substances issued its report in which the world's scientific literature on the health aspects of caffeine was reviewed (81). Although the attention focused on the safety of dietary caffeine, questions were also raised concerning the safety of theobromine due to its structural similarity to caffeine.

To address these questions, the major worldwide producers of chocolate initiated a comprehensive toxicological testing program on theobromine and cocoa. The program included a comprehensive review of the literature (19) as well as studies to determine the acute toxicity (19), genotoxicity (82), teratogenicity (83, 84), carcinogenicity (85) and reproductive effects (86).

Toxicity/carcinogenicity

Median lethal dose (LD_{50}) values of 200 mg/kg for caffeine and 950 mg/kg for theobromine have been determined in the rat (19, 81). The structural differences in these two compounds and the resulting absorption kinetics may account for the variation in LD_{50} values (19). In humans, the fatal oral dose of caffeine is estimated to be 10 g (28). However, deaths from caffeine overdoses are rare, since gastric irritation and vomiting typically occur before toxic amounts are absorbed (87, 88). Reports of toxicity are associated with consumption of caffeine-containing medications (87, 89); there are no reports of acute toxicity from caffeine ingested in foods and beverages (45).

There are even fewer reports on the toxicological effects of theobromine and cocoa. In rats fed up to 5% cocoa in the diet for 104 weeks, no evidence of chronic toxicity or carcinogenicity was observed (85). This cocoa level provided a methylxanthine intake of 60 and 75 mg/kg body weight/day for males and females, respectively, and represented human intakes of greater than 50 times the theobromine consumption from all chocolate sources at the 90th percentile. The incidence of bilateral diffuse testicular atrophy was increased and spermatogenesis was decreased in male rats fed 5% cocoa. These effects were not unexpected based on previous work, which showed that the testes are a target organ for methylxanthine effects (90).

Reproductive/developmental

Several reviews of the reproductive effects of the methylxanthines are available (19, 23, 81, 91). Since the methylxanthines readily cross the placental barrier, exposure begins during gestation. Tests in rats and mice showed birth defects in mice from high doses of caffeine (the human equivalent of 40 cups of coffee at one time); however, there is no scientific evidence linking caffeine with similar abnormalities in humans. Daily caffeine consumption of greater than 300 mg during pregnancy has been associated with reduced birth weights, premature births and spontaneous abortions in some studies; however, study results have been inconsistent and causality has not been established (81, 92–94). The FDA has evaluated the scientific evidence on caffeine and concluded that caffeine does not adversely affect reproduction in humans. Nevertheless, pregnant women are advised to consume caffeine in moderation.

Comprehensive studies of cocoa and theobromine revealed no frank teratogenic or embryotoxic effects, although there was delayed calcification of fetal bone at dietary levels of 0.135% theobromine in rats and 0.125% theobromine in rabbits (83, 84). As mentioned previously, the testes have been identified as a target organ in methylxanthine studies in animals. Dietary concentrations of 0.6% theobromine for 4 weeks resulted in testicular atrophy in rats (95). Due to the link between methylxanthine intake and testicular atrophy, Hostetler *et al.* (86) evaluated the effects of cocoa (which contains both theobromine and caffeine) on reproductive function in rats in a conventional multi-generation exposure study. Continuous exposure to dietary cocoa in rats, providing up to 104 and 126 mg/kg/day of total methylxanthines to males and females, respectively, had no consistent effect on reproductive function over three generations. Based on these results, it is unlikely that consumption of cocoa and cocoa-based products has any potential to adversely affect reproductive capacity in humans.

Fibrocystic breast disease

Fibrocystic breast disease is a catch-all term used to describe fibrocystic changes or benign fibrous lumps in the breast. It is found in more than 50% of women and is believed by some to be a risk factor for breast cancer (37). Women often are advised to abstain from coffee, cola drinks, tea and chocolate to minimize aggravation of this condition. This recommendation is largely based on two studies in which palpable breast nodules disappeared after women abstained from methylxanthine-containing beverages (96, 97). These studies have been criticized for the lack of long-term follow-up. However, epidemiological findings from a number of other investigations do not support this supposed link (98–102) even though at least one case-control study suggested a weak relationship between theobromine consumed in chocolate drinks and candy and a risk of breast cancer (103).

In one prospective study, breast nodules spontaneously disappeared regardless of methylxanthine consumption (104). As others have noted, 'waxing and waning' or fluctuation of symptom severity with time is a characteristic feature of fibrocystic breast disease (104), making it difficult to confirm any causal association between elimination of methylxanthines from the diet and disappearance of symptoms. A large case control study sponsored by the US National Cancer Institute showed no association between caffeine intake and breast cancer (105).

Based on a review of the scientific literature, the Council on Scientific Affairs of the American Medical Association has concluded:

> 'There is currently no scientific basis for associating methylxanthine consumption with fibrocystic disease of the breast. Indeed, it has been suggested that lumpy, fibrous breast tissue in women is normal and represents a response to physiological hormonal variation.' (106)

Discussion

The physiological effects of the methylxanthines are clearly recognized, as evidenced by their history of therapeutic uses and the popularity of coffee and tea as 'stimulating' beverages. Caffeine, for example, has been used in the treatment of infant apnea and migraine headaches, in bronchial and cardiac stimulants, and in over-the-counter products for diuresis and alertness. Additionally, more than 80% of the US adult population consumes caffeine daily from beverages and foods. In general however, although caffeine is recognized for its powerful effects on the CNS, theobromine is virtually inactive.

The precise mechanisms behind the physiological effects of the methylxanthines are complicated, and have been discussed by others (28). Most of the effects exerted by the methylxanthines involve the antagonism of adenosine receptors. However, their role in inhibiting phosphodiesterase and mobilizing calcium is also important. The methylxanthines, especially theophylline, inhibit phosphodiesterase activity necessary to catalyze the conversion of cyclic 3′,5′-adenosine monophosphate (AMP) to 5′-AMP. Both methylxanthines and catecholamines increase cyclic AMP concentrations, but at differing sites of action. This suggests that the methylxanthines could potentiate the effect of the catecholamines. In other situations, the methylxanthines can have opposite effects on cyclic AMP metabolism. For example, adenosine acts to stimulate cyclic AMP synthesis at certain sites, such as the brain, whereas the methylxanthines inhibit the stimulatory effect of adenosine.

Evaluation of the physiological and health effects of the methylxanthines is complicated by methodological weaknesses in research design. Controversy arises over the extrapolation of data from methylxanthine-free animals to methylxanthine-tolerant individuals. In human studies, accuracy in measuring methylxanthine intake is difficult to achieve. Consumption patterns vary from weekday to weekend. Preparation methods and volume of a 'cup' differ widely among consumers, ultimately affecting methylxanthine intake.

Many caffeine studies have focused on coffee consumption, since most dietary caffeine is consumed through coffee. It has been suggested that coffee drinkers may practice a cluster of risky health behaviors, including increased rates of smoking, alcohol use, and saturated fat and cholesterol consumption (37). As a result, caffeine intake may not be the causative agent as implicated in many studies, but actually may be a marker for dietary practices which increase the risk of certain diseases.

In general, notable physiological effects are observed after moderate amounts of 300–500 mg of caffeine, although some individuals experience effects at lower doses. The American Council on Science and Health (ACSH) concluded

> 'that caffeine as generally consumed in foods, beverages, and over-the-counter drugs is not a threat to the health of most Americans. However, some people who consume large amounts of products that contain caffeine may experience health problems, including chronic headaches, sleep disturbances, rapid heart beat, anxiety, and stomach upset.' (107)

These effects can occur at daily consumption levels equal to more than four to five cups of brewed coffee or 15 12-ounce servings of caffeinated soft drinks. Individual response to the physiological effects of methylxanthines depends on a number of factors including rate of absorption, age, amount and frequency of consumption (108). Individuals who typically consume little if any methylxanthines may be more sensitive than those who habitually consume methylxanthines.

Assessment of the physiological effects of methylxanthines consumed in cocoa and chocolate products is difficult due to the relative scarcity of studies specifically focused on cocoa consumption. The available intake data suggests that the average daily consumption of caffeine from cocoa and chocolate-containing foods is very low. Among children, average caffeine consumption of less than 10 mg/day from chocolate foods has been reported (14, 16).

Since theobromine is the predominant methylxanthine in cocoa, the average daily intake of theobromine from cocoa and chocolate-containing foods is much higher and can be estimated as 10–15 times the concentration of caffeine.

However, the physiological effects from theobromine are generally weak to moderate in strength. In addition, the physiological effects of the methylxanthines are greatly reduced when consumed as constituents of cocoa products compared to pure compounds or solutions (25). The difference is partially attributed to reduced gastrointestinal absorption of methylxanthines due to the polyphenol or fat content. As a result of these factors, consumption of cocoa and chocolate foods does not result in any noticeable physiological effects in the majority of individuals.

Summary

The physiological effects of the methylxanthines are widely recognized although the mechanisms are not completely understood. Caffeine exerts its strongest effect on the brain and skeletal muscle, while theobromine is a relatively weak stimulant in comparison. Coffee has been and continues to be a popular beverage for the stimulatory effect of its caffeine content. In contrast, cocoa and chocolate foods contribute very little to the average daily intake of caffeine in both children and adults. While the theobromine content of chocolate foods is at least ten times higher than that of caffeine, theobromine's physiological effects are generally weak. Based on scientific studies, moderate consumption of methylxanthines from any food source does not cause adverse physiological or health effects. Furthermore, due to the low dietary intakes of caffeine from chocolate-containing foods and the weak physiological effects of theobromine, consumption of cocoa and chocolate-containing foods does not result in any noticeable physiological effects in the majority of individuals.

References

1. Spiller, G.A. (Ed.) (1998) *Caffeine*. CRC Press, New York.
2. Apgar, J.L. and Tarka, S.M., Jr (1998) Methylxanthine composition and consumption patterns of cocoa and chocolate products. In *Caffeine*. (Ed. by Spiller, G.A.), pp. 163–192. CRC Press, New York.
3. Kreiser, W. and Martin, R. (1978) Cacao products – high-pressure liquid chromatographic determination of theobromine in cocoa and chocolate products. *J. Assoc. Off. Anal. Chem.* **61**, 1424–1427.
4. Zoumas, B., Kreiser, W. and Martin, R. (1980) Theobromine and caffeine content of chocolate products. *J. Food Sci.* **45**, 314–316.
5. DeVries, J., Johnson, K. and Heroff, J. (1981) HPLC determination of caffeine and theobromine content of various natural and red Dutch cocoas. *J. Food Sci.* **46**, 1968–1969.
6. Craig, W.J. and Nguyen, T.T. (1984) Caffeine and theobromine levels in cocoa and carob products. *J. Food Sci.* **49**, 302–303, 305.

7. Kiefer, B.A. and Martin, R.A. (1987) *Determination of theobromine and caffeine in chocolate products by HPLC*. Hershey Foods Corporation, unpublished data.
8. Hershey Foods Corporation Technical Center, Analytical Research Laboratory, unpublished data.
9. Blauch, J.L., and Tarka, S.M., Jr. (1983) HPLC determination of caffeine in theobromine in coffee, tea, and instant hot cocoa mixes. *J. Food Sci.* **48**, 745–747, 750.
10. Burg, A.W. (1975) How much coffee in the cup. *Tea Coffee Trade J.* **147**, 40–42.
11. Graham, D. (1978) Caffeine – its identity, dietary sources, intake and biological effects. *Nutr. Rev.* **36**, 97–102.
12. Lundsberg, L.S. (1998) Caffeine consumption. In *Caffeine* (Ed. by Spiller, G.A.), pp. 199–224. CRC Press, New York.
13. Barone, J.J. and Roberts, H.R. (1996) Caffeine consumption. *Food Chem. Toxicol.* **34**, 119–129.
14. Morgan, K., Stults, V. and Zabik, M. (1982) Amount and dietary sources of caffeine and saccharin intake by individuals ages 5 to 18 years. *Regul. Toxicol. Pharmacol.* **2**, 296–307.
15. Arbeit, M.L., Nicklas, T.A., Frank, G.C., Webber, L.S., Miner, M.H. and Berenson, G.S. (1988) Caffeine intakes of children from a biracial population: the Bogalusa Heart Study. *J. Am. Diet. Assoc.* **88**, 466–471.
16. Ellison, R.C., Singer, M.R., Moore, L.L., Nguyen, U.D.T., Garrahie, E.J. and Marmor, J.K. (1995). Current caffeine intake of young children: amount and sources. *J. Am. Diet. Assoc.* **95**, 802–804.
17. Cornish, H.G. and Christman, A.A. (1957) A study of the metabolism of theobromine, theophylline, and caffeine in man. *J. Biol. Chem.* **228**, 315–323.
18. Arnaud, M.J. and Welsch, C. (1979) Metabolic pathway of theobromine in the rat and identification of two new metabolites in human urine. *J. Agric. Food Chem.* **27**, 524–527.
19. Tarka, S.M. Jr (1982) The toxicology of cocoa and methylxanthines: a review of the literature. *CRC Crit. Rev. Toxicol.* **9**, 275–312.
20. Arnaud, M.J. (1984) Products of metabolism of caffeine. In *Caffeine* (Ed. by Dews, P.), pp. 3–38. Springer-Verlag, New York.
21. Arnaud, M.J. (1987) The pharmacology of caffeine. *Prog. Drug Res.* **31**, 273–313.
22. Shively, C.A., Tarka, S.M., Jr, Arnaud, M.J., Dvorchik, B.H., Passananti, G.T. and Vesell, E.S. (1985) High levels of methylxanthines in chocolate do not alter theobromine disposition. *Clin. Pharmacol. Ther.* **37**, 415–424.
23. Tarka, S.M., Jr and Shively, C.A. (1987) Methylxanthines. In *Toxicological Aspects of Food* (Ed. by Miller, K.), pp. 373–423, Elsevier, New York.
24. Drouillard, D.D., Vesell, E.S. and Dvorchik, B.H. (1978) Studies on theobromine disposition in normal subjects. *Clin. Pharmacol. Ther.* **23**, 296–302.
25. Czok, G. (1974) Concerning the question of the biological effectiveness of methylxanthines in cocoa products. *Z. Ernaehrungswiss.* **13**, 165–171.
26. Mumford, G.K., Benowitz, N.L., Evans, S.M., *et al.* (1996) Absorption rate of methylxanthines following capsules, cola, and chocolate. *Eur. J. Clin. Pharmacol.* **51**, 319–325.
27. Aldridge, A., Aranda, J.V. and Neims, A.H. (1979) Caffeine metabolism in the newborn. *Clin. Pharmacol. Ther.* **25**, 447–453.

28. Rall, T.W. (1980) Central nervous system stimulants: the xanthines. In *The Pharmacological Basis of Therapeutics*, 6th edn (Ed. by Goodman, L.S. and Gilman, A.), pp. 592–607. Macmillan, New York.
29. Dorfman, L.J. and Jarvik, M.E. (1970) Comparative stimulant and diuretic actions of caffeine and theobromine in man. *Clin. Pharmacol. Ther.* **11**, 869–872.
30. Goldstein, A., Warren, R. and Kaizer, S. (1965) Psychotropic effects of caffeine in man. I. Individual differences in sensitivity to caffeine-induced wakefulness. *J. Pharmacol. Exp. Ther.* **149**, 156–159.
31. Colton, T., Gosselin, R.E. and Smith, R.P. (1968) The tolerance of coffee drinkers to caffeine. *Clin. Pharmacol.Ther.* **9**, 31–39.
32. Goldstein, A., Kaizer, S. and Whitby, O. (1969) Psychotropic effects of caffeine in man. IV. Quantitative and qualitative differences associated with habituation to coffee. *Clin. Pharmacol. Ther.* **10**, 489–497.
33. Fernstrom J.D. and Fernstrom, M.H. (1984) Effects of caffeine on monoamine neurotransmitters in the central and peripheral nervous system. In *Caffeine* (Ed. by Dews, P.), pp. 107–118. Springer-Verlag, New York.
34. Spiller, G.A. (1998) Metabolism and physiological effects of methylxanthines. In *Caffeine* (Ed. by Spiller, G.A.), pp. 225–231. CRC Press, New York.
35. Robertson, D., Frolich J.C., Carr, R.K., *et al.* (1978) Effects of caffeine on plasma renin activity, catecholamines and blood pressure. *N. Engl. J. Med.* **298**, 181–186.
36. Spindel, E.R. and Wurtman, R.J. (1984) Neuroendocrine effects of caffeine in rat and man. In *Caffeine* (Ed. by Dews, P.), pp. 119–128. Springer-Verlag, New York.
37. Lamarine, R.J. (1994) Selected health and behavioral effects related to the use of caffeine. *J. Comm. Health* **19**, 449–466.
38. Green, P.J., Kirby, R. and Suls, J. (1996) The effects of caffeine on blood pressure and heart rate: a review. *Ann. Behav. Med.* **18**, 201–216.
39. Boston Collaborative Drug Surveillance Program (1972) Coffee drinking and acute myocardial infarction. *Lancet* **2**, 1278–1281.
40. Jick, H., Miettinen, O.S., Neff, R.K., Shapiro, S., Heinonen, O.P. and Slone, D. (1973) Coffee and myocardial infarction. *N. Engl. J. Med.* **289**, 63–67.
41. Akinyanju, P. and Yudkin, J. (1967) Effect of coffee and tea on serum lipids in the rat. *Nature* **214**, 426–427.
42. Dawber, T.R., Kannel, W.B. and Gordon, T. (1974) Coffee and cardiovascular disease: observations from the Framingham study. *N. Engl. J. Med.* **291**, 871–874.
43. Yano, K., Rhoads, G.G. and Kagan, A. (1977) Coffee, alcohol, and risk of coronary heart disease among Japanese men living in Hawaii. *N. Engl. J. Med.* **297**, 405–409.
44. Heyden, S., Tyroler, H.A., Heiss, G., Hames, C.G. and Bartel, A. (1978) Coffee consumption and mortality, total mortality, stroke mortality, and coronary heart disease mortality. *Arch. Intern. Med.* **138**, 1472–1475.
45. Bergman, J. and Dews, P.B. (1987) Dietary caffeine and its toxicity. In *Nutritional Toxicology* (Ed. by Hathcock, J.), pp. 199–221. Academic Press, New York.
46. Chou, T.M. and Benowitz, N.L. (1994) Caffeine and coffee: effects on health and cardiovascular disease. *Comp. Biochem. Physiol.* **190C**, 173–189.
47. Gardner, C., Bruce, B. and Spiller, G.A. (1998) Coffee, caffeine, and serum cholesterol. In *Caffeine* (Ed. by Spiller, G.A.), pp. 301–323. CRC Press, New York.

48. Curatolo, P.W. and Robertson, D. (1983) The health consequences of caffeine. *Ann. Intern. Med.* **98**, 641–653.
49. Brouard, C., Moriette, G., Murat, I., *et al.* (1985) Comparative efficacy of theophylline and caffeine in the treatment of idiopathic apnea of premature infants. *Am J. Dis. Child.* **139**, 698–700.
50. Fredholm, B.B. (1984) Effects of methylxanthines on skeletal muscle and on respiration. In *The Methylxanthine Beverages and Foods* (Ed. by Spiller, G.A.), pp. 365–375. Alan R. Liss, New York.
51. Huang, K.C., King, N.B. and Genazzani, E. (1958) Effect of xanthine diuretics on renal tubular transport of PABA. *Am. J. Physiol.* **192**, 373–378.
52. Heaney, R.P. and Recker, R.R. (1982) Effects of nitrogen, phosphorus and caffeine on calcium balance in women. *J. Lab. Clin. Med.* **99**, 46–55.
53. Massey, L.K. and Wise, K.J. (1984) The effect of dietary caffeine on urinary excretion of calcium, magnesium, sodium, and potassium in healthy young females. *Nutr. Res.* **4**, 43–51.
54. Massey, L.K. and Whiting, S.J. (1993) Caffeine, urinary calcium, calcium metabolism and bone. *J. Nutr.* **123**, 1611–1614.
55. Hollingbery, P.W., Bergman, E.A. and Massey, L.K. (1985) Effect of dietary caffeine and aspirin on urinary calcium and hydroxyproline excretion in pre- and post-menopausal women. *Fed. Proc.* **44**, 1149.
56. Massey, L.K. and Berg, T. (1985) Effect of dietary caffeine on urinary mineral excretion in healthy males. *Fed. Proc.* **44**, 1149.
57. Bruce, B. and Spiller, G.A. (1998) Caffeine, calcium, and bone health. In *Caffeine* (Ed. by Spiller, G.A.), pp. 345–356. CRC Press, New York.
58. Lloyd, T., Rollings, N., Eggli, D.F., Kieselhorst, K. and Chinchilli, V.M. (1997) Dietary caffeine intake and bone status of postmenopausal women. *Am. J. Clin. Nutr.* **65**, 1826–1830.
59. Astrup, A., Toubro, S., Cannon, S., Hein, P., Breum, L. and Madsen, J. (1990) Caffeine: a double-blind, placebo-controlled study of its thermogenic, metabolic, and cardiovascular effects in healthy volunteers. *Am. J. Clin. Nutr.* **51**, 759–767.
60. Cohen, S. and Booth, G.H. (1975) Gastric acid secretion and lower-esophageal-sphincter pressure in response to coffee and caffeine. *N. Engl. J. Med.* **293**, 897–899.
61. Smith, B.D. and Tola, K. (1998) Caffeine: effects on psychological functioning and performance. In *Caffeine* (Ed. by Spiller, G.A.), pp. 251–299. CRC Press, New York.
62. Evans, S.M. and Griffiths, R.R. (1992) Caffeine tolerance and choice in humans. *Psychopharmacol.* **108**, 51–59.
63. Mumford, G.K., Evans, S.M., Kaminiski, B.J., *et al.* (1994) Discriminative stimulus and subjective effects of theobromine and caffeine in humans. *Psychopharmacol.* **115**, 1–8.
64. Rapoport, J.L., Berg, C.J., Ismond, D.R., Zahn, T.P. and Neims, A. (1984) Behavioral effects of caffeine in children. *Arch. Gen. Psychiatry* **41**, 1073–1079.
65. Stein, M.A., Krasowski, M., Leventhal, B.L., Phillips, W. and Bender, B.G. (1996) Behavioral and cognitive effects of methylxanthines. *Arch. Pediatr. Adolesc. Med.* **150**, 284–288.

66. Dreisbach, R.H. and Pfeiffer, C. (1943) Caffeine-withdrawal headache. *J. Lab. Clin. Med.* **28**, 1212–1219.
67. Griffiths, R.R., Evans, S.M., Heishman S.J., *et al.* (1990) Low-dose caffeine physical dependence in humans. *J. Pharmacol. Exp. Ther.* **255**, 1123–1132.
68. Strain, E.C., Mumford, G.K., Silverman, K. and Griffiths, R.R. (1994) Caffeine dependence syndrome. *JAMA* **272**, 1043–1048.
69. Hasenfratz, M. and Battig, K. (1994) Acute dose–effect relationships of caffeine and mental performance, EEG, cardiovascular and subjective parameters. *Psychopharmacol.* **114**, 281–287.
70. Koelega, H.S. (1993) Stimulant drugs and vigilance performance: a review. *Psychopharmacol.* **111**, 1–16.
71. Mitchell, V.E., Ross, S. and Hurst, P.M. (1974) Drugs and placebos: effects of caffeine on cognitive performance. *Psychol. Rep.* **35**, 875–883.
72. Clubley, M., Bye, C.E., Henson, T.A., Peck, A.W. and Riddington, C.J. (1979) Effects of caffeine and cyclizine alone and in combination on human performance, subjective effects and EEG activity. *Br. J. Clin. Pharmacol.* **7**, 157–163.
73. Goldstein, A., Kaizer, S. and Warren R. (1965) Psychotropic effects of caffeine in man. II. Alertness, psychomotor coordination, and mood. *J. Pharmacol. Exp. Ther.* **150**, 146–151.
74. Graham, T.E., Rush, J.W.E. and van Soeren, M.H. (1994) Caffeine and exercise: metabolism and performance. *Can. J. Appl. Physiol.* **19**, 111–138.
75. Lamarine, R.J. (1998) Caffeine as an ergogenic aid. In *Caffeine* (Ed. by Spiller, G.A.), pp. 233–250. CRC Press, New York.
76. Costill, D., Dalsky, G. and Fink, W. (1978) Effects of caffeine ingestion on metabolism and exercise performance. *Med. Sci. Sports* **10**, 155–158.
77. Berglund, B. and Hemmingsson, P. (1982) Effects of caffeine ingestion on exercise performance at low and high altitudes in cross-country skiers. *Int. J. Sports Med.* **3**, 234–236.
78. Graham, T. and Spriet, L. (1991) Performance and metabolic response to a high caffeine dose during prolonged exercise. *J. Appl. Physiol.* **71**, 2292–2298.
79. Tarnopolsky, M.A. (1994) Caffeine and endurance performance. *Sports Med.* **18**, 109–125.
80. Ivy, J.L., Costill, D.L., Fink, W.J. and Lower, R.W. (1979) Influence of caffeine and carbohydrate feedings on endurance performance. *Med. Sci. Sports* **11**, 6–11.
81. Life Sciences Research Office of the Federation of American Societies for Experimental Biology and Medicine (1978) *Evaluation of health aspects of caffeine as a food ingredient.* SCOGS-89 Report. FASEB, Washington, DC.
82. Brusick, D., Myhr, B., Galloway, S., Rundell, J., Jagannath, D.R. and Tarka, S.M., Jr (1986) Genotoxicity of theobromine in a series of short-term assays. *Mutat. Res.* **169**, 105–114.
83. Tarka, S.M., Jr, Applebaum, R.S. and Borzelleca, J.F. (1986) Evaluation of the perinatal, postnatal and teratogenic effects of cocoa powder and theobromine in Sprague–Dawley/CD rats. *Food Chem. Toxicol.* **24**, 375–382.
84. Tarka, S.M., Jr, Applebaum, R.S. and Borzelleca, J.F. (1986) Evaluation of the perinatal, postnatal and teratogenic effects of cocoa powder and theobromine in New Zealand white rabbits. *Food Chem. Toxicol.* **24**, 363–374.

85. Tarka, S.M., Jr., Morrissey, R.B., Apgar, J.L., Hostetler, K.A. and Shively, C.A. (1991) Chronic toxicity/carcinogenicity studies of cocoa powder in rats. *Food Chem. Toxicol.* **29**, 7–19.
86. Hostetler, K.A., Morrissey, R.B., Tarka, S.M., Jr, Apgar, J.L. and Shively, C.A. (1990) Three-generation reproductive study of cocoa powder in rats. *Food Chem. Toxicol.* **28**, 483–490.
87. Dimaio, V.J.M. and Garriott, J.C. (1974) Lethal caffeine poisoning in a child. *Forensic Science* **3**, 275–278.
88. Stavric, B. (1988) Methylxanthines: toxicity to humans. Caffeine. *Food. Chem. Toxicol.* **26**, 645–662.
89. Lachance, M.P. (1982) The pharmacology and toxicology of caffeine. *J. Food Safety* **4,** 71–112.
90. Friedman, L., Weinberger, M.A. and Peters, E.L. (1975) Testicular atrophy and aspermatogenesis in rats fed caffeine or theobromine in the presence or absence of sodium nitrite. *Fed. Proc. FASEB.* **34**, 228.
91. Collins, T.F.X. (1979) Review of reproduction and teratology studies of caffeine. *FDA By-Lines* **9**, 352–373.
92. Winick, M. (1998) Caffeine and reproduction. In *Caffeine* (Ed. by Spiller, G.A.), pp. 357–362. CRC Press, New York.
93. Hinds, T.S., West, W.L., Knight, E.M. and Harland, B.F. (1996) The effect of caffeine on pregnancy outcome variables. *Nutr. Rev.* **54**, 203–207.
94. Caan, B., Quesenberry, C.P. and Coates, A.O. (1998) Differences in fertility associated with caffeinated beverage consumption. *Am. J. Public Health* **88**, 270–274.
95. Tarka, S.M., Jr, Zoumas, B.L. and Gans, J.H. (1979) Short-term effects of graded levels of theobromine in laboratory rodents. *Toxic. Appl. Pharmacol.* **49**, 126–149.
96. Minton, J.P., Foecking, M.K., Webster, D.J. and Matthews, R.H. (1979) Caffeine, cyclic nucleotides, and breast disease. *Surgery* **86**, 105–109.
97. Minton, J.P., Abou-Issa, H., Reiches, N. and Roseman, J.M. (1981) Clinical and biochemical studies on methylxanthine-related fibrocystic breast disease. *Surgery* **90**, 299–304.
98. Ernster, V.L., Mason L., Goodson, W.H., III, *et al.* (1982) Effect of caffeine-free diet on benign breast disease: a randomized trial. *Surgery* **91**, 263–267.
99. Heyden, S. and Muhlbaier, L.H. (1984) Prospective study of fibrocystic breast disease and caffeine consumption. *Surgery* **96**, 479–483.
100. Lubin, R., Ron, E., Wax, Y., Black, M., Funaro, M. and Shitrit, A. (1985) A case-control study of caffeine and methylxanthines in benign breast disease. *J. Am. Med. Assoc.* **253**, 2388–2392.
101. Lubin, F. and Ron, E. (1990) Consumption of methylxanthine-containing beverages and the risk of breast cancer. *Cancer Letters* **53**, 81–90.
102. Schairer, C., Brinton, L. and Hoover, R. (1986) Methylxanthines and benign breast disease. *Am. J. Epidemiol.* **124**, 603–611.
103. McLaughlin, C.C., Mahoney, M.C., Nasca, P.C., Metzger, B.B., Baptiste, M.S. and Field, N.A. (1992) Breast cancer and methylxanthine consumption. *Cancer Causes and Control* **3**, 175–178.
104. Heyden, S. and Fodor, J.G. (1986) Coffee consumption and fibrocystic breasts: an unlikely association. *Can. J. Surgery* **29**, 208–211.

105. Schairer, C., Brinton, L. and Hoover, R. (1987) Methylxanthines and breast cancer. *Int. J. Cancer* **40**, 469–473.
106. AMA's Council on Scientific Affairs (1984) Caffeine labeling. *JAMA* **252**, 803–806.
107. Mosher, B.A. (1981) The health effects of caffeine. ACSH, Summit, NJ.
108. Dews, P.B. (1984) Behavioral effects of caffeine. In *Caffeine* (Ed. by Dews, P.), pp. 86–103. Springer-Verlag, New York.

Section IV

Physiological Effects and Health Considerations

Chapter 11

Obesity: Taste Preferences and Chocolate Consumption

Caroline Bolton-Smith and Marion M. Hetherington

This review seeks to document what is known about the relationship between obesity, taste preferences and chocolate consumption. Obesity is a significant health problem in Westernized countries and a growing problem internationally (1). In the USA, it has been estimated to cause 8% of all illness and cost $55 billion per annum (2). Years of multi-disciplinary research into the causation, prevention and treatment of obesity have not yet impacted beneficially on incidence rates. Since chocolate has a popular 'image' of contributing to over-weight, it is important to consider the scientific evidence regarding the role of chocolate in what has been termed *the obesity epidemic*.

In principle this article encompasses all types of chocolate confectionery, chocolate biscuits and other chocolate products, whilst in practice information on chocolate bars and other confectionery predominates. Examining the direct link between chocolate and obesity is not an easy task, for two main reasons. First, there is no nutritional (or indeed known biochemical, physiological or psychological) justification for linking *any single* food item to either the causation or maintenance of obesity in the general populace. Second, and as a consequence of the first, there is virtually no published scientific literature on the direct association between chocolate and obesity. Chocolate is generally dealt with, but seldom separated out in any detail, when high-fat and high-sugar foods or snacking are investigated. This does not mean that chocolate is not widely written about; quite the contrary, since it has been considered to be 'uniquely palatable' (3).

Chocolate – a special place in the diet?

The best of foods, the worst of foods: chocolate occupies a rather ambiguous position in the world of food, with both connotations of reward and high-fat, unhealthy food; luxury and forbidden fruit; treat and temptation. In their study of health information and food preferences, Wardle and Solomons (4) used the phrase 'naughty but nice' to suggest a paradox between familiar, high-fat foods which are rated as well liked compared to newly formulated, low-fat, healthy foods which are rated as less preferred. Chocolate is a food which fits this label well given its long history, beginning with its unique position as a luxurious drink for the Aztec élite to its introduction to the Spanish aristocracy (5) and its current widespread popularity and general availability. Given 17th-century pronouncements on its aphrodisiac qualities, early restrictions of its use to the élite (see Rossner (6) for a summary) and its first appearance in a British medical directory in 1826 as having medicinal uses (7), chocolate appears to confer upon the consumer more than simple calories. Chocolate, therefore, has a long tradition of specialness and in contemporary Western societies, chocolate is offered in gift-giving rituals across a wide spectrum of circumstances – to family and friends, to children, to work associates, and during a variety of religious, family and social occasions. However, in an effort to encourage consumption of chocolate beyond special circumstances or occasions, advertisers promote chocolate as an everyday food item that can be eaten between meals without proportionately reducing caloric intake at meal times, or as a nutritious supplement to meals (7). This means that chocolate occupies different areas of conceptual food space – as an occasional luxury, as a frequent treat, as a snack which will help boost energy and as a yearned-for 'unhealthy' food to be avoided. Much of the public conception about chocolate and 'putting on weight' or being 'overweight' reflects personal subjective experiences, which are strongly influenced by these ambiguities. Therefore it is useful, before going further, to define *obesity* and summarize the factors which are currently thought to cause and contribute to its incidence.

Obesity: definitions and aetiology

Obesity is measured in a number of different ways (8); however the Quetelet index, or body mass index (BMI), which is the ratio of a person's weight (kg) to their height (m^2), is in most common use (9). A number of different cut points have been used for the definition of obesity (9–11), but in general it is adequate to define three stages of obesity: *stage 1* or *overweight*, with a BMI ranging from 25–30; *stage 2* or *clinical obesity*, with a BMI ranging from 30–40; and *stage 3* or *morbid obesity*, with a BMI greater than 40. Other ways of defining obesity which directly measure the percentage body fat become particularly important in sports men and women who may have a high body weight for height due to a high muscle mass (muscle is heavier than fat tissue).

A tendency towards 'central obesity', which may be assessed as waist circumference divided by height, or waist/hip ratios, is more highly related to ill health than is BMI (12). So whilst central obesity may be the more relevant public health measure by which to assess the effects of chocolate consumption, to date the literature is inadequate.

Estimates of the likely genetic component to obesity are in the order of 20–40% depending on the type of study (13). Polymorphisms in the leptin gene may be one of the important genetic variables which regulate appetite. Leptin protein levels have been clearly related to the obese state in animals but less clearly so in humans (14). Despite genetic factors, the important principles are that increasing weight (adiposity) results in movement towards the obese state, and that gain in adiposity (fat mass) results from a *long-term* imbalance between energy intake (energy from food and drink) and energy expenditure (basal metabolic rate + thermic effect of eating + involuntary muscle activity (shivering) + voluntary activity (day-to-day living and specific 'exercise')). A review of the intricacies of food intake, energy balance and body weight control is provided by Doucet and Tremblay (15) (Fig. 11.1).

Weight gain which occurs as a result of prolonged energy imbalance is not all fat tissue: increased lean body mass is required to support the adiposity. Since the basal metabolic rate (BMR) is primarily dependent on body weight, increased weight results in a higher BMR and also a higher energy cost of activity because of the extra weight to move. Thus the total energy intake (TEI) required for weight stability goes up, and selection of energy-dense foods may be required to sustain adiposity. Such selection of energy-dense foods by the obese has been demonstrated (16, 17).

Total energy intake comes from a range of different foods and in a typical Western culture this may be between 15 and 50 different food and drink items per day, with many more being possible. The energy from a food (or drink) is derived from the different macro-nutrient constituents: fat 37 kJ/g (9 kcal/g); carbohydrate (sugars and starch) 16 kJ/g (3.75 kcal/g); protein 17 kJ/g (4 kcal/g); fibre 6 kJ/g (1.5 kcal/g); alcohol 29 kJ/g (7 kcal/g). As fat (whether saturated, mono- or polyunsaturated) provides over twice the energy of carbohydrate and protein per given amount, foods which are fat rich are more calorific, or energy dense, than low-fat foods. This is the basis of why fatty foods are considered 'fattening'. If a diet is predominantly made up of fat-rich foods, then there is a strong possibility of excess energy consumption (in relation to expenditure) due to its low satiety quotient (18), i.e. fat produces a poor satiety response (19–21). This effect has also been called 'passive over-consumption'; however the term 'passive' is not fully appropriate in this context. Fat makes up about 55% of the energy from a chocolate bar, with about 40% from sugar and the remaining 5% from protein.

Fat intake and preferences

There is an association between increased fat content of the diet and obesity in adults (22, 23) and children (24). However, all intake data, and especially from

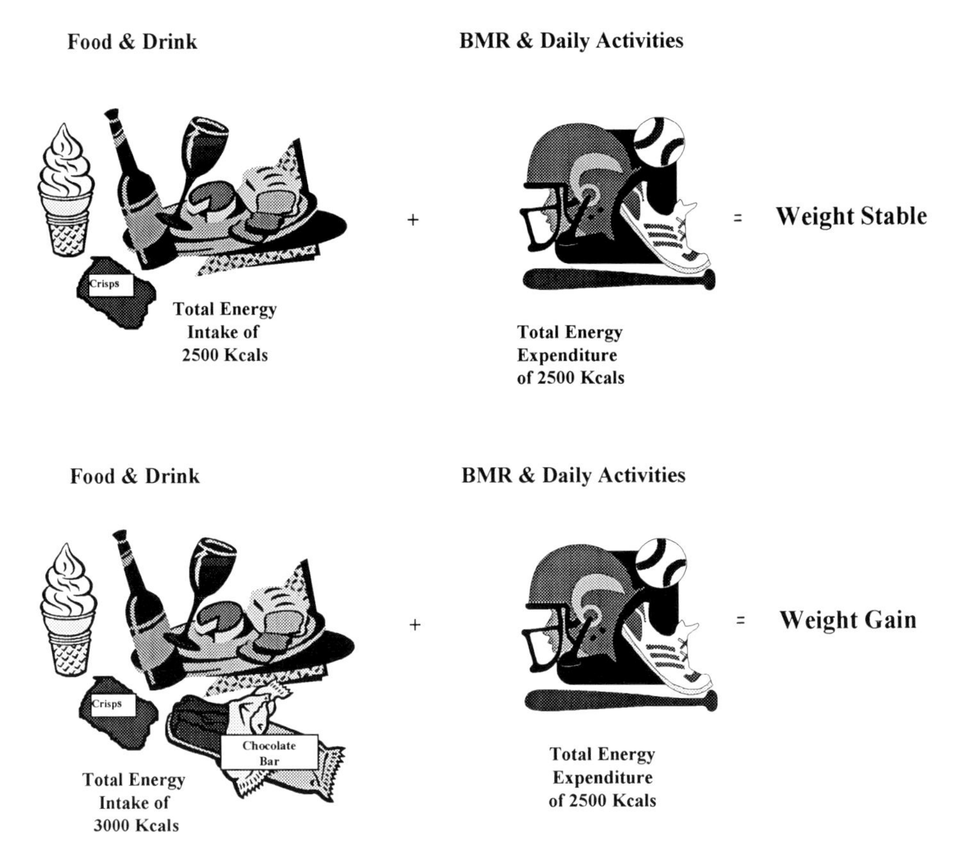

* BMR = basic metabolic rate.

Question: From where do the 500 calories (excess to total energy requirements) which cause weight gain come? The chocolate? The wine? The cheese? The crisps? The ice-cream?
Answer: They do not come from any *one* source – they come from all the food consumed. It is the total number of calories consumed that is important.

Fig. 11.1 Energy imbalance and obesity.

obese subjects, is hampered by under-reporting (25, 26). It has been suggested by Summerbell *et al.* (27) and supported by the recent work of Poppitt *et al.* (28) that omission of snacks from diet records may accentuate the fat content of diets, since snacks tend to be higher in sugars than fats when compared with main meals, and be responsible for the apparent higher proportion of energy from fat in obese subjects. This is unlikely to account for all of the association for, biochemically, dietary fat is used more efficiently and may predispose to the laying down of adipose tissue (29). However, contradictory data have been reported in lean subjects (30).

In determining the association between obesity and fat intake, the role of fat preference is crucial. Laboratory studies of fat preferences in obese consumers reveal a close relationship between taste preferences for fat and body fat (16, 31).

Therefore, one might predict that a high-fat food such as chocolate would be identified as a favourite food by obese consumers. Drewnowski *et al.* (32) asked a large clinical sample of obese men and women to list their ten favourite foods. Both sexes characteristically selected high-fat foods; however, men tended to select high-fat/protein foods (72% of males listed steak as a favourite food) whereas females listed high-fat/high-sweet carbohydrate foods (56% named cakes, cookies or doughnuts as their favourite food). Chocolate was rated by 32% of women and 15% of men in the top ten favourite foods. Thus, there appear to be gender differences in the type of high-fat foods preferred by obese consumers, and even for women, chocolate was not rated as frequently as ice cream (32).

Using taste stimuli containing sucrose and fat mixtures, Drewnowski *et al.* (16) reported that obese individuals demonstrated a strong preference for high-fat samples. The optimal combination of stimulus ingredients was heavy cream (>34% lipid) sweetened with less than 5% sucrose (weight for weight), whereas for normal weight subjects the optimal combination was 10% sucrose and 20% lipid. Since, on a weight basis, chocolate consists of around 30% fat and 50–60% sugar, it does not closely match the preferences of either the obese or normal-weight subjects. Cravings for chocolate amongst obese binge eaters have been noted by Drewnowski *et al.* (33); however, a significant relationship between cravings for salty foods and a high BMI in women has also been noted (34).

In order to explore mechanisms for such chocolate cravings, Drewnowski *et al.* (35) administered the opioid antagonist naloxone to female binge eaters and controls. These subjects were then asked to rate the sugar/fat mixtures and to select from a variety of snacks of different fat and carbohydrate levels. Naloxone reduced preferences for sugar/fat mixtures in both subject groups, and intake of sweet high-fat foods such as chocolate was reduced relative to saline infusions. Thus, in obese binge eaters, cravings for chocolate and other sweet high-fat foods may be mediated by endogenous opioids (35). However, cravings for chocolate are not exclusive to the obese and such observations may be restricted to obese binge eaters rather than obese individuals *per se*.

Taste preferences for fats in a variety of foods were consistently linked to percentage body fat in a sample of normal-weight individuals (31). What is not known, however, is whether such taste preferences develop as a function of adiposity or as a function of habitual dietary patterns (17). For example, Mattes (36) demonstrated that normal-weight subjects placed on a reduced-fat diet over a 12-week period with no exposure to fat sources reported a decline in preference for high-fat foods and a reduction in the preferred fat content of the specific test foods. Thus, fat preference may be influenced by the frequency of eating high-fat foods, exposure to the sensory characteristics of fat sources and habitual intake of foods high in fat. Therefore, the relationship between adiposity and fat preference may be mediated by established eating patterns rather than, or as well as, a propensity to prefer high-fat foods as a function of the need to feed a greater body mass. Similarly, a preference for chocolate cannot be predicted on the basis of body weight or body fat, but rather on liking, intake and pattern of eating across weight categories.

Sugar and obesity

Sugar – the other major energy component of chocolate – has also been implicated in obesity, but on a less sound scientific footing. Carbohydrate intake induces a strong hypothalamic-mediated satiety response (37, 38), i.e. intake is regulated by a feeling of fullness. Sugar-rich foods are necessarily bulkier (less energy dense) than high-fat foods, and also less dense in vitamins and nutrients. This has led to a view that high-sugar foods provide 'empty calories', are unnecessary in the diet, and so when present in the diet they contribute, gratuitously, to excess energy intake and the aetiology of obesity. There is a considerable scientific literature which refutes this viewpoint (22, 29, 39), although the concept is generally sustained through the popular and slimming press. For example, craving for chocolate was reported to be the single most important factor contributing to dietary failure by slimmers in repeated annual surveys (Dr E. Evans, personal communication). This may well be the case if all such highly palatable foods have been excluded from the diet, rather than incorporating 'treats' which allows the natural desire for sweet/palatable foods as part of moderate energy restriction (40).

Chocolate is both a high-fat and high-sugar food. It is contended that the hedonistic (pleasurable) properties of high-fat and high-sugar foods can override regulatory mechanisms and result in excess energy intake and obesity (41). However, when total food intake is considered there is little evidence for a correlation between high-fat and high-sugar diets (42) and fat intake is generally positively associated and sugar intake negatively associated with obesity (22, 43, 44).

Palatability

The palatability of chocolate is a powerful element in its consumption. Chocolate consists of potent olfactory and gustatory properties including a sweet taste, creamy texture and melting mouth-feel. In studies of food preferences, chocolate is rated as highly liked, particularly by young consumers (45, 46). Foods such as chocolate and ice cream obtain highly skewed preference ratings even compared to sweet confectionery (47). These foods seem to combine optimal levels of sweetness and fat content (48). Humans appear to be primed to like sweet tastes, since newborn infants with no prior experience of feeding demonstrate positive affective responses to the presentation of sweet solutions compared to distilled water, sour and bitter solutions (49). In addition, preferences for fat may have, at one time, been adaptive since fat yields a greater amount of energy by weight than either carbohydrate or protein. Thus, chocolate combines two very powerful and possibly primal taste elements.

Morality

Taken together, the symbolism of chocolate and the influence of palatability create a potent mix of sensory, socio-cultural and interpersonal appeal. Not surprisingly, chocolate attracts negative comments from some health professionals. Given that the calories from chocolate are almost entirely derived from sugar and fat, the two most demonized elements of Western diets (50–52), chocolate is targeted as a foodstuff which should be restricted, reduced or eliminated from the diet. However, as sociologists have observed, food morality can reduce the desirability of foods which are good whilst increasing the desirability of foods which are labelled as bad (53). Indeed in a laboratory study of foods labelled as low- or high-fat, Wardle and Solomons (4) found that yoghurt and cheese labelled as low-fat were rated as less pleasant tasting compared to those foods labelled as high-fat. This finding is more interesting given that the fat content was identical across food types. Hence, healthy eating information may not promote selection of foods if low-fat versions are considered less tasty. Chocolate can certainly be considered a food which is 'bad for you', or at least, is not a food to be eaten with abandon. Rogers (54), in his discussion of the phenomenon of 'moreishness' (causing a desire for more) as applied to chocolate, suggests that socially acceptable portion sizes are relatively small for moreish foods, and coupled with their high palatability, produce a conflict in consumers. Chocolate is highly desirable, yet should be eaten with restraint (55). In addition to this, notions of morality around the experience of pleasure support the conceptualization of chocolate as a forbidden indulgence, something which advertizers exploit in their portrayal of chocolate. The marketing appears to be winning over the common public health message to cut back on sweet, calorific, in-between meal snacks since *per capita* consumption of chocolate confectionery is growing (56–58).

Snacking and obesity

Chocolate confectionery and biscuits are commonly eaten as snack foods between traditional 'meal times' and this may be considered the fourth piece of circumstantial evidence for chocolate contributing to obesity.

In a sample of 1213 Americans, 12 years and older, Seligson *et al.* (57) reported that of the 'chocolate-eating occasions', 18% occurred at lunch time, 21% between 12 noon and 4 PM and 22% between 8 PM and midnight. In a French study

(59) of 764 snacks reported by 15–60-year-olds during 7-day food recording, 54% were classified as 'sugars, preserves and confectionery'. Nutritionally, the snacks were higher in sugar and equivalent in fat content to lunch and dinners. There is little direct evidence to support the purported link between excess eating of snacks and obesity and Drummond *et al.* (60) reported an inverse association in men. In the latter study, when the types of snack foods were categorized as chocolate bars, biscuits, cakes, fruit, crisps, soft drinks and sandwiches, 61% reported eating chocolate bars and 73% reported eating crisps during the 7-day recording period. A comparison of those people who reported more or less than four eating occasions per day showed that 69% of both men and women in the frequent-eaters group consumed chocolate bars, whilst 57% of women and 48% of men in the infrequent-eaters group reported eating chocolate bars (Dr S. Drummond, personal communication). It has been suggested that the number of snacks eaten may be involved in the maintenance of obesity, rather than as a causative factor (17).

Chocolate has been found to contribute, on average, only 0.7–1.4% of total daily energy intake (57) but a far higher proportion of energy from all snacks: 18% of the total energy from snacks in 7–8-year-olds and 14% of total snack energy in female students in their twenties (S. Whybrow, personal communication based on data of Ruxton *et al.* (61) and Kirk *et al.* (62), respectively). Summerbell *et al.* (63) reported a lower value of 4% of total snack energy from chocolate in young adults; however, 'snacking' definitions differed, and may have included substantial 'suppers'. Savoury snacks (including potato crisps/corn chips) make up a greater proportion of snacks, have a similar energy density (~530 kcal/100 g) and nutritional composition to chocolate bars (58% energy fat, 38% energy carbohydrate and 4% energy protein) and yet their consumption is not widely perceived as causing obesity. This may be because crisps or chips are seldom viewed as 'addictive' and seldom craved to the same extent as chocolate.

Chocolate addiction

Investigations of so-called chocolate addicts have revealed no particular tendency for such individuals to be obese. Indeed, in a preliminary study of chocolate addiction, the range of BMI in subjects was 16.4–41.0 with a mean of 25.3, suggesting a normal distribution of weight (64). However, when interviewed about their attitudes to chocolate, despite 84% providing positive descriptions of chocolate, 25% believed that chocolate was fattening and 14% described chocolate as 'unhealthy'. This demonstrates the ambiguity of chocolate's identity in consumers who eat chocolate to excess. In a further study of chocolate addicts, Macdiarmid and Hetherington (65) reported that consuming chocolate was accompanied by a significant increase in feelings of guilt. When chocolate addicts were classified according to whether or not they met criteria for binge eating disorder (BED), again no significant difference in BMI was found between binge eaters and non-BED subjects (65).

Obesity and depression

Chocolate is often cited as a high-carbohydrate food which obese carbohydrate cravers select in an attempt to reduce depression (66, 67). However, as Drewnowski *et al.* (35) have suggested, foods which have been targeted by carbohydrate cravers are more accurately described as sugar/fat mixtures (e.g. chocolate candies, chocolate bars, cakes, cookies and ice cream). The self-medicating theory of carbohydrate craving suggests that the obese, premenstrual and individuals suffering from seasonal affective disorder select foods high in carbohydrates in order to increase the availability of tryptophan across the blood–brain barrier, thereby increasing the synthesis of serotonin. However, cravings for foods which are more effective in promoting tryptophan uptake, such as rice, potatoes and pasta, are less likely to be reported than foods which are high in sugar and fat, such as chocolate. Therefore, it remains unclear whether such cravings among obese consumers and others reflects a serotonin deficiency or a simple desire for highly palatable treats to 'cheer themselves up'.

Weight control and chocolate cravings

Chocolate is the single most craved food in studies of food cravings (34, 68–70). Theoretical models used to explain chocolate cravings include abstinence and expectancy models (71). Abstinence models contend that consumers who abstain from consuming a particular food, say for health or weight reasons, are more susceptible to cravings for that particular food. Weight loss interventions have not focused on chocolate *per se*, rather on TEI, activity and, often, snack reduction. In the UK, chocolate is the most frequently cited reason for failure to continue with weight reduction diets (Dr E. Evans, personal communication). In relation to 'sweet' cravings frequently experienced by slimmers who are 'not allowed' sweets and chocolate, the concept of keeping within target caloric intake whilst incorporating 'a-treat-a-day' (often chocolate) has proved successful. If chocolate (in measured amounts) can be a positive aid in weight loss programmes, then this surely invalidates the idea that chocolate *per se* causes or maintains obesity.

Expectancy models propose that cravings for foods such as chocolate develop as a learning process in which sweet foods which are typically consumed at the end of a meal or at a particular time of day induce cravings through habitual intake patterns. Obese consumers who are following a particular dietary plan may be vulnerable to cravings as a function of expectancy or learning. However, cravings are not specific to obese consumers, nor has either of these psychological models been supported by empirical research.

Research on consumers who report food cravings has revealed no specific link to obesity (69), dietary restraint (34, 69, 72) or oestradiol levels (34). However, menstrual cycle (73, 74), pregnancy (75), problem eating (64) and negative affect (69) have been linked to cravings. Therefore, although some obese consumers may experience chocolate cravings, this is no more common than cravings reported by their lean counterparts.

The foregoing commentary suggests that the scientific rationale for a link between obesity, taste preferences and consumption of chocolate products is, at best, tenuous. An additional question remains: 'Is there any evidence for a relationship between chocolate consumption and obesity from population studies?'

Relationships between chocolate consumption and obesity between and within countries

Epidemiological associations between diet and disease have historically (e.g. saturated fat with congenital heart disease (CHD)), and more recently (e.g. antioxidant vitamins with CHD and cancers), provided important indications for causation. Fig. 11.2 reports the currently available information on chocolate

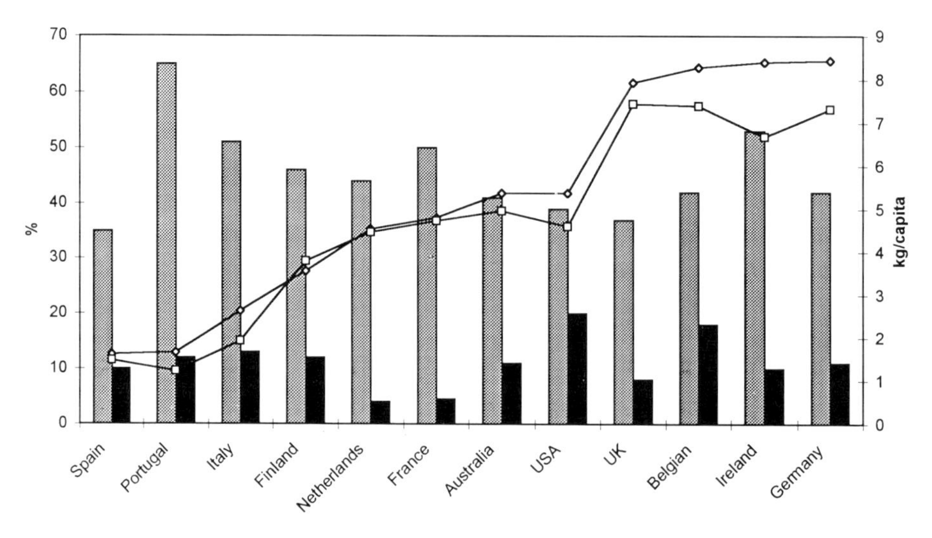

Fig. 11.2 Ecological associations between chocolate consumption and obesity prevalence. Chocolate confectionery consumption (1991 and 1996) with prevalence of over-weight (BMI 25–30) and obesity (BMI >30) in 12 countries (mid-1980s data). ▩ = % overweight; ■ = % obese; ◇ = chocolate consumption in 1996; ☐ = chocolate consumption in 1991.

consumption as kg per capita (56) and obesity prevalence taken from published national data (76–83) in different countries. Whilst the quality of the data is very variable, there appears to be no association between consumption in the 1990s and over-weight and obesity prevalence (data largely derives from the mid-1980s) for the countries shown. Consumption data for the same countries in the 1980s is incomplete (56), but where present also shows no relationship to obesity prevalence.

A stronger epidemiological technique is to look for relationships over time. However, nationally representative data on obesity prevalence is reported only sporadically, such that frequently only two or three time points exist and this makes any assessment of trends very tenuous. Alternatively, as for the National Health and Nutrition Examination Survey (NHANES) in the USA, they span several years and so cannot be matched with *per capita* consumption data (1). Table 11.1(a) does show a trend of increasing *per capita* consumption of chocolate products (56) and prevalence of over-weight and obesity for Australian women but far less so for men (84), and similar relationships for UK men and women (85–87). Table 11.1(b) illustrates the changes in obesity prevalence across the Scottish MONICA (MONItoring trends and determinants in CArdiovascular disease) surveys and the percentage who reported frequent (more than three per week) chocolate confectionery consumption (88, 89 and Bolton-Smith, unpublished observations).

However, such coincident trends in the rising prevalence of obesity and *per capita* chocolate confectionery consumption in the countries examined can be taken to imply neither causation, nor a lack of it. Whilst there *could* be a negative role for chocolate, the trend is far more likely to be due to the wider shifts from conventional three meals a day to snacking or grazing eating patterns, and the ready availability of foods such as chocolate confectionery and other snack foods (ease of access has been reported as a primary determinant of food choice in school children (90)). Similar trends over time occur for the consumption of semi-skimmed milk and low-fat spreads in these countries and these foods are unlikely to be causally associated with increasing obesity prevalence.

Unpublished data from the sequential cross-sectional Scottish MONICA surveys (88) in North Glasgow of 1984, 1989, 1992 and 1995 showed a weak inverse relationship between BMI and the frequency of chocolate confectionery consumption for the total population of 6236 men and women aged 18–75 ($r = -0.026$, $P = 0.041$), and no significant correlation between chocolate consumption and BMI within eight of the nine sex and BMI groups (<25, 25–30, >30). The exception was for the male obese (BMI >30) where there was a weak positive correlation ($r = 0.074$, $P = 0.019$).

In a more detailed analysis of the relationship between chocolate confectionery and body weight status in the 1992 Scottish MONICA data set, total energy intake and the percentage energy from sugar, but not fat, increased sig-

Table 11.1 Trends in obesity prevalence and consumption of chocolate products.
(a) Australia and UK.

Australia	1980	1983/5	1989
Men			
% overweight	40.6	40.0	44.1
% obese	9.3	9.1	11.5
Women			
% overweight	20.2	22.0	25.1
% obese	8.0	10.5	13.2
Consumption of chocolate products (kg/capita)	3.99	4.42	5.01
UK	1980	1983/6	1991
Men			
% overweight	35	39	46
% obese	6	8	12
Women			
% overweight	23	22	29
% obese	8	12	16
Consumption of chocolate products (kg/capita)	6.5	7.6	7.79

Source: data from references (56, 84–87, Bolton-Smith, unpublished observations).

(b) North Glasgow, Scottish MONICA populations.

	1984	1989	1992	1995
Men				
% overweight	40	41	42	42
% obese	11	11	14	20
% consuming chocolate + sweets more than three times/week	23	12	21	25
Women				
% overweight	34	33	32	33
% obese	14	15	20	20
% consuming chocolate + sweets more than three times/week	15	18	25	29

Source: data from references (88, 89, Bolton-Smith, unpublished observations).

nificantly across low (less than one per week) to high (more than three per week) chocolate confectionery groups. No significant relationships were seen between consumption and BMI group for men, whilst for women the proportion of low consumers was greatest and the proportion of high consumers was lowest in the obese (BMI > 30) group. A similar trend was seen with consumption of sweet biscuits (including chocolate coated), while the opposite trends were seen for consumption of ice cream and yoghurts as a single group. This illustrates the need for care in interpreting cross-sectional data, for while one can conclude that low chocolate confectionery and high ice cream + yoghurt consumption is *associated* with obese women, it is clearly inappropriate to suppose that low chocolate and high yoghurt consumption *causes* obesity. Changes consequent upon the obese state are more likely and misreporting of the true intake could also contribute to such trends.

The same data set has shown a lack of association between obesity prevalence and chocolate confectionery consumption by socio-economic groups (91), and New and Grubb (92) also reported no relationship between BMI and chocolate confectionery consumption in 994 middle-aged Scottish women.

An analysis of UK national trends by Prentice and Jebb (93) found that increasing inactivity was associated with increasing prevalence of obesity and not dietary factors. Chocolate bars are often marketed as a quick energy boost for the 'active', and indeed, both men and women who were most active at work reported consuming significantly more chocolate confectionery, whilst there were no significant differences in BMI between the activity groups (89). The BMI of women who were most active during leisure time was significantly higher than that of less active women, but their reported consumption of chocolate did not differ (89).

Whilst the existing data which has allowed the comparison between the obese state and chocolate product consumption is sparse, with the provisos regarding data completeness and bias, there appears to be no evidence of a direct relationship between the two.

Summary

Chocolate is a complex food from a psychological point of view. It occupies a rather unusual place in the food domain as an everyday snack and occasional extravagance. It is considered a forbidden pleasure by a society which increasingly encourages lower fat and sugar intakes for better health. The data support neither greater palatability or craving for chocolate, nor higher consumption of chocolate confectionery in the obese. Thus, in the absence of any longitudinal data, there is no evidence to implicate chocolate consumption *per se* in the aetiology of obesity.

It is perhaps plausible that chocolate could be contributing to weight maintenance in the obese, by providing a highly palatable, readily available and energy-dense snack food, in a similar manner to crisps, although supporting hard data do not exist. It is undoubtedly viewed by slimmers as a cause (or is it an excuse?) of slimming failure.

Overall, there is no available evidence for a causative link between the consumption of chocolate and obesity. While further, focused research would clarify the situation, it seems unlikely that the conclusion would differ.

References

1. Flegal, K.M., Carroll, M.D., Kuczmarski, R.J. and Johnson, C.L. (1998) Over-weight and obesity in the United States; prevalence and trends, 1960–1994. *Int. J. Obes.* **22**, 39–47.
2. Colditz, G. (1992) Economic costs of obesity. *Am. J. Clin. Nutr.* **55**, S503–S507.
3. Macdonald, H. (1993) Flavour development from cocoa bean to chocolate bar. *The Biochemist* **15** (2), 3–5.
4. Wardle, J. and Solomons, W. (1994) Naughty but nice: a laboratory study of health information and food preferences in a community sample. *Health Psychol.* **13**, 180–183.
5. Nielson, N. (1995) *Chocolate*. Trevi, Stockholm.
6. Rossner, S. (1997) Chocolate – divine food, fattening junk or nutritious supplementation? *Eur. J. Clin. Nutr.* **51**, 341–345.
7. James, A. (1990) The good, the bad and the delicious: the role of confectionery in British society. *Sociol. Rev.*, **38**, 666–688.
8. Sichieri, R., Everhart, J.E. and Hubbard, V.S. (1992) Relative weight classifications in the assessment of under-weight and over-weight in the United States. *Int. J. Obes.* **16**, 303–312.
9. Garrow, J.S. and Webster, J. (1985) Quetelet's index (W/H^2) as a measure of fatness. *Int. J. Obes.* **9**, 147–153.
10. RCP (1983) Royal College of Physicians Obesity Report. *J. Royal Coll. Physicians* **17**, 3–58.
11. Kumanyika, S., Antipatis, V., Jeffery, R., Morabia, A., Ritenbaugh, L. and James, W.P.T. (1998) The International Obesity Task Force: its role in public health prevention. *Appetite* **31**, 426–428.
12. Ashwell, M., LeJeune, S. and McPherson, K. (1996) Ratio of waist circumference to height may be a better indicator of need for weight management. *BMJ* **312**, 377.
13. Bouchard, C. (1997) Human variation in body mass: evidence for a role of the genes. *Nutr. Rev.* **55**, S21–S27.
14. Trayhurn, P. (1996) New insights into the development of obesity: obese genes and the leptin system. *Proc. Nutr. Soc.* **55**, 783–791.
15. Doucet, E. and Tremblay, A. (1997) Food intake, energy balance and body weight control. *Eur. J. Clin. Nutr.* **51**, 846–855.
16. Drewnowski, A., Brunzell, J.D., Sande, K., Iverius, P.H. and Greenwood, M.R.C. (1985) Sweet tooth reconsidered: taste responsiveness in human obesity. *Physiol. Behav.* **35**, 617–622.
17. Mela, D.J. (1996) Eating behaviour, food preferences and dietary intake in relation to obesity and body-weight status. *Proc. Nutr. Soc.* **55**, 803–816.
18. Green, S.M., Delargy, H.J., Joanes, D. and Blundell, J.E. (1997) A satiety quotient: a formulation to assess the satiating effect of food. *Appetite* **29**, 291–304.
19. Rolls, B.J., Hetherington, M.M. and Burley, V.J. (1988) The specificity of satiety: the influence of different macro-nutrient content on the development of satiety. *Physiol. Behav.* **43**, 145–153.
20. Rolls, B.J., Kim-Harris, S., Fischman, M.W., Foltin, R.W., Moran, T.H. and Stoner, S.A. (1994) Satiety after pre-loads with different amounts of fat and carbohydrate: implications for obesity. *Am. J. Clin. Nutr.* **60**, 476–487.

21. Cotton, J.R., Burley, V.J., Weststrate, J.A. and Blundell, J.E. (1994) Dietary fat and appetite – similarities and differences in the satiating effect of meals supplemented with either fat or carbohydrate. *J. Hum. Nutr. Diet.* **7**, 11–24.
22. Bolton-Smith, C. and Woodward, M. (1994) Dietary composition and fat to sugar ratios in relation to obesity. *Int. J. Obes.* **18**, 820–828.
23. Mela, D.J. (1993) Consumer perception, acceptance and consumption of dietary fats. *Int. Food Ingredients* **3**, 34–40.
24. Tucker, L.A., Seljaas, G.T. and Hager, R. (1997) Body fat percentage of children varies according to their diet composition. *J. Am. Diet. Assoc.* **97**, 981–986.
25. Lissner, L., Habicht, J.P., Strupp, B.J., Levitsky, D.A., Haas, J.D. and Roe, D.A. (1989) Body composition and energy intake: do overweight women over-eat and under-report? *Am. J. Clin. Nutr.* **49**, 320–325.
26. Lightman, S.W., Pisarska, K., Berman, E.L., *et al.* (1992) Discrepancy between self-reported and actual caloric intake and exercise in obese subjects. *N. Engl. J. Med.* **327**, 1893–1898.
27. Summerbell, C.D., Moody, R.C., Shankgs, J., Stocks, M.J. and Geissler, C. (1996) Relationship between feeding pattern and body mass index in 220 free-living people in four age groups. *Eur. J. Clin. Nutr.* **50**, 513–519.
28. Poppitt, S.D., Swann, D., Black, A.E. and Prentice, A.M. (1998) Assessment of selective under-reporting of food intake by both obese and non-obese women in a metabolic facility. *Int. J. Obes.* **22**, 303–311.
29. Flatt, J.P. (1978) The biochemistry of energy expenditure. In *Recent Advances in Obesity II* (Ed. by Bray, G.S.), pp. 211–228. Newmann, London.
30. Bobbioni-Harsch, E., Habicht, F., Lehmann, T., James, R.W., Rohner-Jeanrenaud, F. and Golay, A. (1997) Energy expenditure and substrates oxidative patterns after glucose, fat or mixed load in normal weight subjects. *Eur. J. Clin. Nutr.* **51**, 370–374.
31. Mela, D.J. and Sacchetti, D.S. (1991) Sensory preferences for fats in foods: relationships to diet and body composition. *Am. J. Clin. Nutr.* **53**, 908–915.
32. Drewnowski, A., Kurth, C., Holden-Wiltse, J. and Saari, J. (1992) Food preferences in human obesity: carbohydrates versus fats. *Appetite* **18**, 207–221.
33. Drewnowski, A., Kurth, C.L. and Rahaim, J.E. (1991) Taste preferences in human obesity: environmental and familial factors. *Am. J. Clin. Nutr.* **54**, 635–641.
34. Rodin, J., Mancuso, J., Granger, J. and Nelbach, E. (1991) Food cravings in relation to body mass index, restraint and estradiol levels: a repeated measures study in healthy women. *Appetite* **17**, 177–185.
35. Drewnowski, A., Krahn, D., Demitrack, M.A., Nairn, K. and Gosnell, B.A. (1992) Taste responses and preferences for sweet high-fat foods: evidence for opioid involvement. *Physiol Behav.* **51**, 371–379.
36. Mattes, R.D. (1993) Fat preference and adherence to a reduced-fat diet. *Am. J. Clin. Nutr.* **57**, 373–381.
37. Hoebel, B.G. (1997) Neuroscience and appetite behaviour research: 25 years. *Appetite* **29**, 119–133.
38. Blundell, J.E., Green, S. and Burley, V.J. (1994) Carbohydrates and human appetite. *Am. J. Clin. Nutr.* **59**, S728–S734.
39. Stubbs, R.J., Mugatroyd, P.R., Goldberg, G.R. and Price, A.M. (1993) Carbohydrate balance and the regulation of day-to-day food intake in humans. *Am. J. Clin. Nutr.* **57**, 897–903.

40. Mela, D.J., Rogers, P.J., Shepherd, R. and Mackie, H.J.H. (1993) Real people, real foods, real eating situations – real problems and real advantages. *Appetite* **19**, 69–73.
41. Green, S.M. and Blundell, J.E. (1996) Effect of fat- and sucrose-containing foods on the size of eating episodes and energy intake in lean dietary restrained and unrestrained females: potential for causing over-consumption. *Eur. J. Clin. Nutr.* **50**, 625–635.
42. Gibney, M.J., Sigman-Grant, M., Stanton, J.L. and Keast, D.R. (1995) Consumption of sugars. *Am. J. Clin. Nutr.* **62** (Suppl.), 178S–194S.
43. Dreon, D.M., Frey-Hewitt, B., Ellsworth, N., Williams, P.T., Terry, R.B. and Wood, P.D. (1988) Dietary fat:carbohydrate ratio and obesity in middle-aged men. *Am. J. Clin. Nutr.* **47**, 995–1000.
44. Tucker, L.A. and Kano, M.J. (1992) Dietary fat and body fat: a multivariate study of 205 adult females. *Am. J. Clin. Nutr.* **56**, 616–622.
45. Tuorila, H. and Pangborn, R.M. (1988) Prediction of reported consumption of selected fat-containing foods. *Appetite* **11**, 81–95.
46. Tuorila-Ollikainen, H. and Mahlamaki-Kultanen, S. (1985) The relationship of attitudes and experiences of Finnish youths to their hedonic response to sweetness in soft drinks. *Appetite* **6**, 115–124.
47. Tuorila, H. (1992) Preferences related to fat-containing foods. In *Dietary Fats: Determinants of Preference, Selection and Consumption* (Ed. by Mela, D.J.), pp. 27–41. Elsevier, London.
48. Drewnowski, A. and Greenwood, M.R.C. (1983) Cream and sugar: human preferences for high-fat foods. *Physiol. Behav.* **30**, 629–633.
49. Steiner, J.E. (1977) Facial expressions of the neonate infant indicating the hedonics of food-related chemical stimuli. In *Taste and Development: the Genesis of Sweet Preference* (Ed. by Weiffenbach, J.M.), pp. 173–188. DHEW, Bethesda.
50. Mintz, S. (1985) *Sweetness and Power: the Place of Sugar in History*. Viking, New York.
51. Yudkin, J. (1986) *Pure White and Deadly*. Penguin, London.
52. Klein, R. (1996) *Eat Fat*. Picador, London.
53. Mead, M. (1980) A perspective on food patterns. In *Issues in Nutrition in the 1980s* (Ed. by Tobias, L.A. and Thompson, P.J.) Wadsworth, Monterey.
54. Rogers, P.J. (1994) Mechanisms of moreishness and food craving. In *Pleasure: the Politics and the Reality* (Ed. by Warburton, D.M.), pp. 38–49. Wiley, Chichester.
55. Drewnowski, A. (1992) Sensory preferences and fat consumption in obesity and eating disorders. In *Dietary Fats: Determinants of Preference, Selection and Consumption* (Ed. by Mela, D.J.), pp. 59–77. Elsevier, London.
56. IOCCC (1997) *International Statistical Review of the Cocoa, Chocolate and Sugar Confectionery Industries 1996*. International Office of Cocoa, Chocolate and Sugar Confectionery, Brussels.
57. Seligson, F.H., Krummel, D.A. and Apgar, J.L. (1994) Patterns of chocolate consumption. *Am. J. Clin. Nutr.* **60**, S1060–S1067.
58. Ministry of Agriculture, Fisheries and Food (1997) *The National Food Survey 1996*. Her Majesty's Stationery Office, London.
59. Francois, P.L., Calamassi-Tran, G., Hebel, P., Renault, C. and Lebreton, V.J.L. (1996) Food and nutrient intake outside the home of 629 French people of fifteen years and over. *Eur. J. Clin. Nutr.* **50**, 826–831.

60. Drummond, S.E., Crombie, N.E., Cursiter, M.C. and Kirk, T.R. (1998) Evidence that eating frequency is inversely related to body weight status in male, but not female, non-obese adults reporting valid dietary intakes. *Int. J. Obes.* **22**, 105–112.
61. Ruxton, C.H.S., Kirk, T.R. and Belton, N.R. (1996) Energy and nutrient intakes in a sample of 136 Edinburgh 7–8-year-olds: a comparison with United Kingdom dietary reference values. *Br. J. Nutr.* **75**, 151–160.
62. Kirk, T.R., Burkill, S. and Cursiter, M. (1997) Dietary fat reduction achieved by increasing consumption of a starch food – an intervention study. *Eur. J. Clin. Nutr.* **51**, 455–461.
63. Summerbell, C.D., Moody, R.C., Shanks, J., Stock, M.J. and Geissler, C. (1995) Sources of energy from meals versus snacks in 220 people in four age groups. *Eur. J. Clin. Nutr.* **49**, 33–41.
64. Hetherington, M.M. and Macdiarmid, J.I. (1993) Pleasure and excess: liking for and over-consumption of chocolate. *Physiol. Behav.* **57**, 27–35.
65. Macdiarmid, J.I. and Hetherington, M.M. (1995) Mood modulation by food: an exploration of affect and cravings in 'chocolate addicts'. *Br. J. Clin. Psychol.* **34**, 129–138.
66. Wurtman, J.J. (1984) The involvement of brain serotonin in excessive carbohydrate snacking by obese carbohydrate cravers. *J. Am. Diet. Assoc.* **84**, 1004–1007.
67. Wurtman, R.J., Wurtman, J.J., Growdon, J.H., Henry, P., Lipscomb, A. and Zeisel, S.H. (1981) Carbohydrate craving in obese people: suppression by treatments affecting serotoninergic transmission. *Int. J. Eating Disord.* **1**, 2–15.
68. Weingarten, H.P. and Elston, D. (1991) Food cravings in a college population. *Appetite* **17**, 167–175.
69. Hill, A.J., Weaver, C.F.L. and Blundell, J.E. (1991) Food craving, dietary restraint and mood. *Appetite* **17**, 187–197.
70. Rozin, P., Levine, E. and Stoess, C. (1991) Chocolate craving and liking. *Appetite* **17**, 199–212.
71. Weingarten, H.P. and Elston, D. (1990) The phenomenology of food cravings. *Appetite* **15**, 231–246.
72. Lappalainen, R., Sjoden, P.-O., Hursti, T. and Vesa, V. (1990) Hunger/craving responses and reactivity to food stimuli during fasting and dieting. *Int. J. Obes.* **14**, 679–688.
73. Tomelleri, R. and Grunewald, K.K. (1987) Menstrual-cycle and food cravings in young college women. *J. Am. Diet. Assoc.* **87**, 311–315.
74. Bowen, D.J. and Grunberg, N.E. (1990) Variations in food preferences and consumption across the menstrual-cycle. *Physiol. Behav.* **47**, 287–291.
75. Schmidt, H.J. and Beauchamp, G.K. (1990) Biological determinants of food preferences in humans. In *Diet and Behaviour* (Ed. by Krasnegor, N.A., Miller, G.D. and Simopoulos, A.), pp. 33–47. Springer-Verlag, New York.
76. Ament, A. *et. al.* (1991) Overgewicht vormt forse kostenpost in gezondheidszorg (CBS-Gezondheidsenquete). *Voeding* **52**, 293–295.
77. National Heart Foundation of Australia and Australian Institute of Health (1989) *Risk Factor Prevalence Study*. Survey No. 3. National Health Foundation of Australia, Sydney.
78. Deutsche Gesellschaft für Ernahrung (1992) *Ernahrungsbericht*. Druchkerie Henrich, Frankfurt.

79. Direccion General de Prevencion y Promocion de Salut (1994) *Encuesta de Nutricion de la Communidad de Madrid*. Documento Tecnico de Salud Publica No. 18. Spanish Government Publication, Madrid.
80. INDI (1990) *Irish National Nutrition Survey*. Irish Nutrition and Dietetic Institute, Sligo.
81. Laurier, D., Guiguet, M., Chau, N.P., Wells, J.A. and Valleron, A.-J. (1992) Prevalence of obesity: a comparative survey in France, the United Kingdom and the United States. *Int. J. Obes.* **16**, 565–572.
82. Tavani, A., Negri, E. and Lavecchia, C. (1994) Determinants of body-mass index – a study from Northern Italy. *Int. J. Obes.* **18**, 497–502.
83. Office for National Statistics (1997) *The Health of Adult Britain 1841–1994. Volume 1* (Ed. by Charlton, J. and Murphy, M.). Government Statistical Service, London.
84. Bennett, S.A. and Magnus, P. (1994) Trends in cardiovascular risk factors in Australia. *Med. J. Aust.* **161**, 519–527.
85. Gregory, J., Foster, K., Tyler, H. and Wiseman, M. (1990) *The Dietary and Nutritional Survey of British Adults*. Her Majesty's Stationery Office, London.
86. Knight, I. (1994) *The Heights and Weights of Adults in Great Britain*. Her Majesty's Stationery Office, London.
87. Colhoun, H. and Prescott-Clarke, P. (eds) (1996) *Health Survey for England 1994. A survey carried out on behalf of the Department of Health*. Vol. 1. Her Majesty's Stationery Office, London.
88. The WHO MONICA Project (1989) A worldwide monitoring system for cardiovascular diseases: cardiovascular mortality and risk factors in selected communities. *World Health Statistics* **41**, 27–149.
89. Bolton-Smith, C. and McCluskey, M.-K. (1997) The relationship between confectionery consumption, physical activity and body mass index in a Scottish population. *Proc. Nutr. Soc.* **56**, 158A
90. Meiselman, H.L., Hedderley, D., Standdon, S.L., Pierson, B.J. and Symonds, C.R. (1994) Effect of effort on meal selection and meal acceptability in a student cafeteria. *Appetite* **23**, 43–55.
91. Wreiden, W., McCluskey, M.-K. and Bolton-Smith, C. (1995) Social status differences in biscuit, cake and confectionery consumption in North Glasgow. *Proc. Nutr. Soc.* **54**, 200A.
92. New, S.A. and Grubb, D.A. (1996) Relationship of biscuit, cake and confectionery consumption to body mass index and energy intake in Scottish women. *Proc. Nutr. Soc.* **55**, 122A.
93. Prentice, A.M. and Jebb, S.A. (1995) Obesity in Britain: gluttony or sloth? *BMJ* **311**, 437–439.

Chapter 12

Chocolate Consumption and Glucose Response in People with Diabetes

Janette C. Brand Miller

Diabetes is a disorder of carbohydrate metabolism characterised by high blood glucose (sugar) levels. Untreated, the disease produces serious complications including blindness, renal failure, nerve damage and coronary heart disease. Thus one of the main goals in the management of diabetes is to maintain before-meal and after-meal blood glucose levels as close as possible to normal. Other goals include normalising blood lipid levels, blood pressure and weight status. These objectives apply to both types of diabetes, type 1 (insulin-dependent, juvenile onset diabetes) and type 2 (non-insulin-dependent, adult onset diabetes). Historical changes in diets for people with diabetes have been so marked that the reader can be forgiven for being confused. In the past, the prescribed diets were low in carbohydrate (i.e. starch and sugars) because dietary carbohydrate was considered the main factor responsible for high blood glucose levels (1). This had the unintended consequence of increasing fat intake, especially saturated fat, and people with diabetes often died from premature coronary heart disease as a result.

In the 1970s, the importance of avoiding excessive consumption of fat led to the liberalisation of dietary carbohydrate intake to as much as 55% of energy, up from a low of 10–20%. However, in the 1990s, concerns about the potentially undesirable effects of very high-carbohydrate diets on certain plasma lipid fractions (triglyceride and high-density lipoprotein (HDL) cholesterol) led to more flexible recommendations regarding the proportion of carbohydrate and fat in the diabetic diet (2). The distribution of calories between fat and carbohydrate can vary and be individualised based on the overall goals of treatment (degree of reduction in blood glucose, blood lipids and weight loss needed).

Which carbohydrates – starch or sugars?

For the greater part of the 20th century, the most widely held belief about the dietary treatment of diabetes has been that simple sugars should be avoided and replaced with complex carbohydrates (starch, dietary fiber) (2). Chocolate, of course, was restricted because of its high sucrose content (50–55% by weight). The terms *simple* and *complex* carbohydrates are, in fact, no longer recommended because this chemical distinction means little in clinical terms (3). The assumption that sugars would be more rapidly digested and absorbed than starches, and therefore aggravate high blood glucose levels to a greater degree, was found to be incorrect. Similarly, the reasoning that the large molecular size of starches would render them more slowly digested and absorbed was also wrong in most cases. There is now abundant scientific proof that modern starchy foods have a greater impact on blood glucose and insulin levels than most foods containing sugar, whether naturally occurring or refined (4, 5).

While it is now recognised by the experts that the simple-versus-complex carbohydrate distinction has little clinical significance, refined sugars continue to be discouraged in the diets of both diabetic and non-diabetic individuals (6). Arguments about 'empty' calories, weight gain and dental caries continue to be put forward to justify the restriction on sugars. Even these reasons, however, do not stand up to scientific scrutiny. In the past decade, many large, well-designed, carefully controlled studies show that sugar is not the villain we thought it was (7). The role of sugar in Western diets is discussed further below.

Why restrict chocolate?

The high sugar content of chocolate is the main reason why avoidance has been recommended for people with diabetes, yet the blood glucose responses to chocolate and chocolate products, as we shall see, are universally low even when matched for carbohydrate content. Certain types of chocolate have also been accused of eliciting high insulin responses to the ingested carbohydrate, resulting in rebound hypoglycemia (low blood sugars levels) followed by fatigue, hunger and dizziness (8). While some chocolate products do appear to generate disproportionately high insulin responses (9), this is not a universal finding and certainly does not indicate increased risk of hypoglycemic symptoms (10). Foods such as chocolate that are high in both fat and carbohydrate are energy dense and may encourage 'passive over-consumption' of calories (11), but concerns about compromised blood glucose control are scientifically unfounded.

The glycemic index (GI) of foods

How can we compare the glucose response to chocolate with that of other foods? In 1981, the GI of foods was introduced by Jenkins *et al.* (4) as a scientifically based method of assessing and comparing the blood glucose response (glycemic response) to carbohydrate-containing foods.

It compares foods on an equal-carbohydrate basis and expresses the glycemic impact as a percentage of that seen with a standard food (usually glucose or white bread). Over 600 individual foods have since been tested for their GI (12). Table 12.1 shows the GI of a range of common foods.

One of the most important messages from GI research is that most modern

Table 12.1 The glycemic index of foods using a scale where glucose = 100. The values represent averages reported in the literature (12).

Food	GI
Breads	
White bread	70
Wholemeal bread	69
Pumpernickel bread	41
Dark rye bread	76
Sourdough bread	57
Heavy mixed grain bread	30–45
Breakfast cereals	
Cornflakes	84
Rice Krispies	82
Cheerios	83
Puffed Wheat	80
All Bran	42
Porridge	46
Legumes	
Lentils	28
Soybeans	18
Baked beans (canned)	48
Dairy foods	
Milk (full fat)	27
Milk (skim)	32
Ice cream (full fat)	61
Yogurt (low fat, fruit)	33
Soft/sport drinks	
Fanta	68
Gatorade	78
Fruits	
Apple	38
Orange	44
Peach	42
Banana	55
Watermelon	72
Snack foods	
Mars bar	68
Jelly beans	80
Chocolate bar	49

starchy foods are rapidly digested and absorbed, giving high blood glucose responses and correspondingly high GI scores. The foods with a high GI (>70 on a scale where glucose has a value of 100) include white and wholemeal bread, potatoes, most rices, and the majority of breakfast cereals (12). The starch in these foods has been fully gelatinised during processing and is therefore quickly digested and absorbed in the small intestine to glucose.

In contrast, there are some starchy foods that undergo less gelatinisation during cooking or contain factors which slow down starch digestion and absorption, such as soluble fibre. Porridge, pasta, legumes and cooked 'whole' intact grains have low GI values (< 55). The GI concept has been widely applied in nutritional science and shown to be clinically useful in the management of diabetes (13). Diets based on low-GI foods produce improvements in blood glucose and lipid control in people with diabetes (14).

Glycemic index of foods containing refined sugars

Where do sugary foods fall on the GI scale? In a large study of foods containing sugars, GI values ranged from very low to very high (5). The median GI, however, was only 56 (glucose = 100). The GI of the foods containing naturally occurring sugars only was similar to that of foods containing added sugars (median 53 versus 58 respectively, $P = 0.08$). Similarly, the median insulin index of the foods containing natural sugars was not significantly different from that of the foods containing added sugars (56 versus 61, $P = 0.16$). In addition, there was no significant difference in responses between cakes, muffins, cookies and crackers made with or without added sugars. Foods where added sugar produced substantially higher values included canned peaches, milk, yogurt and soft drinks. However, even in these instances the GI values of the foods containing added sugar were less than those of starchy staples such as bread.

The explanation for the low-to-moderate blood glucose responses to sugary foods lies in the fact that sugars in foods are usually a mixture of sucrose, lactose, glucose, fructose or galactose. Only the glucose component (free and that derived from hydrolysis of sucrose and lactose) has a marked effect on blood glucose levels. Fructose and galactose have little effect on either glycemic or insulin responses (10). Thus, a 50 g carbohydrate portion of a sugary food will produce fewer 'glucose equivalents' than a 50 g portion of a starch that is hydrolysed entirely to glucose, resulting in less effect on blood glucose.

It is therefore a scientifically proven fact that foods containing sugar, whether naturally occurring or refined, give glycemic and insulin responses that are similar to or lower than those of many common starchy foods in the Western diet. Indeed, the strongest determinant of whole diet GI in 342 individuals with non-insulin-dependent diabetes mellitus (NIDDM) was the intake of simple sugars – the *higher* the intake of simple sugars, the *lower* the overall GI of the diet (15).

In most medium-to-long-term studies, diets high in sucrose (versus diets high in starch) have been found to be compatible with good blood glucose and lipid control (10). While there are lingering concerns that extremely large amounts of sugars may have undesirable effects on blood triglycerides, this is not the case for the typical amounts of sucrose found in western diets (16). The most recent recommendations for people with diabetes in the USA (2) state: 'Scientific evidence has shown that the use of sucrose as part of the meal plan does not impair blood glucose control in individuals with type I or type II diabetes.'

Glucose and insulin responses to chocolate products in diabetic subjects

There is remarkably little information in the literature concerning blood glucose and insulin responses to chocolate products *in people with diabetes*. Peters *et al.* (17) compared the responses in type 1 diabetic patients to an isocaloric meal in which chocolate cake was substituted for baked potato. The glucose response and urinary excretion of glucose were no different between the two meals. However, potato has one of the highest GI values (GI = 70–90) and the chocolate cake appeared to be no better.

Gee *et al.* (18) compared the glucose and insulin responses to conventional sucrose-sweetened chocolate with fructose- and isomalt-based chocolates in people with type 2 diabetes. Isomalt is a sweet disaccharide that has no glycemic effect when consumed as a pure compound. Fructose when consumed alone produces only one-third of the glycemic and insulin effects of sucrose (10). All three chocolates provoked a sustained rise in blood glucose, the highest levels occurring after the conventional chocolate. The isomalt chocolate gave one-third less glycemia while the fructose-based product was not significantly different to that of the conventional chocolate. Insulin responses correlated with the level of glycemia, but there were no statistically significant differences between any of the formulations. These results give us no comparison of chocolate with other foods but nonetheless have interesting implications. They indicate that it is not just the sucrose in conventional chocolate that is responsible for all the glycemia or insulin demand. The isomalt-based chocolate might have been expected to produce little glycemia because the remaining carbohydrate was a small amount of lactose derived from milk. The high insulin response to all three chocolate products also indicates that the fat and protein in chocolate have marked stimulatory effects on insulin secretion. It is possible, therefore, that certain amino acids and fatty acids in chocolate provoke exceptional insulin responses.

Responses to chocolate products in healthy subjects

Since there is an excellent correlation between responses in diabetic and non-diabetic subjects (19), studies in healthy subjects may be used as a guide to the likely responses in people with diabetes. Most studies report that ingestion of a chocolate bar causes a substantially lower rise in plasma glucose and insulin than equivalent amounts of other carbohydrates (20–22). Shively *et al.* (9) conducted one of the most comprehensive studies in normal subjects of glucose and insulin responses to various snack foods, including chocolate, potato chips, granola bars and peanut butter cups. They found that plasma glucose responses to the snacks, on both an isocaloric or equivalent carbohydrate basis, were uniformly lower than that after a glucose load. In contrast, insulin responses to the snacks exhibited more variability, the milk chocolate bar giving higher responses than those predicted by the level of glycemia. In fact, the insulin response to the chocolate bar was as high as that seen after the glucose load. There was no evidence, however, of rebound hypoglycemia.

Recently blood glucose and insulin responses were determined after consumption of 10 different chocolate products in 12 healthy volunteers with normal glucose tolerance (23). Their ages ranged from 21–39 years (mean ± SD, 31 ± 6) and their body mass index (BMI) from 19 to 24 (22 ± 2) kg/m^2. The subjects took 50 g available carbohydrate (starch + sugars) portions of each test food in random order on separate mornings after a 10-hour overnight fast. The nutrient composition and weight of the 50 g carbohydrate portion of the foods is shown in Table 12.2. The reference food (white bread) was tested twice during the study. Foods were consumed over 12 min with sufficient water to bring the total meal volume to 600 ml. Finger-prick capillary blood samples were taken at 0 (fasting), 15, 30, 45, 60, 90 and 120 min after the meal began. Blood glucose was assayed using the glucose hexokinase method and plasma insulin concentration was determined by radioimmunoassay using commercial kits.

To determine the GI of the foods according to standard methodology (24), the area under the incremental plasma glucose curve (AUC) was calculated geometrically using Simpson's rule, with the fasting value as the baseline. The AUC for the test food was then expressed as a percentage of the AUC for the reference food (the average AUC for the two bread tests) for that individual (= GI in that individual). GI values were multiplied by 0.7 so that a scale on which the glucose = 100 could be used to express the final results. The mean (± SE) GI of each food was calculated as the average GI of all 12 individuals. The insulin index (II) was calculated in an analogous manner using the area under the plasma insulin curve. In instances where an individual gave a GI or II outside two standard deviations from the mean of the group, the result was excluded from further analysis.

The glycemic index of chocolate products

The GI and II of the foods tested are shown in Table 12.3. Figures 12.1 and 12.2 illustrate the glucose and insulin responses to the chocolate bar.

The findings indicate that most of the chocolate products produced relatively

Table 12.2 The composition, normal serving size and weight of the 50 g carbohydrate portion of the test foods based on manufacturers' data.

Food	Energy (kcal)	Protein (%)	Fat (%)	Carbohydrate (%)	Normal serving size (g)	Weight of 50 g carbohydrate portion (g)
White bread (control)	244	8.9	2.4	43.8	60	114.2
Candy-coated chocolate peanut pieces	507	10.1	26.3	60.9	49.3	82.1
Chocolate-covered caramel and nougat bar	460	3.3	18.0	70.5	61.0	71.0
Chocolate-covered caramel and nougat bar[1] ('lite' formulation)	382	2.3	11.2	76.4[1]	44.5	65.5
Chocolate-covered caramel cookie bar	477	5.1	23.9	63.0	58.7	79.4
Chocolate-covered peanut and caramel bar	477	8.5	23.8	61.3	58.7	81.6
Chocolate granola bar	432	7.2	16.2	68.3	27.8	73.2
Energy bar 1 (cocoa-based)	354	15.4	3.1	69.2	65.0	72.3
Energy bar 2 (cocoa-based)	353	7.7	4.6	72.0	65.2	69.4
Energy drink[2] (chocolate milk)	88	3	2.5	13	100 ml	385 ml
Milk chocolate bar	542	5.4	32.5	59.6	36.9	83.9

[1] Includes polydextrose.
[2] Per 100 ml.

low glucose responses, the average GI being 49 and average II being 59. The candy-coated chocolate peanut pieces gave the lowest GI (GI = 33) while the chocolate granola bar and the chocolate-covered caramel and nougat bar had the highest (GI = 62). The latter value was comparable to that obtained by Jenkins *et al.* in 1981 for a similar product (Mars Bar) sold in Canada (4).

One reason for the relatively low GI values is the high fat content of the products. Fat in food is well known to slow gastric emptying, thereby producing lower glucose responses, but importantly, insulin responses are not always correspondingly low (25). Hence, high-fat foods such as chocolate may appear to be very desirable if GI were the only criterion for food selection (it shouldn't be).

One would normally expect the GI and II to be roughly similar since the blood glucose rise is the major stimulus for insulin secretion. Plasma glucose and insulin responses have correlated well in studies of sugary foods (5, 10) and starchy foods (26). Surprisingly, among this group of foods there was no positive correlation between glucose and insulin ($r = -0.3$, NS, Fig. 12.2). In fact, in three cases the II

Table 12.3 The glycemic and insulin indices of chocolate products.

Chocolate product	Glycemic index* (mean ± SE)	Insulin index* (mean ± SE)
Candy-coated chocolate peanut pieces	33 ± 3	57 ± 17
Chocolate-covered caramel and nougat bar	62 ± 8	61 ± 9
Chocolate-covered caramel and nougat bar ('lite' formulation)	45 ± 5	46 ± 11
Chocolate-covered caramel cookie bar	44 ± 6	46 ± 6
Chocolate-covered peanut and caramel bar	41 ± 5	88 ± 20
Chocolate granola bar	62 ± 8	48 ± 13
Energy bar 1 (cocoa-based)	58 ± 5	55 ± 11
Energy bar 2 (cocoa-based)	49 ± 8	49 ± 14
Energy drink (chocolate milk)	46 ± 4	56 ± 15
Milk chocolate bar	45 ± 8	87 ± 31

* Glucose = 100.

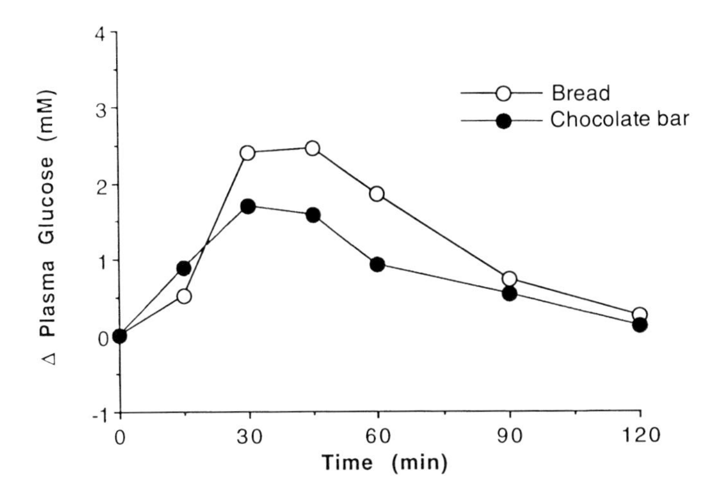

Fig. 12.1 Plasma glucose response to milk chocolate in ten healthy subjects.

was double that predicted (chocolate bar, candy-coated chocolate peanut pieces, chocolate-covered peanut and caramel bar). These unexpected findings therefore corroborate other reports that chocolate products sometimes produce disproportionately high insulin responses (10).

While the fat content might seem the obvious reason for the disparity between glucose and insulin, it is not the entire explanation. Foods with a similarly high fat content, such as potato chips and peanut butter, do not produce as much insulinemia as certain chocolate products (9). Furthermore, Brand Miller *et al.* (5) found that some *low-fat* chocolate products (<2% fat w/w) displayed the same disparity between glucose and insulin. For example, Kellogg's Coco-pops (a

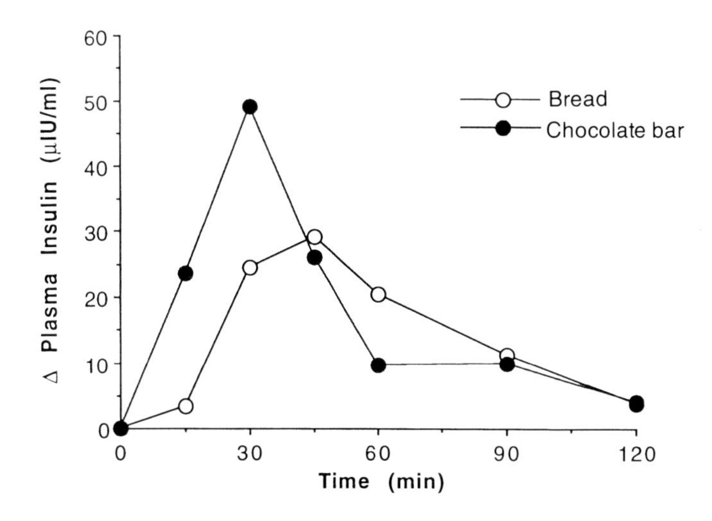

Fig. 12.2 Plasma insulin response to milk chocolate in ten healthy subjects.

chocolate-coated puffed rice breakfast cereal) gave a GI of 77 but an II almost twice as high as expected (II = 124), the highest value seen in that study of 39 common foods). Similarly, the GI of a low-fat chocolate milk was 34 while the II was 69. In these instances, the high insulin response cannot be explained by large amounts of fat.

The quantity and quality of the protein may be another explanation for the high insulin response to some chocolate products. Like carbohydrate, protein also elicits an insulin response (27) and cocoa powder contains significant amounts of protein (18.5 g/100 g) (20). In healthy subjects, the degree of insulinemia (AUC) after consuming a 'pure' protein load is about 30% of that seen with the same amount of glucose (27). However, the amount of protein per 50 g carbohydrate portion was similar in most of the products that we tested. It is possible that cocoa is a source of non-nutritive substances that stimulate insulin secretion, but given the variation in insulin responses among similar chocolate products, other food ingredients may be responsible for the effect (28). Further research in this area is needed.

While a small number of chocolate-containing products appear to raise insulin levels higher than anticipated, their small contribution of calories to the diet suggests that they are not a significant health threat.

Significance of high insulin responses

Individual variation in insulin responses to foods is determined to a large degree by the level of *insulin resistance* in the individual (29). Insulin resistance is defined as a state where greater than normal insulin levels are required to elicit a quantitatively normal glucose response. The term is often used interchangeably with *diminished insulin action* or *decreased insulin sensitivity*. The acute metabolic response to insulin resistance is hyperinsulinemia. People who are older, overweight or with a family history of diabetes show higher insulin responses than insulin-sensitive individuals. Underlying insulin resistance can be worsened by continuously raised insulin levels and by consumption of high-saturated fat (30) and high-GI diets (31, 32). The long-term result of insulin resistance and compensatory hyperinsulinemia is a variety of clinical and biochemical abnormalities, which combine to significantly increase the risk of NIDDM and coronary heart disease. Any food which stimulates much higher insulin responses than expected and makes a substantial contribution to energy intake might be considered undesirable.

Sugar is not the villain we thought

Sugars play a unique role in human diets. They satisfy our instinctual desire for sweetness and contribute 20–22% of the energy in modern Western diets. Revised estimates of honey intake in prehistoric times indicate that amounts eaten may have approached the current levels of added sugars (33). Added sugars have many functional roles in foods, which extend beyond their sweetness, including preservative, textural and flavour-modifying qualities. Unfortunately, sugars have a ‘bad reputation’ especially in respect of dental caries, which is no longer deserved. Since it is now clear that neither sugary foods in general nor chocolate itself produce exceptionally high glucose responses, other arguments are often raised to justify a restriction on sugar. However, even in these areas, new research dispels much of the old dietary dogma.

In the past, refined sucrose was believed to cause diabetes, excessive weight gain, heart disease, micro-nutrient deficiencies and even hyperactivity in children. A wealth of new research has shown most of these assumptions to be false (7). We now know that the average intake of refined sugars is much lower than original estimates, i.e. we eat 40–70 g/day, representing 10–12% total calorie intake (34, 35). This is considered an acceptable level and within a healthy eating regime. More importantly, it is now clear that sugar intake correlates *inversely* with fat intake (i.e. the higher the sugar as a percentage of energy, the lower the fat and vice versa) and that high-sugar diets are associated with lower body weight (35, 36, 37). Thus, lowering sugar intake may be counterproductive, increasing fat intake and the likelihood of over-weight. Lastly, moderate intake of sugars is associated with the highest intakes of micro-nutrients (35, 38). Lower intakes of sugar are often associated with diets of poorer nutritional quality.

These findings have been confirmed repeatedly in large, well-designed studies in many countries all around the world; in adults, children and adolescents, and in people with diabetes. While dental caries are still associated with high sucrose consumption in non-industrialised countries, refined starch is also implicated (39). The use of fluoridated water supplies and toothpastes in Western countries has dramatically reduced the problem of tooth decay and made sugar intake less important.

Conclusion

People with diabetes have long suffered from the widespread belief that normal everyday pleasures such as chocolate are to be eaten sparingly or not at all. It can be concluded that the majority of chocolate products give glycemic and insulin responses that are lower than those shown for white or wholemeal bread. Some products, however, stimulate disproportionately more insulin relative to the level of glycemia and may be undesirable in large amounts in the diet of people with insulin resistance syndrome (i.e. individuals who are overweight, hypertensive, with impaired glucose tolerance or NIDDM).

Although the subjects studied were often normal non-diabetic individuals, the findings are likely to apply to individuals with diabetes. Hence, chocolate products should not be blamed for high blood sugar levels, nor should they be recommended for the treatment of clinical hypoglycemia (low blood sugar) in insulin-dependent diabetes. Chocolate products are usually very energy-dense foods and should therefore play a small but nonetheless pleasurable and guilt-free role (e.g. one daily 'indulgence') in the diet of people with type 2 diabetes. Active individuals with type 1 diabetes can afford to include more generous amounts of chocolate as long as blood glucose and lipid control remain within desirable levels.

References

1. American Diabetes Association (1988) Report of the American Diabetes Association's Task Force on Nutrition. *Diabetes Care* **11**, 127–128.
2. American Diabetes Association (1994) Nutrition recommendations and principles for people with diabetes mellitus. *Diabetes Care* **17**, 519–522.
3. FAO/WHO (1998) Joint FAO/WHO Expert Consultation. *Carbohydrates in Human Nutrition.* Paper 66, 14–18 April 1997, FAO Food and Nutrition. FAO, Rome.
4. Jenkins, D.J.A., Wolever, T.M.S. and Taylor, R.H. (1981) Glycemic index of foods: a physiological basis for carbohydrate exchange. *Am. J. Clin. Nutr.* **34**, 362–366
5. Brand Miller, J., Pang, E. and Broomhead, L. (1995) The glycaemic index of foods containing sugars: comparison of foods with naturally occurring versus added sugars. *Br. J. Nutr.* **73**, 613–623.
6. Waldron, S., Swift, P.G.F., Raymond, N.T. and Botha, J.L. (1997) A survey of the dietary management of children's diabetes. *Diabetic Med.* **14**, 698–702.

7. Anderson, G.H. (1997) Sugars and health: a review. *Nutr. Res.* **17**, 1485–1498.
8. Anon (1981) Exploding exercise myths. *Curr. Health* November, 20–21.
9. Shively, C.A., Apgar, J.L. and Tarka, S.M. (1986) Postprandial glucose and insulin responses to various snacks of equivalent carbohydrate content in normal subjects. *Am. J. Clin. Nutr.* **43**, 335–342.
10. Wolever, T.M.S. and Brand Miller, J. (1995) Sugar and blood glucose control. *Am. J. Clin. Nutr.* **62** (Suppl.), 212S–227S.
11. Green, S.M., Burley, V.J. and Blundell, J.E. (1994) Effect of fat- and sucrose-containing foods on the size of eating episodes and energy intake in lean males: potential for causing over-consumption. *Eur. J. Clin. Nutr.* **48**, 547–555.
12. Foster-Powell, K. and Brand Miller, J. (1995) International tables of glycemic index. *Am. J. Clin. Nutr.* **62**, 871S–893S.
13. Brand, J.C., Colagiuri, S., Crossman, S., Allen, A., Roberts, D.C.K. and Truswell, A.S. (1991) Low-glycemic index foods improve long-term glycemic control in NIDDM. *Diabetes Care* **14**, 95–101.
14. Brand Miller, J. (1994) The importance of glycemic index in diabetes. *Am. J. Clin. Nutr.* **59** (Suppl.), 747S–752S.
15. Wolever, T.M.S., Nguyen, P., Chiasson, J., *et al.* (1994) Determinants of diet glycemic index calculated retrospectively from diet records of 342 individuals with non-insulin-dependent diabetes mellitus. *Am. J. Clin. Nutr.* **59**, 1265–1269.
16. Frayn, K.N. and Kingman, S.M. (1995) Dietary sugars and lipid metabolism in humans. *Am. J. Clin. Nutr.* **62**, 242S–249S.
17. Peters, A.L., Davidson, M.B. and Eisenberg, K. (1990) Effect of isocaloric substitution of chocolate cake for potato in type 1 diabetic subjects. *Diabetes Care* **13**, 888–892.
18. Gee, J.M., Cooke, D., Gorick, S., *et al.* (1991) Effects of conventional sucrose-based, fructose-based and isomalt-based chocolate on postprandial metabolism in non-insulin-dependent diabetics. *Eur. J. Clin. Nutr.* **45**, 561–566.
19. Jenkins, D.J.A., Wolever, T.M.S., Taylor, R.H., *et al.* (1980) Rate of digestion of foods and postprandial glycaemia in normal and diabetic subjects. *BMJ* **2**, 14–17.
20. Nguyen, N.U., Henriet, M.T., Dumoulin, G., Widmer, A. and Regnard, J. (1994) Increase in calciuria and oxaluria after a single chocolate bar load. *Horm. Metab. Res.* **26**, 383–386.
21. Akgun, S. and Ertel N.H. (1980) A comparison of carbohydrate metabolism after sucrose, sorbitol and fructose meals in normal and diabetic subjects. *Diabetes Care* **3**, 582–585.
22. Macdonald, I., Keyser, A. and Pacy, D. (1980) Some effects in man of varying the load of glucose, sucrose, fructose or sorbitol on various metabolites in blood. *Am. J. Clin. Nutr.* **31**, 1305–1311.
23. Brand Miller, J.C., Wang, B., McNeil, Y. and Swan, V. (1997) The glycaemic index of more breads, breakfast cereals and snack products. *Proc. Nutr. Soc. Aust.* **21**, 144.
24. Wolever, T.M.S., Jenkins, D.J.A., Jenkins, A.L. and Josse, R.G. (1991) The glycemic index: methodology and clinical implications. *Am. J. Clin. Nutr.* **54**, 846–854.
25. Collier, G. and O'Dea, K. (1983) The effect of co-ingestion of fat on the glucose, insulin and gastric inhibitory polypeptide responses to carbohydrate and protein. *Am. J. Clin. Nutr.* **37**, 941–944.
26. Holt, S.H.A., Brand Miller, J.C., Petocz, P. (1996) Inter-relationships among post-

prandial satiety, glucose and insulin responses and changes in subsequent food intake. *Eur. J. Clin. Nutr.* **50**, 788–797.

27. Krezowski, P.A., Nuttall, F.Q., Gannon, M.C. and Bartosh, N.H. (1986) The effect of protein ingestion on the metabolic response to oral glucose in normal individuals. *Am. J. Clin. Nutr.* **44**, 847–856.
28. Paul, A.A. and Southgate, D.A.T. (1978) *McCance and Widdowson's The Composition of Foods*. HMSO, London.
29. DeFronzo, R.A. and Ferrannini, E. (1991) Insulin resistance A multi-faceted syndrome responsible for NIDDM, obesity, hypertension, dyslipidemia, and atherosclerotic cardiovascular disease. *Diabetes Care* **14**, 173–194.
30. Storlien, L.H., Baur, A.L., Kriketos, A.S., *et al.* (1996) Dietary fats and insulin action. *Diabetologia* **39**, 621–631.
31. Higgins, J.A., Brand Miller, J.C. and Denyer, G.S. (1996) Development of insulin resistance in the rat is dependent on the rate of glucose absorption from the diet. *J. Nutr.* **126**, 596–602.
32. Salmeron, J., Manson, J.E., Stampfer, M.J., Colditz, G.A., Wing, A.L. and Willet, W.C. (1997) Dietary fiber, glycemic load and risk of non-insulin-dependent diabetes mellitus in women. *JAMA* **277**, 472–477.
33. Allsop, K.A. and Brand Miller, J. (1996) Honey revisited: the role of honey in preindustrial diets. *Br. J. Nutr.* **75**, 513–520.
34. Baghurst, K.I., Record, S.J., Syrette, J.A., Crawford, D.A. and Baghurst, P.A. (1989) Intakes and sources of a range of dietary sugars in various Australian populations. *Med. J. Aust.* **151**, 512–518.
35. Sigman-Grant, M., Stanton, J., Keast, D.R. and Gibney, M. (1995) Consumption of sugars in the United States and in the European Union. *Am. J. Clin. Nutr.* **62**, 178S–194S.
36. Bolton-Smith, C. and Woodward, M. (1994) Dietary composition and fat to sugar ratios in relation to obesity. *Int. J. Obes.* **18**, 820–828.
37. Baghurst, K.I., Baghurst, P.A. and Record, S.J. (1992) Demographic and nutritional profiles of people consuming varying levels of added sugars. *Nutr. Res.* **12**, 1455–1465.
38. Gibson, S.A. (1993) Consumption and sources of sugars in the diets of British schoolchildren: are high-sugar diets nutritionally inferior? *J. Hum. Nutr. Diet.* **6**, 355–371.
39. Navia, J.M. (1994) Carbohydrates and dental health. *Am. J. Clin. Nutr.* **59** (Suppl.), 719S–727S.

Chapter 13

Chocolate and Dental Health

Martin E.J. Curzon

Chocolate has long been associated in the public's mind, with confectionery, as being a cause of tooth decay (usually known as dental caries). This probably dates back to the 19th century when honey ceased to be the main source of sweetness in the diet and was replaced by sugar, as sucrose. There was also a view that chocolate and confectionery were associated with a hedonistic aspect of life and therefore somewhat reprehensible. Recent research indicates that our use of sucrose is probably at about the same level as our use of honey over many periods of our history (1). Cheap sugar from cane and then beet meant a changeover from honey – not easy to store and transport – to sugar. This also gave rise to the growth of confectionery at the same time that chocolate gained favour. Hence the association of chocolate with confectionery. The extensive sale of confectioneries and other refined carbohydrates, such as biscuits, cakes and baked goods, was associated with a rapid rise in the prevalence of dental caries. Therefore in the late 19th century the dental profession, followed by the lay public, came to associate dental caries with the use of sugar and confectionery, including chocolate. This association was widely accepted and became 'common knowledge', although there was no scientific evidence for it.

Thus, chocolate has been considered as one of the main aetiological agents in the development of dental caries. There has been a considerable amount of 'third-party blaming' in this approach to chocolate, as with all confectionery, in that it is always easier to blame someone or something else, rather than one's own lack of oral health care. A review of the many surveys to determine any relationship of confectionery to dental caries has shown that the eating of such snack foods is not the principal cause of dental caries and that sugar consumed in other forms, such as with starches, is probably of greater importance (2). However, the evidence for a direct relationship of chocolate to dental caries is thin. While most dental textbooks on the prevention of dental caries include chocolate as a cariogenic food (that is, inducible to the development of dental caries), there is little evidence for this. This chapter assesses the experimental research for any relationship of chocolate to dental caries. The approach taken is to consider the published research for a cariogenic role of chocolate under a number of headings related to the different experimental methods used in cariogenicity testing.

Cariogenic load

When assessing the cariogenicity of a food, consideration must be given to the cariogenic load of the diet. Each food will add or subtract from this load depending on the properties of the food itself. In the case of chocolate, factors that are likely to increase or decrease caries based on its composition are important. These factors will comprise the chocolate itself (cocoa), milk or dairy products added to make the chocolate, sugar and other chemicals such as casein or fluoride, and further ingredients such as nuts, nougat or fudge, etc.

These compositional factors may also affect the retentiveness of the chocolate, as the longer a carbohydrate based food is retained in the mouth the more likely it is to induce caries. It is clear, therefore, that the assessment of the cariogenicity of chocolate is not a simplistic one.

Research methods used on chocolate and dental caries

Research on the relationship of nutrition, food and diet to dental caries is based upon a number of different types of study. These fall under the following headings:

- Human observational studies
- Human intervention studies
- Intra-oral plaque pH studies using pH electrodes
- Artificial mouths
- *In vitro/in vivo* studies
- Animal studies
- Other studies – food retention and acid production

Human observational studies

These are defined as studies assessing the prevalence of dental caries in a free-living population in relation to lifestyle and, for our purposes, the types of foods or diets used. They are by necessity rather crude, as it is very difficult to determine the actual diets used. As dental caries usually takes over 2 years to develop, so point dietary analyses are difficult to significantly equate with a disease process taking several years.

The majority of such studies were carried out many years ago and long before the remit of the past 25 years – the brief for this publication. However, because there have been so few recent studies it is appropriate to briefly mention some of the earlier, more important ones.

The work on chocolate in human observational studies is often associated with sugar. Nonetheless, many studies commenced in the 1930s focusing on retrospective diet diaries of adults or children related to the development of dental caries. The most extensive of these retrospective studies was published by Nizel and Bibby (3) using dental caries data for recruits to the US armed forces during the Civil War (1861–65), World War I (1917–18) and World War II (1942–45). Mapping dental caries by state and relating this data to food consumption and dietary patterns, the authors discovered that caries was not, as expected, related to sugar or confectionery use. The low-caries states of the south-west and south-central USA used far higher levels of soft drinks (soda pop) and confectionery, which would have included chocolate, while the high-caries states used higher levels of white flour.

The many human observational studies on diet and dental caries rarely mention chocolate and, if they do, it is included with confectionery. Those studies that have found a relationship of caries to a particular food, or groups of foods, have reported caries as significantly related to 'sugars', 'carbohydrates', 'frequency of sugar' or some such term. Chocolate is never mentioned as a particular food.

Further important observational studies were those of World War II, where dental caries was shown to drop after the introduction of wartime rationing. The lower caries incidence was ascribed to the restriction of sugar and confectionery. Again, there was no mention in these studies of chocolate *per se*. However, *all foods* were restricted at that time and eating was reduced to mealtimes only. What is more, the extraction rate of flour was put back up to 100% so that wholemeal flour became the norm, rather than white. With all these restrictions of availability of foods, including chocolate, it was not surprising that there was a reduction in dental caries prevalence.

Other research in more recent years has not shown any significant relationship of dental caries to chocolate or confectionery use. The Michigan (USA) study by Burt and Ismail (4) followed over 400 children for 2 years to determine if dental caries was related to the type and form of the diet. Only very weak relationships of sugar and caries were found affecting those few children that did develop dental caries. The authors concluded that in Western societies, where fluoride was extensively used, sugars and related products in the diet have little relationship to caries except in the most susceptible individuals.

Another study in Northumbia (England) gave very similar results (5). Out of 405 children followed for 2 years, only 31 developed dental caries and the sugars-to-caries relationship, while statistically significant, could only explain 4% of the variance.

The human observational studies therefore give no direct evidence of a significant relationship of chocolate to dental caries.

Human intervention studies

These are the classic type of studies used in the pharmaceutical industry in which a group(s) of people are observed closely over a period of time under controlled circumstances to determine the development or absence of disease, and whether a drug is effective or not. While they are carried out every day in medicine, for many years they have not been allowed in dental research on foods and diets. One cannot deliberately instigate dental caries, purely for experimental purposes. For this reason, there has been only one proper study. Known as the Vipeholm Study (6), it remains the only intervention study involving chocolate in a human population. It plays a major role in any consideration of diet and dental caries.

The Vipeholm project involved over 436 inmates of an institution for the mentally retarded. Over a study period of several years, groups of adult subjects were given dietary supplements, either as between-meals snacks or added to meals. The caries incidence was monitored and relationships between caries increment and dietary supplement assessed. All three of the candy-eating (confectionery) groups developed more cavities than the no-candy groups. The greatest incidence was found in the group that used toffees most frequently. The group of subjects receiving additional chocolates between meals had a lower incidence (Fig. 13.1). The chocolate group had a lower caries increment than the group of inmates using caramels between meals. The highest caries increment occurred in those subjects using 24 caramels between meals. The chocolate group therefore ranked midway between the highest and lowest caries groups.

There have been no other intervention studies in which chocolate was specifically used in a test group of individuals. In the Turku xylitol study (7), some 200

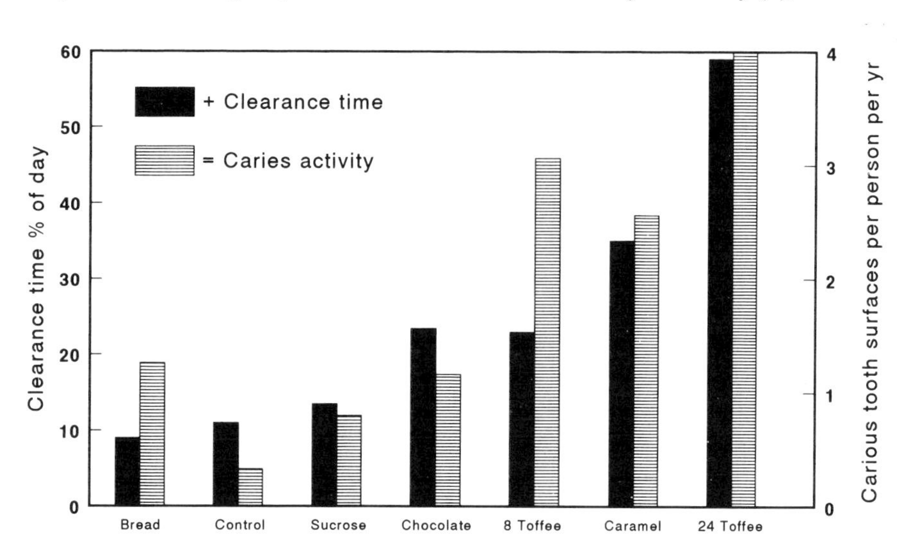

Fig. 13.1 Relationship of dental caries increment and clearance time of snack foods used by various groups of subjects in the Vipeholm Study (6).

adults were given controlled diets in which the refined carbohydrate was either sugar, fructose or a sugar substitute, xylitol. The total diet for these people was controlled and therefore the sugar component was either kept standard or substituted, which would have included chocolate as well. The results showed a reduction in caries increment in the xylitol group. However, chocolate does not appear to be a factor in this study.

Intra-oral plaque pH studies

These involve the measurement of pH (*acidity*) in plaque (the accumulated adhesive mass of bacteria, food debris and mucins that collects on teeth) on or in between the teeth. The ability of a food to encourage or create an acid plaque (*acidogenicity*) is related to a food's ability to initiate or encourage the development of dental caries. The tests commonly use a pH microelectrode which can be touched on the plaque *in situ* (touch method), placed within a dental appliance (indwelling method), or where plaque samples are scraped off and tested outside of the mouth (harvesting method).

This is an area of study in which chocolate, or chocolate products, have been more widely tested. The method lends itself very well to the evaluation of specific foods but, of course, cannot test diets. In these studies, a group of volunteers is used to sequentially test various foods for their ability to promote an acidogenic plaque. The maximum drop in pH is recorded as well as the time taken for the plaque to be buffered by the saliva and return to normal. By always using a set of two foods, sucrose and sorbitol (10% solutions), as positive and negative controls, respectively, reasonable comparisons can be made. Using sucrose as the main comparator, and given an arbitrary score of 1.0, each test food can be given a Cariogenic Potential Index (CPI) score.

Large numbers of foods, snack foods and confectioneries have been tested and tables of pH scores for comparison as least or most acidogenic have been devised (Table 13.1) (8, 9).

This comparison of two separate studies shows that chocolate falls in the middle of scores and is therefore only of moderate cariogenicity. What is also interesting is the effect of added ingredients to chocolate. Thus, the chocolate peanuts ranked lower in the acidogenic potential scores than sweet chocolate.

Artificial mouths

Artificial mouths recreate in a bench-top device the conditions that occur in the human mouth. These closed systems use extracted teeth with natural or artificial plaque, a food source and appropriate bacteria to produce caries-like lesions in the enamel. By using several different foods, such as chocolate, a comparison can

Table 13.1 Plaque pH scores to assess the acidogenicity of various snack foods and candies in two separate studies.

Grading	Study	
	Rugg-Gunn *et al* (9)	Edgar and Rugg-Gunn (8)*
Least acidogenic	Sugarless gum	Caramels
	Peanuts	Sugared gum
	Bread & butter	*Chocolate*
	Potato crisps/chips	Liquorice
	Ice lolly/pops	Sugarless gum
	Sugared coffee	
	Chocolate peanuts	
	Chocolate-fondant-caramel bar	
	Apple	Orange
	Cream-filled biscuit	Jellies
	Ice cream	
	Sweet chocolate	
	Chocolate biscuit	
	Chewing gum	Rock candy
	Plain biscuit	Clear mints
	Orange drink	Sourballs
	Toffee	Fruit gums
↓	Sugared coffee	Fruit
Most acidogenic	Boiled sweet/candies	Lollipops

* Candies tested out of 54 different foods.

be made on relative cariogenicity. They have not been used to assess the cariogenicity of chocolate.

In vivo/in vitro methods

These seek to combine an intra-oral environment with the effects of foods outside of the mouth. They are therefore known as intra-oral cariogenicity tests (ICT). Slabs of human sterile enamel are placed into an intra-oral, denture-like device. These become colonized by oral bacteria and plaque. The device(s) can be removed from the mouth and immersed in a food slurry, such as chocolate, at various times during the day. Over a period of several weeks, caries-like lesions develop in the enamel slabs, which can be measured for an index of cariogenicity. As with artificial mouths, the ICT method has not been used to assess chocolate, although it could readily do so.

Animal studies

These use an animal model, usually the laboratory rat, to produce *in vivo* dental caries. By controlled feeding experiments, groups of rats can be fed a nutritional diet as well as varying numbers of snacks as experimental foods. After a suitable period of time, 35 days in the case of rats, the animals are sacrificed and the number of dental caries lesions/cavities counted to assess a comparative cariogenicity or CPI of the foods tested. As with other methods described above, it is usual to use sucrose as the control against which foods are tested.

Imfeld *et al.* (10), using an animal model, recorded the CPI of sucrose at 1.0, apple at 0.6, a chocolate biscuit at 0.7 and a low cariogenic food at 0.4.

A major study of a number of snack foods (11) showed that chocolate gave a CPI value of 0.72, compared with sucrose with a score of 1.0. Morrisey *et al.* (12) showed that four out of five of the products tested containing chocolate were in the relatively lower half of the cariogenicity table. Reynolds and Black (13) noted that chocolate fed to rats was cariogenic but its effect could be reduced by the addition of calcium caseinate or milk products. An extensive study (14) included 22 snack foods in a series of experiments using rats, but also included milk chocolate as one of two control groups along with sucrose. The results (Table 13.2) showed that chocolate had a comparative CPI of 0.8, which was similar to the finding of Bowen *et al.* (11).

A study by Navia and Lopez (15) had also shown that chocolate or chocolate

Table 32.2 Cariogenic Potential Indices for various snack foods compared with sucrose (given a score of 1.0).

Snack food	CPI
Gelatin dessert	0.4
Corn chips	0.4
Peanuts	0.4
Bologna	0.4
Yoghurt	0.4
Pretzels	0.5
Potato chips (crisps)	0.6
Saltines/crackers	0.6
Natural snack mix*	0.6
Cornstarch	0.7
Rye crackers	0.7
Fried cake	0.7
Graham crackers	0.8
Milk chocolate	*0.8*
Sponge cake-filling	0.8
Bread	0.9
Sucrose	*1.0*
Granola cereal	1.0
French fries	1.1
Bananas	1.1
Cupcakes	1.2
Raisins	1.2

*Mixture of nuts, raisins and dried fruits.

products were only moderately cariogenic by comparison with sugar. More recently, Grenby and Mistry (16) reported that caries scores in rats were nearly 30% higher in animals fed plain (dark) chocolate rather than those on a milk chocolate regimen.

Food retention, acid production and enamel dissolution

This method measures the amount of a food retained in the mouth after eating (clearance time), the amount of acid inherent in a food (or produced by incubation with saliva) or the amount of human enamel dissolved per unit time when enamel is incubated with a slurry of food and saliva. These measures have been used to assess cariogenicity, since a food that is retained in the mouth for a long time provides a substrate for bacteria to produce acid, which will be more likely to dissolve human tooth enamel.

The early studies by Bibby and Mundorff (17) showed that chocolate products, particularly those containing added ingredients such as nuts, etc., cleared from the mouth quickly. When the foods used in the Vipeholm Study (6) were compared for rates of clearance (Fig. 13.1), it can be seen that the chocolate group of subjects recorded lower clearance times than the toffee and caramel groups in which the higher caries occurred. Chocolate would therefore be classified as a low cariogenic food, as its lower rate of retention would give less opportunity for oral bacteria to ferment and produce less amounts of acid.

Studies by Bibby and Mundorff (17) have shown that tooth enamel lost to acid dissolution is not related to acid formed (Table 13.3). Some foods have low acid but dissolve enamel, such as white flour, while others, such as milk chocolate, have a high acid formation but dissolve little enamel.

Foods may have a varying degree of inherent acidity and therefore may also be more or less conducive to acid production by oral cariogenic bacteria. The amount of acid produced intra-orally can be measured and used as yet another

Table 13.3 Acid formation and enamel dissolution by some foods and candies.

Food	Acid formed (0.05 M NaOH ml)	Enamel dissolved (mg × 10)
Wholewheat bread	11.0	2.0
White bread	8.4	4.0
White flour	4.9	10.0
Graham flour	5.1	5.0
Cornflakes	4.3	5.0
All Bran, cereal	16.2	1.0
Chocolate coconut bar	11.1	11.0
Milk chocolate	13.2	1.0

Source: after Bibby and Mundorff (17).

indicator of cariogenicity. However, just because a food is acidogenic does not mean that tooth enamel is attacked in a direct relationship to the quantity of acid produced. This is because there may be other components in the food that buffer any acid produced. Milk products, as noted above, calcium caseinate and other ingredients may have such a buffering effect. The lower CPI of milk chocolate, noted by Grenby and Mistry (16) may be explained by this.

Cocoa fractions and caries inhibition

Stralfors (18) showed that various fractions of cocoa had an inhibitory effect on dental caries in hamsters. *In vitro* observations made by Ferguson and Jenkins (19) also showed that extracts of commercial cocoa had a marked effect in reducing acid solubility of tricalcium phosphate, similar to enamel. Later, s'Gravenmade and Jenkins (20), also using *in vitro* tests, indicated that there is a cariostatic agent in cocoa. The moderate CPI of chocolate may also be explained, therefore, by an inherent cariostatic effect of cocoa used in the manufacture of chocolate.

Conclusion

Although chocolate has been associated in the lay public's mind, as well as within much of the dental profession, with the promotion of dental caries, the experimental evidence for such an association is thin. The limited human studies have shown chocolate to have only a moderate cariogenic potential which has been confirmed in a number of animal experiments. Studies on the effects of chocolate and chocolate products on intra-oral plaque pH measurements also indicate that chocolate is only moderately related to acidogenicity. The moderate cariogenicity of chocolate is probably related to the cocoa fraction of chocolate, which contains chemicals inhibitory to oral bacteria. In addition, added ingredients such as nuts may reduce the clearance time and will therefore render the products even less potentially cariogenic. As chocolate also usually includes dairy products, which are cariostatic, milk chocolate in particular has further protection.

Whilst not entirely free of a cariogenic potential, chocolate when used in moderation should not be considered as playing a major role in the aetiology of dental caries. Chocolate, however, is not always eaten as chocolate but more often as a chocolate bar with other ingredients. Some of these combinations may be more or less cariogenic, but the detailed research has yet to be carried out to determine which is which. Such research would be valuable in formulating chocolate bars with a low-to-moderate cariogenic potential that could be recommended as part of a dental caries preventive programme.

References

1. Allsop, K.A. and Brand Miller, J. (1996) Honey revisited: a reappraisal of honey in pre-industrial diets. *Br. J. Nutr.* **75**, 513–521.
2. Bibby, B.G. (1990) *Food and the Teeth.* Vantage Press, New York.
3. Nizel, A. and Bibby, B.G. (1944) Geographic variations in caries prevalence in soldiers. *J. Am. Dent. Assoc.* **31**, 1618–1626.
4. Burt, B.A. and Ismail, A.I. (1986) Diet, nutrition and food cariogenicity. *J. Dent. Res.* **65**, 1475–1484.
5. Rugg-Gunn, A.J. and Hackett, Ö. (1987) Relative cariogenicity of starches and sugars in a 2-year longitudinal study of 405 English school children. *Caries Res.* **21**, 464–473.
6. Gustaffsson, B.E., Quesnel, C.E., Lanke, L.S., *et al.* (1952) The Viepholm Dental Caries Study. *Acta Odont. Scand.* **11**, 232–264.
7. Scheinin, A. and Makinen, K.K. (1975) The Turku sugar studies. *Acta Odont. Scandin.* **33**, (Suppl. 70), 1–349.
8. Edgar, W.M. and Rugg-Gunn, A.J. (1975) Acid production in plaques after eating snack foods: modifying factors in foods. *J. Am. Dent. Assoc.* **90**, 418-425.
9. Rugg-Gunn, A.J., Edgar, W.M. and Jenkins, G.N. (1978) The effect of eating some British snacks upon the pH of human dental plaque. *Br. Dent. J.* **145**, 95–100.
10. Imfeld, T., Schmid, R., Lutz, F. and Guggenheim, B. (1991) Cariogenicity of Milch-snitte and apple in program-fed rats. *Caries Res.* **25**, 352–358.
11. Bowen, W.H., Amsbaugh, S.M., Monell-Torrens, S., Brunelle, J., Kuzmiak-Jones, H. and Cole, M.F. (1980) A method to assess the cariogenic potential of foodstuffs. *J. Am. Dent. Assoc.* **100**, 677–681.
12. Morrisey, R.B., Burkholder, B.D. and Tarka, S.M. (1984) The cariogenic potential of various snack foods. *J. Am. Dent. Assoc.* **109**, 589–591.
13. Reynolds, E.C. and Black, C.L. (1987) Reduction of chocolate's cariogenicity by supplementation with sodium caseinate. *Caries Res.* **21**, 445–451.
14. Mundorff, S.A., Featherstone, J.D., Bibby, B.G., Curzon, M.E.J., Eisenberg, A.D. and Espeland, M.A. (1990) Cariogenic potential of foods. Caries in the rat model. *Caries Res.* **24**, 344–355.
15. Navia, J.M. and Lopez, H. (1983) Rat caries assay of reference foods and sugar-containing snacks. *J. Dent. Res.* **62**, 893–898.
16. Grenby, T.H. and Mistry, M. (1995) Precise control of the frequency and amount of food provided for small laboratory animals by a new electronic metering technique, used to evaluate the cariogenic potential of chocolate. *Caries Res.* **29**, 418–423.
17. Bibby, B.G. and Mundorff, S.A. (1975) Enamel demineralization by snack foods. *J. Dent. Res.* **54**, 461–470.
18. Stralfors, A. (1966) Inhibition of hamster caries by cocoa: the effect of whole and defatted cocoa and the absence of activity in cocoa fat. *Arch. Oral Biol.* **11**, 149–161.
19. Ferguson, D.B. and Jenkins, G.N. (1967) Potential caries-reducing substances in pecan shells and cocoa. *J. Dent. Res.* **46**, 103.
20. s'Gravenmade, E.J. and Jenkins, G.N. (1986) Isolation, purification and some properties of a potentially cariostatic factor in cocoa that lowers enamel solubility. *Caries Res.* **20**, 433–436.

Chapter 14

Food Allergy, Intolerance and Behavioral Reactions

Steve L. Taylor and Susan L. Hefle

Lucretius once said, 'One man's food is another man's poison.' Food allergies and intolerances, also known collectively as food sensitivities, are food-related illnesses that affect some consumers but not others. While Lucretius was not referring to food allergies and intolerances, he could have been.

Several distinct types of illnesses are included among the food sensitivities (1, 2). True food allergies are those illnesses that involve an abnormal or heightened response of the body's immune system to specific food components. Several different types of true food allergies exist which have different immunological mechanisms. Food intolerances, on the other hand, do not involve immunological mechanisms. Again, several different types of food intolerances exist. The distinction between the mechanisms involved in true food allergies and food intolerances has some practical importance. Individuals with true food allergies can tolerate very little of the offending food in their diets, while individuals with food intolerances can usually tolerate modest amounts of the offending food or food component in their diet. Additionally, true food allergies tend to manifest with more severe symptoms than food intolerances.

Consumers, and even some medical professionals, frequently fail to distinguish among the various types of food sensitivity (3). Consumers tend to label any sort of unexplainable, post-eating discomfort as a food allergy. Surveys show that up to 20% of consumers believe that they or a member of their family has one or more food allergy (4). In fact, the prevalence of true food allergy is much lower, perhaps in the range of 1–2% of the population (5, 6), although infants and young children have a higher rate of prevalence than the rest of the population (7). Thus, many people do not appreciate the exquisite degree of sensitivity and the seriousness of the symptoms experienced by some individuals with true food allergies.

True food allergies

As noted above, true food allergies involve abnormal immunological responses of the body to substances in foods. The allergens are typically naturally occurring proteins that are present in certain foods. However, the true food allergies can be divided into two categories: antibody-mediated or cell-mediated reactions. The antibody-mediated food allergies are usually mediated by allergen-specific immunoglobulin E (IgE) antibodies, while cell-mediated food allergies are mediated by sensitized lymphocytes (2).

IgE-mediated food allergies

IgE-mediated food allergies are also referred to as immediate hypersensitivity reactions. Symptoms develop within a few minutes to a few hours after the inadvertent ingestion of the offending food. In IgE-mediated food allergies (Fig. 14.1), susceptible individuals will form allergen-specific IgE antibodies upon exposure to an allergen (8). The IgE attaches itself to mast cells in various tissues of the body and basophils in the blood. Mast cells and basophils are specialized cells in the body that contain a host of physiologically active molecules (mediators) including histamine, prostaglandins and leukotrienes. The process of IgE

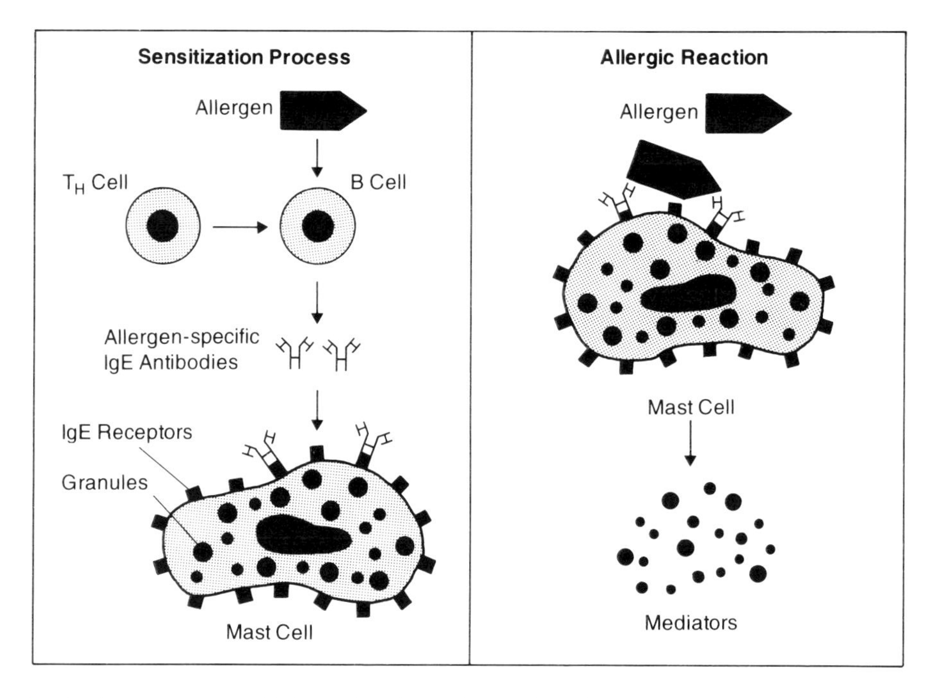

Fig. 14.1 Mechanism of IgE-mediated allergic reactions to food.

production and attachment to mast cell and basophil membranes is known as sensitization.

No symptoms occur during the sensitization process. However, upon a second exposure of the sensitized individual to the same allergen, the allergen cross-links IgE molecules on the surface of the mast cell or basophil membrane, stimulating the cells to degranulate and release the mediators mentioned above. These mediators circulate throughout the body and interact with receptors in various tissues, leading to the development of a wide range of possible symptoms. In IgE-mediated allergies, the interaction of very small amounts of the allergen with the membrane-bound IgE results in the release of massive quantities of histamine and the other mediators. Hence, consumers with IgE-mediated food allergies have a very low tolerance for the offending food.

Allergies to various environmental substances (pollens, mold spores, animal danders, dust mites, etc.), bee venom and certain drugs (e.g. penicillin) also occur through this same IgE-mediated mechanism. The only difference is the source, nature, route of exposure and identity of the allergen.

Individuals with IgE-mediated food allergies can tolerate very little of the offending food in their diet. Recent studies have shown that the tolerance dose for peanut among peanut-allergic individuals ranges from 2 mg to >50 mg for development of objective symptoms and down to as little as 100 μg for development of mild subjective symptoms (9). The severity of the symptoms will vary with the amount consumed. While milligram levels of allergenic foods might not provoke reactions in all sensitive individuals, serious reactions have been provoked in some allergic individuals by ingestion of milligram levels of peanuts and tree nuts (10, 11). Some individuals are clearly more sensitive than others (9).

A variety of different symptoms can be involved with IgE-mediated food allergy (2). Different individuals will experience different symptoms on exposure to the same offending food. Gastrointestinal symptoms (nausea, vomiting and diarrhea) are very common because the gastrointestinal tract is the organ of initial insult in food allergies. Cutaneous symptoms (hives, itching, eczema and swelling) are also fairly common; eczema is especially common in infants. Respiratory symptoms (rhinitis, asthma) are uncommon symptoms with food allergies but asthma is a very serious, potentially life-threatening manifestation (12). Anaphylactic shock is the most frightening symptom associated with food allergies, although fortunately it is fairly rare. Anaphylactic shock involves multiple effects of the mediators on numerous tissues with gastrointestinal, cutaneous, respiratory and cardiovascular symptoms. Death can occur within 15 min of the onset of anaphylactic shock. Deaths have been documented from inadvertent consumption of the offending food by those with extraordinary degrees of allergic sensitivity (12, 13).

Allergies can develop at any age but they develop most commonly in infancy or early childhood (7). Most infants outgrow their food allergies within a matter of months or years (14). Some food allergies are much more commonly outgrown than others. Allergies to cows' milk, eggs and soybeans are more likely to be outgrown than allergies to peanuts, tree nuts, fish and crustaceans. Peanut and tree nut allergies are virtually never outgrown. The basis for this development of tolerance to the offending food is not fully understood. Some foods tend to provoke more severe allergic reactions than others. Peanuts and tree nuts also seem more likely to elicit severe reactions than milk and eggs.

The most common allergenic foods on a worldwide basis are cows' milk, eggs, peanuts, soybeans, tree nuts, wheat, crustaceans and fish (15). Allergies to cows' milk, eggs, soybeans and wheat are much more common among infants than among adults because these food allergies are frequently outgrown. Other countries may experience different patterns of food allergy. In Japan, for example, peanut allergy is rare, while soybean and rice allergies are much more common than in North America. This is probably the result of the popularity of the food in the various countries. The above list of eight foods or food groups, sometimes referred to as the *Big Eight*, probably accounts for 90% of all food allergies. It is possible to develop an allergy to any food that contains protein; over 160 other foods have been reported on at least rare occasions to cause food allergy (16). The most common allergenic foods are foods with high protein contents that are frequently consumed. However, beef, pork and poultry meat are rarely allergenic despite their high protein contents and frequent consumption.

While chocolate is often regarded by consumers as a commonly allergenic food, only rare allergies have been documented to chocolate itself (16, 17). Of course, milk, egg, peanut and tree nuts are often mixed with chocolate to produce confectionery items, and these components are frequently offending foods which probably accounts for the confusion.

The allergens in these foods are naturally occurring proteins (18). While foods contain tens of thousands of proteins, only a few are allergenic. These food allergens tend to be resistant to heat processing and digestion (19). Thus, the allergens in peanuts, tree nuts, milk and eggs are likely to survive chocolate processing.

Effective treatments do not exist for IgE-mediated allergies. While antihistamines can be taken to relieve some of the symptoms, such treatments are not preventive. Epinephrine, also known as adrenaline, can be a life-saving drug for individuals susceptible to anaphylactic shock and such individuals should carry a self-injectable form of epinephrine at all times (12).

Specific avoidance diets are the only effective method of preventing food allergy (20). If allergic to peanuts, for example, simply do not eat peanuts. The construction of safe and effective avoidance diets is difficult for numerous reasons including:

- The exquisite sensitivity of most food-allergic individuals (individuals must avoid even traces of the offending food).

- The lack of labeling for some foods (restaurant items and individual candies are examples).
- The numerous and often confusing terms used to describe these foods and ingredients derived from them (casein and whey are not known as milk components to many consumers).
- The existence of cross-reacting foods. For example, milk from all species and eggs from all species tend to be cross-reactive (21, 22); however, peanuts are legumes, but most peanut-allergic individuals can eat other legumes without difficulty (23).

Cell-mediated food allergies

Celiac disease is an example of a cell-mediated type of food allergy. Cell-mediated food allergies are sometimes called delayed hypersensitivity reactions because symptoms develop 24–72 hours after ingestion of the offending food (24). Celiac disease is associated with a cell-mediated, localized inflammatory reaction in the intestinal tract (25). Wheat, rye, barley, triticale, spelt and kamut are the offending foods in celiac disease (15, 26). Oats were once thought to be associated with celiac disease as well, but recent evidence discounts their role (27). The illness manifests as a malabsorption syndrome, because the localized inflammatory process destroys the absorptive functions of epithelial cells lining the small intestine. The symptoms include body wasting, anemia, bone pain and diarrhea; in children, failure to thrive is another noteworthy symptom (28, 29). The risk of death from celiac disease is quite low (30); untreated celiac disease is associated with considerable discomfort. Individuals who suffer from celiac disease for prolonged time periods are at increased risk for development of T-cell lymphoma (29). The prolamine protein fraction of wheat, rye and barley has been implicated in the causation of celiac disease, although the precise protein segments have not been conclusively identified (31). With wheat, these specific fractions are the gluten, or more specifically the gliadin, fraction. Individuals must totally avoid the gluten fraction of wheat, rye, barley, triticale, spelt and kamut in their diet (32). Ingredients made from these grains must also be avoided if the ingredients contain protein residue (32). While the tolerance for the offending grains has not been conclusively demonstrated, a few isolated studies have concluded that levels of 10 mg of gliadin per day will be tolerated by most patients with celiac disease (33). The prevalence of celiac disease is likely underestimated but has been estimated to be as high as 1 in every 250 people in some European groups (28).

Celiac disease is likely not a major concern to chocolate manufacturers, because wheat, rye, barley and the other responsible grains are not incorporated into most chocolate products. However, celiac sufferers also try to avoid ingredients derived from these grains (32). Only ingredients containing protein residues from these grains would be expected to elicit symptoms. However, celiac sufferers will often avoid products such as rye-based alcohol, wheat starch and malted products made from barley. Proof of the role of these grain-based ingredients in the causation of celiac disease is lacking.

Food intolerances

Several types of food intolerances exist. The major categories include anaphylactoid reactions, metabolic food disorders and idiosyncratic reactions (2). The anaphylactoid reactions are not discussed further here because there is no conclusive proof that such reactions occur related to foods. Metabolic food disorders occur in individuals who are unable to normally metabolize a food component. Lactose intolerance is the best example, and is an example that has relevance to chocolate manufacturing (see below). Idiosyncratic reactions are those illnesses that occur through unknown mechanisms. By definition, relatively little is known about idiosyncratic reactions. In many cases, the cause-and-effect relationship with specific foods or food components has not been well established. The role of sugar in hyperkinetic and other abnormal behaviors in children is discussed briefly below as an example of an idiosyncratic reaction. Individuals with these various forms of intolerances can often tolerate some of the offending food or ingredient in their diets.

Lactose intolerance

Lactose intolerance is a metabolic disorder caused by a deficiency of the enzyme β-galactosidase (lactase) in the intestine (34). As a result, lactose-intolerant individuals cannot properly metabolize lactose (milk sugar). Lactase hydrolyzes lactose into its constituent monosaccharides galactose and glucose, which can then be absorbed for energy. Without the action of lactase, the lactose cannot be digested and absorbed from the small intestine. The lactose passes into the colon where large numbers of bacteria exist and metabolize the lactose to carbon dioxide, hydrogen and water. The resulting symptoms are bloating, flatulence, abdominal cramping and frothy diarrhea (2, 34). While lactose intolerance is treated with a dairy product-avoidance diet, many individuals with lactose intolerance can tolerate several grams or more of lactose in their diets.

Few chocolate confectionery products would contain sufficient lactose to cause symptoms in lactose-intolerant individuals. Lactose can be a significant component of several ingredients used in chocolate manufacturing, including milk and whey. Little concern exists about the minuscule levels of lactose present in other dairy-based ingredients such as caseinates or butter. Since lactose-intolerant consumers can tolerate some lactose in their diets (often as much as 10 g or more), the amount of lactose in most confectionery products would be insufficient to cause reactions in most lactose-intolerant consumers. Few problems would be anticipated from the use of shared equipment between lactose-containing and lactose-free formulations.

Sugar and aberrant behavior

The intake of sugar or sucrose, a common component of chocolate confectionery products, has been linked to several forms of aberrant behavior especially in young children and adolescents. In young children, sugar intake has been suggested as a cause of hyperkinesis, more correctly known as attention-deficit hyperactivity disorder (ADHD) (35). In adolescents, a hypothesis was raised that sugar contributed to aggressive or delinquent behavior (35). A critical examination of the data from multiple studies suggests that few individual hyperactive children respond adversely to a sucrose challenge (35). The most carefully designed and controlled studies of the effect of sucrose on hyperactive behavior found no convincing evidence of any association (35). Some evidence does exist to suggest that both juvenile delinquents and adult criminals display anomalies in carbohydrate metabolism that may be related to their antisocial behavior (35, 36). However, carefully controlled challenge studies of incarcerated juvenile delinquents with sucrose demonstrated that sucrose had no deleterious effects on behavior (35, 37). In fact, the ingestion of sucrose improved behavioral measures in this study (37). Thus, despite popular opinion to the contrary, sugar-sweetened confections are not directly involved in the provocation of aberrant behavior among consumers.

Issues for the confectionery industry

The food industry including the confectionery industry bears considerable responsibility for the prevention of allergic reactions, especially IgE-mediated, true food allergies. Since individuals with IgE-mediated food allergies must specifically avoid their offending foods and can tolerate very little of the offending food, the food industry must properly and completely declare the presence of allergenic foods on the ingredient labels of their packages. Allergic consumers are avid label readers. They also frequently call food companies seeking additional information and must be given accurate answers.

If food products inadvertently become contaminated with even small amounts of an allergenic food which is unlabeled, the consequences can be serious indeed. Individuals with IgE-mediated food allergies can react adversely to the ingestion of traces of the offending food, sometimes with devastating consequences. If larger amounts of the offending food are present, severe reactions can certainly occur.

Given the huge volumes of food products manufactured daily, there are few examples of contamination that have been noteworthy. Nonetheless, serious episodes have occurred including several in the confectionery industry. In Europe, a consumer inadvertently ingested a chocolate product containing unlabeled hazelnuts. Although only 6 mg of hazelnut was estimated to have been ingested, the consumer experienced a serious asthmatic reaction (11). More serious by far, several years ago in Canada, a consumer died after the ingestion of a chocolate-coated wafer contaminated with undeclared peanut, although the level of peanut in the contaminated product remains uncertain. The peanut residues in this product apparently arose from the use of shared production lines and inadequate flushing between peanut-containing and peanut-free products.

Immunoassays currently in various stages of development will allow the detection of residues of specific allergenic foods in other products (38). There are presently such immunoassays available for the detection of peanut, milk, egg and almond residues in other foods. The sensitivity of these immunoassays approaches 1 p.p.m. Unfortunately, satisfactory analytical methods for the detection of residues of other commonly allergenic foods do not yet exist. The best approach to the assessment of the adequacy of cleaning procedures is to test samples of the first product made on the shared equipment after flushing or clean-up following the processing of products containing major allergenic foods such as peanuts.

With respect to ingredients, any ingredient derived from an allergenic food that contains protein could elicit allergic responses. Refined oils do not contain protein residues and are probably safe (39–41), although some countries have chosen to require source labeling of oils. The source labeling of hydrolyzed proteins is quite important. Even extensively hydrolyzed proteins have triggered allergic reactions in exquisitely sensitive individuals (42, 43). In the USA, the term *hydrolyzed vegetable protein* is no longer allowed; the source of the protein must be declared. Flavoring ingredients, both natural and artificial, can occasionally contain small amounts of allergenic proteins (44). In such situations, the allergenic source of the protein in the flavoring should somehow be declared on the ingredient statement. While flavor suppliers are understandably reluctant to release their formulations, they will often divulge if the formulation contains proteins from any of the most common allergenic foods. Lecithin is obtained from soybeans or eggs, both of which are commonly allergenic foods. Lecithin may contain trace residues of protein, but many soybean- and egg-allergic individuals do not avoid lecithin. However, the allergenicity of lecithin has not been tested. When lecithin is directly added to food products as part of the formulation, the source of the lecithin should be included on the label in the view of the authors.

Conclusion

Food allergies affect only a small percentage of consumers. However, some of these consumers can experience life-threatening reactions to the inadvertent consumption of even trace amounts of the offending food. Thus, the food industry must be alert to the avoidance of allergen cross-contact and the proper labeling of allergenic ingredients. Highly specific and sensitive tests are being developed for the detection of allergen cross-contact that will be useful in the establishment of hazard analysis of critical control points (HACCP) approaches to the control of allergen cross-contact.

References

1. Anderson, J.A. (1986) The establishment of a common language. *J. Allergy Clin. Immunol.* **78**, 140–143.
2. Lemke, P.J. and Taylor, S.L. (1994) Allergic reactions and food intolerances. In *Nutritional Toxicology* (Ed. by Kotsonis, F.N., Mackey, M. and Hjelle, J.) , pp. 117–37. Raven Press, New York.
3. Taylor, S.L. (1987) Allergic and sensitivity reactions to food components. In *Nutritional Toxicology*, Vol. 2 (Ed. by Hathcock, J.N.), pp. 173–198. Academic Press, Orlando.
4. Sloan, A.E. (1986) A perspective on popular perceptions of adverse reactions to foods. *J. Allergy Clin. Immunol.* **78**, 127–132.
5. Niestijl Jansen, J.J., Kardinaal, A.F.M., Huijbers, G.H., Vlieg-Boestra, B.J., Martens, B.P.M. and Ockhuizen, T. (1994) Prevalence of food allergy and intolerance in the adult Dutch population. *J. Allergy Clin. Immunol.* **93**, 446–456.
6. Young, E., Stoneham, M.D., Petruckevitch, A., Barton, J. and Rona, R. (1994) A population study of food intolerance. *Lancet* **343**, 1127–1130.
7. Bock, S.A. (1987) Prospective appraisal of complaints of adverse reactions to foods in children during the first three years of life. *Pediatrics* **79**, 683–688.
8. Mekori, Y.A. (1996) Introduction to allergic diseases. *Crit. Rev. Food Sci. Nutr.* **36S**, 1–18.
9. Hourihane, J.O'B., Kilburn, S.A., Nordlee, J.A., Hefle, S.L., Taylor, S.L. and Warner, J.O. (1997) An evaluation of sensitivity of peanut-allergic subjects to very low doses of peanut protein: a randomized double-blind, placebo-controlled food challenge study. *J. Allergy Clin. Immunol.* **100**, 596–600.
10. McKenna, C. and Klontz, K.C. (1997) Systemic allergic reaction following ingestion of undeclared peanut flour in a peanut-sensitive woman. *Ann. Allergy Asthma Immunol.* **79**, 234–236.
11. Yman, I.M., Eriksson, A., Everitt, G., Yman, L. and Karlsson, T. (1994) Analysis of food proteins for verification of contamination or misbranding. *Food Agric. Immunol.* **6**, 167–172.
12. Sampson, H.A., Mendelson, L. and Rosen, J.P. (1992) Fatal and near-fatal anaphylactic reactions to food in children and adolescents. *N. Engl. J. Med.* **327**, 380–384.
13. Yunginger, J.W., Sweeney, K.G., Sturner, W.Q., *et al.* (1988) Fatal food-induced anaphylaxis. *JAMA* **260**, 1450–1452.
14. Bock, S.A. (1982) The natural history of food sensitivity. *J. Allergy Clin. Immunol.* **69**, 173–177.

15. FAO (1995) *Report of the FAO Technical Consultation on Food Allergies*. Food and Agriculture Organization, Rome.
16. Hefle, S.L., Nordlee, J.A. and Taylor, S.L. (1996) Allergenic foods. *Crit. Rev. Food Sci. Nutr.* **36S**, 69–89.
17. Fries, J.H. (1978) Chocolate: a review of published reports of allergic and other deleterious effects, real or presumed. *Ann. Allergy* **41**, 195–207.
18. Bush, R.K. and Hefle, S.L. (1996) Food allergens. *Crit. Rev. Food Sci. Nutr.* **36S**, 119–163.
19. Taylor, S.L. and Lehrer, S.B. (1996) Principles and characteristics of food allergens. *Crit. Rev. Food Sci. Nutr.* **36S**, 91–118.
20. Taylor, S.L., Bush, R.K. and Busse, W.W. (1986) Avoidance diets – how selective should we be? *N. Engl. Reg. Allergy Proc.* **7**, 527–532.
21. Dean, T.P., Adler, B.R., Ruge, F. and Warner, J.O. (1993) *In vitro* allergenicity of cows' milk substitutes. *Clin. Exp. Allergy* **23**, 205–210.
22. Langeland, T. (1983) A clinical and immunological study of allergy to hens' egg white. VI. Occurrence of proteins cross-reacting with allergens in hens' egg white as studied in egg white from turkey, duck, goose, sea gull, and in hen egg yolk, and hen and chicken sera and flesh. *Allergy* **38**, 399–412.
23. Bernhisel-Broadbent, J. and Sampson, H.A. (1989) Cross-allergenicity in the legume botanical family in children with food hypersensitivity. *J. Allergy Clin. Immunol.* **83**, 435–440.
24. Sampson, H.A. (1991) Immunologic mechanisms in adverse reactions to foods. *Immunol. Allergy Clin. North Am.* **11**, 701–716.
25. Strober, W. (1986) Gluten-sensitive enteropathy: a nonallergic immune hypersensitivity of the gastrointestinal tract. *J. Allergy Clin. Immunol.* **78**, 202–211.
26. Marsh, M.N. (1992) Gluten, major histocompatibility complex, and the small intestine. A molecular and immunobiologic approach to the spectrum of gluten sensitivity ('celiac sprue'). *Gastroenterology* **102**, 330–354.
27. Janatuinen, E.K., Pikkarainen, P.H., Kemppainen, T.A., *et al.* (1995) A comparison of diets with and without oats in adults with celiac disease. *N. Engl. J. Med.* **333**, 1033–1037.
28. Troncone, R., Greco, L. and Auricchio, S. (1996) Gluten-sensitive enteropathy. *Pediatr. Clin. North Am.* **43**, 355–373.
29. Marsh, M.N. (1997) Transglutaminase, gluten and celiac disease: food for thought. *Nature Med.* **3**, 725–726.
30. Logan, R.F.A., Rifkind, E.A., Turner, I.D., and Ferguson, A. (1989) Mortality in celiac disease. *Gastroenterology* **97**, 265–271.
31. Cornell, H.J. (1996) Coeliac disease: a review of the causative agents and their possible mechanisms of action. *Amino Acids* **10**, 1–19.
32. Hartsook, E.I. (1984) Celiac sprue: sensitivity to gliadin. *Cereal Foods World* **29**, 157–158.
33. Hekkens, W.T.J.M. and van Twist de Graaf, M. (1990) What is gluten-free – levels and tolerance in the gluten-free diet. *Nahrung* **34**, 483–487.
34. Suarez, F.L. and Savaiano, D.A. (1997) Diet, genetics, and lactose intolerance. *Food Technol.* **51** (3), 74–76.
35. Gans, D.A. (1991) Sucrose and delinquent behavior: coincidence or consequence? *Crit. Rev. Food Sci. Nutr.* **30**, 23–48.

36. Kruesi, M.J.P., Rapoport, J.L., Cummings, E.M., *et al.* (1987) Effects of sugar and aspartame on aggression and activity in children. *Am. J. Psychiatr.* **144**, 1487–1490.
37. Bachorowski, J., Newman, J.P., Nichols, S.L., Gans, D.A., Harper, A.E. and Taylor, S.L. (1990) Sucrose and delinquency: behavioral assessment. *Pediatrics* **86**, 244–253.
38. Taylor, S.L. and Nordlee, J.A. (1996) Detection of food allergens. *Food Technol.* **50** (5), 232–234, 238.
39. Taylor, S.L., Busse, W.W., Sachs, M.I., Parker, J.L., and Yunginger, J.W. (1981) Peanut oil is not allergenic to peanut-sensitive individuals. *J. Allergy Clin. Immunol.* **68**, 372–375.
40. Hourihane, J.O'B., Bedwani, S.J., Dean, T.P. and Warner, J.O. 1997. Randomised, double-blind, crossover challenge study of allergenicity of peanut oils in subjects allergic to peanuts. *BMJ* **314**, 1084–1088.
41. Bush, R.K., Taylor, S.L., Nordlee, J.A., and Busse, W.W. (1985) Soybean oil is not allergenic to soybean-sensitive individuals. *J. Allergy Clin. Immunol.* **76**, 242–245.
42. Saylor, J.D. and Bahna, S.L. (1991) Anaphylaxis to casein hydrolysate formula. *J. Pediatr.* **118**, 71–74.
43. Schwartz, R.H., Keefe, M.W., Harris, N. and Witherly, S. (1989) The spectrum of IgE-mediated acute allergic reactions to cows' milk in children as determined by skin testing with cows' milk protein hydrolysate formulas. *Pediatr. Asthma Allergy Immunol.* **3**, 207–215.
44. Taylor, S.L. and Dormedy, E.S. (1998) The role of flavoring substances in food allergy and intolerance. *Adv. Food Nutr. Res.* **42**, 1–44.

Chapter 15

Chocolate and Headache: Is There a Relationship?

Lisa Scharff and Dawn A. Marcus

Migraine is a common disorder, affecting approximately 17% of women and 6% of men in the USA (1). Exactly what factors migraine patients perceive as triggering their pain has been examined by several researchers (2–4) with varying conclusions. Stress and the menstrual cycle are by far the most commonly cited migraine triggers (4), and both of these factors have been empirically demonstrated to be related to migraine onset (5). Various foods have also been indicated as triggers by migraine patients, although studies have found inconsistencies in the percentage of migraineurs who implicate specific foodstuffs (varying from 0% (6) to 75% (7)). Chocolate tends to be the most frequently identified food as a headache trigger, with approximately 23–75% of chronic headache sufferers indicating that eating chocolate is associated with headache onset (3, 7). The evidence that food in general, and chocolate specifically, triggers migraine is mixed. This chapter reviews such evidence, with special attention to chocolate.

The rationale behind foods as headache triggers

Various foods have been linked to headache, and there have been multiple studies investigating this relationship (7–9). Many migraine sufferers are educated about the potential link between food and headache, and are provided with a list of implicated foods by their physicians. A relatively small percentage of headache patients actually alter their diets after receiving this information, and many of them report that dietary changes do not affect their headache at all. One study (10) reported that 75% of a group of 130 headache sufferers were aware of the potential connection between food and headache; however, only about 37% had attempted dietary changes, and approximately half of those patients reported any degree of improvement. Because of the unenthusiastic response of many migraine sufferers to the suggestion of dietary changes, the percentage of headache patients that actually suffers from 'dietary' headache is unknown.

Assuming that at least some individuals experience food-triggered headaches, two main theories have been presented to account for this effect. The first is that vasoactive amines in food affect blood vessels in the brain, causing the vasodilatation associated with migraine pain. According to the vascular theory of migraine, a migraine episode is associated with changes in the vascular system, with a period of vasoconstriction (a decrease in blood flow to the brain, often associated with an 'aura'), followed by a reactive vasodilatation which is associated with the sensation of throbbing pain. Vasoactive amines in foods such as chocolate are thought to trigger vascular reactions in those who are prone to migraine or sensitive to the effects of that particular amine. The second, more controversial hypothesis is that these patients suffer from food allergies and that migraine pain is an allergic reaction.

Vasoactive amines and headache

The connection between vasoactive amines and headache was initially identified in patients taking monoamine oxidase (MAO) inhibitors. MAO is commonly thought to absorb vasoactive substances in the gut. Individuals taking MAO inhibitors for depression reported severe headaches after ingesting foods containing tyramine, a vasoactive substance commonly found in foods such as aged cheese, processed meats and red wine. (11)

A variety of amines have been implicated in the development of headache, most commonly tyramine, histamine and β-phenylethylamine (BPEA). Chocolate, cheeses (especially Camembert, Cheddar and Parmesan), peanuts, meats (especially pork and venison) and alcohol are known to contain these amines. Each of these foods has also been identified as causally related to headache activity.

Chocolate is known to contain relatively large amounts of BPEA (12). BPEA is metabolized by MAO, and headache caused by chocolate may be related to a deficient metabolism for BPEA. Sandler *et al.* (12) reported reduced oxidative capacity of MAO for BPEA in migraineurs, and identified a headache occurrence rate of 50% in migraine headache patients exposed to BPEA, compared to 6% of those receiving a placebo. Glover *et al.* (13) also observed a reduction in MAO activity during migraine attacks. Thus, some migraine sufferers may not fully metabolize BPEA, leaving this chemical in their systems longer and strengthening its effect on the vascular system. Table 15.1 groups the various vasoactive amines described above.

When BPEA is metabolized, it is in part converted into tyramine, another vasoactive chemical commonly identified as a migraine trigger (19–22). Tyramine itself has also been detected in chocolate products in varying amounts (14, 18).

Table 15.1 Vasoactive amines and headache-triggering agents in chocolate.

Chemical	Amount typically found in chocolate			
Phenylethylamine (BPEA)	0.4–6.6 μg/g	(14);	0.38–6.4 μg/g	(15)
Theobromine	Average 2 mg/g	(16);	1.36–5.65 mg/g	(17)
Caffeine	Average 0.2 mg/g	(16);	0.116–0.617 mg/g	(17)
Tyramine	3.8–12 μg/g	(14)		
Spermine	1.63 μg/g	(18)		

A variety of studies have investigated the role of ingested tyramine as a trigger for headache. All of the studies supporting a causal relationship between tyramine and migraine were generated from a single laboratory (Hanington *et al.* (11, 19, 20, 23)), and one series of their research reports describe updates of an ongoing study, with each subsequent report including subjects and data from previous reports.

These researchers included a 'dietary headache' group comprised of only migraine sufferers who avoided specific foods because of a recognized connection between eating those foods and migraine activity. Subjects in this dietary headache group (n = 11 in the last report of this study) were compared to a non-dietary headache group (n = 9) and a group of headache patients in which the relationship between food and headache was unclear (n = 7). Subjects were orally administered either 100 mg of tyramine (equivalent to $3\frac{1}{2}$ oz of cheese with a high tyramine content) or a lactose placebo on a varying number of occasions. The dietary headache group was more likely to report headaches after ingesting the tyramine that after ingesting the lactose, and this group was also significantly more likely to report a headache after tyramine than subjects in either of the other two groups. In a later study (23), the same researchers reported that 35 dietary migraine sufferers were more likely to develop headaches after 125 mg of tyramine in comparison to lactose and a non-dietary control group of 27 migraine sufferers.

Other controlled studies of tyramine and headache have failed to replicate the findings of Hanington *et al.*, even when the same subject selection criteria of food sensitivity and avoidance were applied (8, 21, 22, 26, 27). Shaw *et al.* (22) reported that none of the nine dietary migraine subjects in their study experienced a headache after ingestion of 200 mg of tyramine. Moffett *et al.* (21) reported no differences in headache occurrence in a group of dietary headache sufferers and non-dietary headache sufferers after ingestion of 125 mg of tyramine or placebo.

Although there is some evidence that BPEA and tyramine contribute to migraine headache in some patients, there is some dispute concerning the contribution of dietary as opposed to endogenously produced BPEA. Karoum *et al.* (26) measured BPEA and tyramine in human blood, cerebrospinal fluid and urine, as well as in rat brains before and after ingestion of foods including chocolate, and found no alteration of either BPEA or tyramine. The authors concluded that the dietary contribution of these amines did not directly contribute in a significant way to their quantity in the body, but rather that these amines are produced from endogenous sources. This failure of ingested foods to alter physiological levels of BPEA and tyramine may explain why some studies fail to demonstrate changes in headache after food ingestion.

Chemicals found in chocolate other than BPEA and tyramine have also been implicated as causing headaches, although not as frequently. Caffeine has also been linked to headache, and theobromine – a caffeine metabolite – is typically present in chocolate and cocoa products (27). It is the theobromine rather than the caffeine content in chocolate that is postulated as a headache trigger (28). Recently, another vasoactive amine present in chocolate, spermine, has been identified as an important transmitter of pain (29). The role of spermine specifically as a trigger for headache is unknown.

Headache as an expression of food allergy

The proposed relationship between food allergy and headache is controversial. Some researchers have asserted that migraine itself may be an expression of food allergy (30, 31). The majority of researchers investigating this possible link, however, have dismissed the possibility that the vast majority of food-induced migraines are caused by food allergies (7, 32).

One common link between allergy and food-induced migraine is that vasoactive amines (most notably histamine) are released during an allergy attack, and increased levels of plasma histamine have been identified during migraine episodes (33), as well as in migraine patients when they ingest implicated trigger foods (34). Studies that have examined RAST or skin prick tests of allergies in comparison to identified food triggers of migraine, however, have rarely identified any link between positive allergy findings and food-induced headache. Schuller *et al.* (35) have reviewed the research literature regarding this subject, and found that studies supporting the association between food allergy and migraine are far outnumbered by studies that have not supported this association. If there were indeed a link between positive food allergy findings and clinical symptoms such as headache, chocolate would indeed demonstrate this association. One study found that 67% of a group of allergy patients tested positive for chocolate allergy, and the vast majority of these patients regularly ate chocolate (36).

Restrictive diet studies

Anecdotally, dietary restriction is associated with headache improvement in one-third to two-thirds of migraineurs. Several studies have attempted to investigate the relationship between headache and diet by conducting prospective studies of various diets, with some eliminating foods rich in vasoactive amines (8), and others eliminating foods that may cause allergic reactions (30, 37, 38). The results of these studies have been mixed and have suffered from multiple methodological problems.

One problem with most diet restriction studies has been the lack of an adequate control group. Most studies (39, 40) have used the patient's usual diet as a baseline control, rather than offering a 'placebo' diet to control for expectation effects. These studies generally show positive effects in the form of decreased headaches on the restrictive diet. In contrast, Salfield *et al.* (41) placed subjects on both restrictive and control diets, and showed equal headache improvement on either diet. Medina and Diamond (8) also reported no significant difference in headache activity when subjects were placed on tyramine-rich or tyramine-free diets. It would appear from the results of these two studies that any type of dietary manipulation might result in headache improvement, regardless of the type of diet used.

A second problem with the majority of dietary challenge and restriction studies has been the wide interval from food ingestion to time of headache, with studies attributing headaches to foods consumed up to 72 hours before the headache (42). In addition, these studies have failed to control additional potential triggers such as menstruation, stress and excessive analgesic use. One case study (42) reported a patient with near daily headache and excessive aspirin use that was hospitalized for ulcer disease. Treatment included discontinuation of aspirin and dietary restriction, which resulted in a resolution of the chronic daily headache. This patient would now be recognized as suffering from analgesic overuse or drug rebound headache, and improvement in headache would be attributed to elimination of excessive aspirin. At the time, however, his headache improvement was attributed to the dietary restrictions.

Investigations of chocolate as a trigger of headache

Three studies have investigated the relationship between chocolate and headache directly through placebo-controlled challenge studies. In one such study, Gibb *et al.* (43) selected 20 subjects from a headache clinic who believed chocolate was a trigger of their migraines and divided them into chocolate (n = 12) and carob 'placebo' (n = 8) groups. Subjects ate 40 g samples and were contacted by telephone 32 hours after ingestion. Five of the individuals who ate chocolate and none of the individuals who ate the placebo reported the development of a headache within 24 hours of eating the sample, a significant difference (P = 0.051).

In contrast to the findings of Gibb *et al.*, two published studies have failed to find an association between chocolate and headache. Moffet *et al.* (44) selected

25 migraine sufferers from a pool of advertisement respondents based on their reports that even a small amount of chocolate triggered migraine. Study participants were mailed a chocolate sample and a carob sample (both 44 g) 2 weeks apart, and were given questionnaires to complete and return 48 hours after eating the sample. Eleven subjects failed to respond with a headache to either sample. Eight subjects (32%) reported headaches after the chocolate but not the placebo and five (20%) reported headaches after the placebo but not the chocolate. The remaining subject reported headaches after both samples. There was no difference in headache occurrence after either sample. In a second study, the researchers repeated the same procedure with 15 of the original 25 subjects and again found no difference in reported headache after chocolate or placebo.

A recent study conducted by the authors (18) also demonstrated the lack of a relationship between chocolate and headache in a large sample of women with migraine or tension-type headaches. Study subjects were not selected because of dietary sensitivity; however, a subgroup of subjects who reported that chocolate was a trigger for their headaches was identified. After following a vasoactive amine elimination diet, 63 subjects ingested two 60 g chocolate samples and two 60 g carob samples in random order on four different occasions, with no samples administered during the menstrual week. Results indicated that chocolate was no more likely to trigger a headache than carob, even in subjects who were convinced that chocolate was a trigger. Subjects were also unsuccessful at guessing which samples were chocolate during the trials.

Neither migraine nor tension-type headaches were triggered by chocolate ingestion. The same numbers of headaches were reported after ingestion of chocolate as were reported after ingestion of carob. In subjects who reported a mild headache before eating the sample, headaches tended to worsen throughout the following 12 hours regardless of what type of sample was eaten. In addition, failure to adhere to the diet and consume other postulated headache trigger foods along with the chocolate samples did not result in an increase in headache activity. A separate evaluation was performed using only the 17% of women who had identified themselves as sensitive to chocolate as a trigger for headache. Even in these women, chocolate ingestion did not result in headache.

The results of studies investigating the relationship between chocolate ingestion and headache are conflicting, and one possible explanation for this is the subject selection criteria that were used in each of them. The Gibb *et al.* (43) study used selection criteria similar to the Hanington *et al.* studies (11, 19, 20, 23). In order to be included in the study, subjects had to both suspect chocolate as a trigger and actively avoid eating chocolate. The Moffet *et al.* (44) study included patients who believed chocolate was a trigger, but did not impose the criterion of chocolate avoidance. The authors' study (18) included a general headache sample, with special attention paid to a subgroup of subjects who implicated chocolate as a trigger, but who also did not report avoidance of eating chocolate.

The distinction between suspecting chocolate as a headache trigger and avoiding eating chocolate due to that suspicion seems a minor one, but may be vital in terms of the conflicting study findings. Hanington has reported that, although 73% of a sample of 500 headache sufferers implicated chocolate as a trigger for them (7), only 5% of the patients in a headache clinic met the voluntary diet restriction criterion. Thus, the results of all of the studies reviewed here lead to the conclusion that chocolate is not a significant trigger of migraine, and that many migraine sufferers believe chocolate triggers their headache when in fact it does not.

Explanations for the perceived causal relationship between chocolate and headache

If chocolate does indeed cause headaches, it is in a very limited percentage (5% or less) of headache sufferers. Given this finding, it is curious that many times that percentage of headache sufferers believe that chocolate is a trigger for them. Some of these individuals may have encountered educational materials identifying chocolate as a headache trigger and assumed that chocolate caused their headaches without a personal experience of that relationship. Others may have had an experience that led them to believe chocolate caused their headache when this was not the case. As chocolate is the most commonly craved food in the USA (45), there are ample opportunities to make false associations. This may be particularly true for women, who are three times more likely than men to suffer from migraine (1) and also crave chocolate more frequently than men (46). Sweet craving itself has been reported to be a prodromal symptom of migraine (47). Thus, craving and consuming chocolate may be a symptom rather than a cause of migraine.

The gender difference in headache prevalence is often explained by hormonal differences. Estrogen fluctuation associated with the onset of menses has been identified as a frequent migraine trigger, occurring regularly in about 60% of women (48). Headache diaries have demonstrated this association reliably (49). Interestingly, the onset of menses has also been associated with an increase in carbohydrate and chocolate craving (46, 50). Women may also become more sensitive to the effects of vasoactive chemicals such as tyramine and BPEA in the perimenstrum (51). Thus, the occurrence of a menstrual headache may be erroneously associated with chocolate ingestion, the craving of which is also associated with menses onset. Additionally, chocolate may actually contribute as a trigger, but only during the perimenstrum.

In addition, stress has been endorsed as a headache-triggering factor in almost three-quarters of headache sufferers (2), and has been reliably demonstrated to be associated with headache onset (5). Stress is also often associated with sweet craving. In addition, fasting or skipping meals has been linked to headache and is endorsed by almost half of chronic headache suffers (2). A candy bar from the office vending machine is often used as a meal replacement in individuals who are skipping balanced meals, and are also likely to be experiencing a stressful day. Inadequate nutrition and stress rather than the chocolate bar itself may be responsible for headache onset.

Conclusion

From the results of the studies reviewed above, it can certainly be concluded that chocolate probably doesn't cause headaches in the majority of headache sufferers, and certainly not an average candy bar-sized amount of chocolate. It is possible that many of the individuals who believe chocolate triggers their migraines may have experienced a coincidental association between chocolate and another, more reliable headache trigger, such as stress or menstruation. Another possibility that has been suggested is that migraine may be associated with chocolate through a conditioned taste aversion (52). In other words, simply experiencing one migraine after eating chocolate may initiate a conditioned response of migraine after eating chocolate.

There are factors that have not been investigated in the relationship between chocolate and headache that may shed additional light on the association. First, all of the studies that have been conducted using chocolate as a potential trigger of headache have used an average candy bar-sized sample. The results, therefore, cannot be generalized to the chocolate 'binge', or the ingestion of larger amounts of chocolate or BPEA. In addition, it is possible that the relationship between chocolate and migraine may be more than one of simple cause and effect. Combinations of several factors may be necessary to trigger a headache: both stress and chocolate, for example, as opposed to just one or the other. Nonetheless, the vast majority of evidence suggests we dismiss the hypothesis that chocolate is a significant migraine trigger.

References

1. Stewart, W. and Lipton, R. (1994) Migraine epidemiology in the United States. In *Headache Classification and Epidemiology* (Ed. by Olsen, J.), pp. 239–246. Raven Press, New York.
2. Scharff, L., Turk, D.C. and Marcus, D.A. (1995) Triggers of headache episode and coping responses of headache diagnostic groups. *Headache* **35**, 397–403.
3. Van den Bergh, V., Amery, W.K. and Waelkens, J. (1987) Trigger factors in migraine: a study conducted by the Belgian Migraine Society. *Headache* **27**, 191–196.
4. Drummond, P.D. (1985) Predisposing, precipitating and relieving factors in different categories of headache. *Headache* **25**, 16–22.
5. Kohler, T. and Haimerl, C. (1990) Daily stress as a trigger of migraine attacks: results of thirteen single-subject studies. *J. Consult. Clin. Psychol.* **58**, 870–872.

6. Osterhaus, S. and Passchier, J. (1992) Perception of triggers in young nonclincial school students with migrainous headaches and with tension headaches. *Percept. Motor Skills* **75**, 284–286.
7. Hanington, E. (1980) Diet and migraine. *J. Hum. Nutr.* **34**, 175–180.
8. Medina, J.C. and Diamond, S. (1978) The role of diet in migraine. *Headache* **18**, 31–34.
9. McQueen, J., Loblay, R.H., Swain, A.R., Anthony, M. and Lance, J.W. (1988) A controlled trial of dietary modification of migraine. In *New Advances in Headache Research* (Ed. by Rose, F.C.). Smith-Gordon, London.
10. Guarnieri, P., Radnitz, C.L. and Blanchard E.B. (1990) Assessment of dietary risk factors in chronic headache. *Biofeedback Self-Regul.* **15**, 15–25.
11. Hanington, E. (1969) The effect of tyramine in inducing migrainous headache. In *Background to Migraine: Second Migraine Symposium, 1967* (Ed. by Smith, R.), pp. 10–18. Heinemann, London.
12. Sandler, M., Youdim, M.B.H. and Hanington, E. (1974) A phenylethylamine oxidizing defect in migraine. *Nature* **350**, 335–337.
13. Glover, V., Sandler, M., Grant, E., *et al.* (1977) Transitory decrease in platelet monoamine oxidase activity during migraine attacks. *Lancet* **1**, 391–393.
14. Hurst, W.J. and Toomey, P.B. (1981) High-performance liquid chromatographic determination of four biogenic amines in chocolate. *Analyst* **106**, 394–402.
15. Schweitzer, J.W., Friedhoff, A.J. and Schwartz, R. (1975) Chocolate, β-phenylethylamine and migraine re-examined. *Nature* **257**, 256–257.
16. Shively, C.A. and Tarka, S.M. (1984) Methylxanthine composition and consumption patterns of cocoa and chocolate products. In *The Methylxanthine Beverages and Foods: Chemistry, Consumption and Health Effects* (Ed. by Spiller, G.), pp. 149–178. Alan R. Liss, New York.
17. Craig, W.J. and Nguyen, T.T. (1984) Caffeine and theobromine levels in cocoa and carob products. *J. Food Sci.* **49**, 302–305.
18. Marcus, D.A., Scharff, L., Turk, D.C. and Gourley, L.M. (1997) A double-blind provocative study of chocolate as a trigger of headache. *Cephalalgia* **17**, 855–862.
19. Hanington, E. (1967) Preliminary report on tyramine headache. *BMJ* **2**, 550–551.
20. Hanington, E. and Harper, A.M. (1968) The role of tyramine in the aetiology of migraine and related studies on the cerebral and extracerebral circulations. *Headache* **8**, 84–97.
21. Moffett, A., Swash, M. and Scott, D.F. (1972) Effect of tyramine in migraine: a double-blind study. *J. Neurol. Neurosurg. Psychiatry* **35**, 496–499.
22. Shaw, S.W., Johnson, R.H. and Keogh, N.J. (1978) Oral tyramine in dietary migraine sufferers. In *Current Concepts in Migraine Research* (Ed. by Green, R.), pp. 31–39. Raven Press, New York.
23. Hanington, E., Horn, M. and Wilkinson, M. (1970) Further observations on the effects of tyramine. In *Background to Migraine: Third Migraine Symposium, 1969* (Ed. by Cochrane, A.L.), pp. 113–119. Heinemann, London.
24. Ryan, R.E. (1974) A clinical study of tyramine as an etiological factor in migraine. *Headache* **14**, 43–48.
25. Forsythe, W. and Redmond A. (1974) Two controlled trials of tyramine in children with migraine. *Dev. Med. Child Neurol.* **16**, 794–799.
26. Karoum, F., Nasrallah, H. and Potkin, S. (1979) Mass fragmentography of phene-

thylamine, *m*- and *p*-tyramine and related amines in plasma, cerebrospinal fluid, urine, and brain. *J. Neurochem.* **33**, 201–212.
27. Zoumas, B.L., Kreiser, W.R. and Martin, R.A. (1980) Theobromine and caffeine content of chocolate products. *J. Food Sci.* **45**, 314.
28. Miller, J.L., Jr (1960) Caffeine, chocolate and withdrawal headache. *Northwest Med.* **59**, 502–504.
29. Kolhekar, R., Meller S. and Gebhart, G. (1994) *N*-methyl-D-aspartate receptor-mediated changes in thermal nociception: allosteric modulation at glycine and polyamine recognition sites. *Neuroscience* **63**, 925–936.
30. Egger, J., Carter C.M., Wilson, J., Turner, M.W. and Soothill, J.F. (1983) Is migraine food allergy? A double-blind controlled trial of oligoantigenic diet treatment. *Lancet* **1**, 865–868.
31. Martelletti, P., Sutherland, J., Anastasi, E., DiMario, U. and Giacovazzo, M. (1989) Evidence for an immune-mediated mechanism in food-induced migraine from a study on activated T cells IgG4 subclass, anti-IgG antibodies and circulating immune complexes. *Headache* **29**, 664–670.
32. Pradelier, A., Weinman, S., Launay, J.M., Baron, J.F. and Dry, J. (1983) Total IgE, specific IgE and prick tests against foods in common migraine – a prospective study. *Cephalalgia* **3**, 231–234.
33. Maimart, M., Pradalier, A., Launay, J.M., Dreux, C. and Dry, J. (1987) Whole blood and plasma histamine in common migraine. *Cephalalgia* **7**, 39–42.
34. Olsen, C.G., Vaughn, T.R. and Ledoux, R.A. (1989) Food-induced migraine: search for immunologic mechanisms. *J. Allergy Clin. Immunol.* **83**, 238.
35. Schuller, D.E., Cadman, T.E. and Jeffereys, W.H. (1996) Recurrent headaches: what every allergist should know. *Ann. Allergy Asthma Immunol.* **76**, 219–230.
36. Maslansky, L. and Wein, G. (1971) Chocolate allergy: a double-blind study. *Conn Med.* **35**, 5–9.
37. Mansfield, L.E., Vaughn, T.R. and Waller, S.F. (1985) Food allergy and adult migraine, double-blind and mediator confirmation of an allergic etiology. *Ann. Allergy.* **55**, 126–129.
38. Monro, J., Brostoff, J., Carini, C. and Zilkha, K. (1980) Food allergy in migraine: study of dietary exclusion and RAST. *Lancet* **2**, 1–4.
39. Hasselmark, L., Malgren, R. and Hannerz, J. (1987) Effect of a carbohydrate-rich diet, low in protein-tryptophan, in classic and common migraine. *Cephalalgia* **7**, 87–92.
40. McQueen, J., Loblay, R., Swain, A., Anthony, M. and Lance, J.W. (1988) A controlled trial of dietary management in migraine. Abstract from Australian Headache Society 2nd annual meeting 1988. *Cephalalgia* **8**, 295.
41. Salfield, S.A.W., Wardley, B.L. and Houlsby, W.T. (1987) Controlled study of exclusion of dietary vasoactive amines in migraine. *Arch. Dis. Child.* **62**, 458–460.
42. Gettis, A. (1987) Serendipity and food sensitivity: a case study. *Headache* **27**, 73–75.
43. Gibb, C.M., Davies, P.T.G., Glover, V., Steiner, T.J., Rose, F.C. and Sandler, M. (1991) Chocolate is a migraine-provoking agent. *Cephalalgia* **11**, 93–95.
44. Moffet, A.M., Swash M. and Scott, D.F. (1974) Effect of chocolate in migraine: a double-blind study. *J. Neurol. Neurosurg. Psychiatry* **37**, 445–448.
45. Weingarten, H.P. and Elston, D. (1991) Food cravings in a college population: a questionnaire study. *Appetite* **15**, 167–175.

46. Rozin, P., Levine, E. and Stoess, C. (1991) Chocolate craving and liking. *Appetite* **17**, 199–212.
47. Blau, J.N. and Diamond, S. (1984) Dietary factors in migraine precipitation: the physician's view. *Headache* **25**, 184–187.
48. Nattero, G. (1982) Menstrual headache. In *Advances in Neurology*, Vol. 33 (Ed. by Critchley, M.). Raven Press, New York.
49. Johannes, C.B., Linet, M.S., Stewart, W.F., Celentano, D.D., Lipton, R.B. and Szklo, M. (1995) Relationship of headache to phase of the menstrual cycle among young women: a daily diary study. *Neurology* **45**, 1076–1082.
50. Bowen, D.J. and Grunberg, N.E. (1990) Variations in food preference and consumption across the menstrual cycle. *Physiol. Behav.* **47**, 287–291.
51. Ghose, K. and Turner, P. (1977) The menstrual cycle and the tyramine pressor response test. *Br. J. Clin. Pharmacol.* **4**, 500–502.
52. Jessup, B. (1978) The role of diet in migraine: conditioned taste aversion. *Headache* **18**, 228.

Chapter 16

The Role of Chocolate in Exercise Performance

Louise M. Burke

Modern sports nutrition guidelines offer athletes an important tool in the achievement of their sporting potential, with goals and strategies being developed from the outcomes of rigorous scientific research. Sports nutrition encompasses issues in the everyday or training diet, which promote the health of the athlete and optimize the benefits of training (Table 16.1). Special strategies can be undertaken before, during and after exercise to enhance performance and recovery; these are particularly relevant to the competition setting (Table 16.1). While the importance of these issues is most evident in the case of élite athletes, many apply equally well to the much larger number of committed recreational athletes and people undertaking regular exercise. Therefore, this chapter refers to all such people as *athletes*.

The purpose of this chapter is to review the role of chocolate, and chocolate bars in particular, in the achievement of optimal sports performance. It is difficult to assess the role of a single food, or even a group of foods in the nutrition of athletes. Generally, a menu based on food variety is promoted to ensure adequate nutrient intake and eating pleasure. Table 16.1 illustrates the diversity of sports nutrition goals which vary between athletes and sports, and the period of the sporting calendar. Furthermore, individuals may be able to use a number of foods, or combinations of foods, to meet a specific nutritional goal, according to their personal preference and previous experience.

Since carbohydrate is the primary nutrient supplied by a chocolate bar, this review focuses on the acute intake of carbohydrate and exercise performance. Investigations of carbohydrate intake before, during and after exercise are examined, and studies in which chocolate bars have served as the carbohydrate feeding are indicated.

To consider whether the results of carbohydrate-feeding studies can be applied to chocolate, issues such as the effects of adding fat and protein to carbohydrate, and differences between liquid and solid feedings, are reviewed.

Table 16.1 The goals of sports nutrition.

Everyday nutrition goals	For competition, the athlete must
Continue to enjoy food and the pleasure of sharing meals. Keep healthy – especially by looking after the increased needs for some nutrients resulting from a heavy training programme. Get into ideal shape for their sport – achieve a level of body mass, body fat and muscle mass that is consistent with good health and good performance. Refuel and rehydrate well during each training session so that they perform at their best out of each session. Practise any intended competition eating strategies so that beneficial practices can be identified and fine-tuned. Enhance adaptation and recovery between training sessions by providing all the nutrients associated with these processes. Eat for long-term health by paying attention to community nutrition guidelines.	In the case of weight-classed sports, achieve the weigh-in target without sacrificing fuel stores and body fluid levels. 'Fuel up' with adequate body carbohydrate stores prior to the event. Minimize dehydration during the event by using opportunities to drink fluids before, during and after the event. Supply additional carbohydrate during events >1 hour in duration or other events where body carbohydrate stores become depleted. Achieve pre- and during-event strategies without causing gastrointestinal discomfort or upsets. Promote recovery after competition, particularly in sports played as a series of heats and finals, or as a tournament.

Source: adapted from Burke and Read (1).

Of equal importance with the *science* of sports nutrition is its *practice*. Athletes are often required to meet nutrient intake targets that are beyond the constraints of appetite, gastrointestinal comfort or access to food. Whether chocolate bars are a practical food choice in typical sporting situations, and whether they assist in the achievement of other goals of sports nutrition, is also discussed.

Carbohydrate intake and exercise performance

The availability of carbohydrate for oxidation by the muscle and central nervous system is a critical factor in the performance of prolonged sessions (>60–90 min) of submaximal or intermittent high-intensity exercise, and is a permissive factor in the performance of brief high-intensity exercise (for a review, see (2)). However, the total body stores of carbohydrate are limited, and are often substantially less than the fuel requirements of the exercise programmes of many athletes. Carbohydrate intake before, during and in the recovery after exercise provides a variety of options for increasing body carbohydrate availability.

'Fuelling up' before exercise

Before competition, an athlete should ensure that liver and muscle glycogen stores are able to support the anticipated fuel needs of the event. For sports events lasting less than 60 min, muscle glycogen stores, which have been normalized to the resting levels of trained athletes, are considered adequate (3). In the absence of muscle damage, muscle glycogen levels can be restored by 24–36 hours of high carbohydrate intake, in conjunction with a reduction in exercise volume and intensity (4). Thus, 'fuelling up' for most events simply consists of high carbohydrate eating and tapered training on the day before competition. Athletes who compete in events longer than ~90 min may improve their performance by maximizing muscle glycogen stores over the 3–6 days prior to their competition via an exercise-diet programme known as glycogen (or carbohydrate) loading. 'Loaded' glycogen stores permit the athlete to continue exercising at their optimal pace for a longer time, postponing the onset of fatigue and decreasing the time taken to complete a set task of prolonged duration (3).

The original carbohydrate loading protocol, developed by Scandinavian scientists in the late 1960s, used extremes of diet and exercise to first deplete then supercompensate glycogen stores (5). Recent research suggests that trained athletes do not need to undertake a severe depletion phase to subsequently 'load' glycogen stores. It appears that similar increases in muscle glycogen can be achieved simply by tapering training and consuming a daily carbohydrate intake of ~8–10 g/kg body mass over the 72 hours prior to an event (6). The most important dietary factor in glycogen storage is the amount of carbohydrate consumed (4) (see 'Carbohydrate and post-exercise recovery' below). Muscle glycogen storage increases in relation to dietary carbohydrate intake, reaching a threshold above which no further storage occurs at a carbohydrate intake of ~500–600 g/day or 8–10 g/kg body mass/day (7).

In theory, a variety of carbohydrate-rich menus may be used by athletes to 'fuel up' before events. However, studies have reported that in real life, athletes do not have sufficient practical nutrition knowledge to achieve such carbohydrate intakes and may require dietary counselling (8). Dietary patterns that may limit total carbohydrate intakes include reliance on three meals per day instead of more frequent food intake opportunities, emphasis on high fibre and bulky carbohydrate-rich foods which may cause gastrointestinal fullness before fuel intake targets are reached, and the failure to recognize sugar and sugary foods as suitable carbohydrate-rich sources (8). Education should promote the advantages of frequent meals and snacks and compact, dietary forms of carbohydrate such as sugar-rich foods and carbohydrate-containing drinks.

A high carbohydrate intake should be a daily routine for athletes who need to promote recovery between prolonged training sessions. In the situation of everyday nutrition, athletes need to choose foods that balance their fuel requirements with total nutrient needs and general energy balance. However, even in competition preparation when these other nutritional goals are of lower priority, it appears advantageous to eat a variety of carbohydrate-rich foods to meet carbohydrate intake goals. It has been noted from a review of several laboratory studies of carbohydrate loading that when athletes are fed 'loading' diets in which the additional carbohydrate is supplemented from a single food source, the increase in glycogen above the athletes' control diet appears less than that attained from a high-carbohydrate diet of mixed origin (9). Whether this is a real effect requires further investigation; it is possible that reliance on a single food in a diet can lead to an underestimation of total carbohydrate intake, due to the compounding of inaccurate food composition data or malabsorption of the food.

Chocolate bars and 'fuelling up'

Chocolate bars offer a compact and enjoyable source of carbohydrate for an athlete who has high carbohydrate requirements. They may provide a useful part of a 'fuelling up' programme, particularly as a portable snack that requires minimal preparation or storage facilities. However, the value of a varied food intake should be promoted, particularly in the everyday eating patterns of athletes in heavy training.

Pre-exercise carbohydrate intake

Studies have shown that eating a large, carbohydrate-rich meal (> 200 g carbohydrate) in the 4-hour period before exercise promotes endurance (10) and enhances the performance of a time trial undertaken at the end of a prolonged exercise session (11, 12). Pre-exercise meals may enhance carbohydrate availability during situations of endurance exercise by increasing muscle and liver glycogen stores (13), or by providing a source of glucose in the gut for later release. This may be particularly important for exercise undertaken in the morning after an overnight fast, when liver glycogen stores are likely to be depleted.

Guidelines for pre-exercise meals need to take into account gastrointestinal comfort as well as the potential for enhancing body carbohydrate stores. In some cases, an athlete may need to eat prior to a prolonged work-out or competition in order to prevent hunger during the subsequent exercise session. However, in all cases the athlete must ensure that such pre-exercise intake does not cause excessive fullness, or result in gastrointestinal discomfort and upsets during exercise. Excessive stomach fullness caused by the ingestion of large volumes of fluid before intense exercise is not only associated with gastrointestinal problems, but also with impairment in exercise performance (14). This presumably also applies to the excessive intake of solid foods, and to the intake of meals close to the onset of exercise. There appears to be considerable inter-individual variability in tolerance to the size and timing of pre-event intake.

The general advice regarding pre-event eating is that a carbohydrate-rich meal or snack should be consumed 2–4 hours prior to the exercise, and that the fat, protein and fibre content of these meals should be kept to modest levels. There is evidence that for some athletes, the consumption of significant amounts of fat, protein and fibre in pre-race meals may increase the risk of experiencing gastrointestinal problems during the event (15). The 'ideal' pre-exercise meal should be chosen according to the situation and the individual athlete, and be based on their previous experience and experimentation during training. In general, 3–4 hours may be needed for the digestion of a large, carbohydrate-rich, low-fat meal. On the other hand, a smaller, carbohydrate-rich snack or drink may be consumed without gastric problems 30–90 min prior to exercise. This may be a more suitable strategy for athletes who are competing in early morning events or as a 'top up' when there is a long gap between the last meal and the exercise session.

Chocolate bars and carbohydrate intake in the hour before exercise

A potential disadvantage of carbohydrate intake prior to exercise is that it stimulates a rise in plasma insulin concentrations, which in turn suppresses fat metabolism and increases carbohydrate oxidation during the subsequent exercise (16–18). Although metabolic alterations may persist even when pre-exercise meals are eaten 4 hours before exercise (13), this effect is mostly likely to occur when carbohydrate is consumed within the hour before exercise.

An insulin-mediated increase in carbohydrate oxidation during exercise may lead to a decrease in plasma glucose levels and/or an accelerated rate of muscle glycogen utilization. Many studies have noted that the feeding of glucose or carbohydrate-rich foods 30–60 min prior to submaximal exercise causes a small dip in blood glucose levels after the exercise begins (17–22). Furthermore, Foster *et al.* (18) reported that feeding 75 g of glucose 30 min prior to exercise reduced the time able to cycle before exhaustion at 80% of maximal oxygen uptake (VO_{2max}). The reduction in endurance was attributed to hypothesized acceleration in muscle glycogenolysis (18).

The negative publicity surrounding this study has lead to generalized warnings to avoid carbohydrate intake during the hour prior to endurance exercise. This advice still persists in many athletic circles, despite the evidence from at least a dozen subsequent studies that carbohydrate feeding in the hour prior to exercise enhances, or at least fails to affect, work capacity and performance of prolonged, moderate-intensity exercise (for reviews see (4, 23)). In at least one study, chocolate bars provided the pre-exercise meal choice and did not alter the endurance of cyclists riding at 70% VO_{2max} compared with a placebo feeding, despite the transient lowering of blood glucose concentrations after 15 min of exercise (19).

One strategy to reduce the risk of a detrimental response to pre-exercise feedings is to choose carbohydrate sources that produce a minimal glycaemic and insulinaemic response. These include fructose (17, 20, 21) or carbohydrate-rich foods with a low glycaemic index (GI). Thomas *et al.* (22) reported that a low-GI, carbohydrate-rich food (lentils) eaten 60 min prior to cycling increased the time taken to reach fatigue at ~70% VO_{2max} compared with the ingestion of an equal amount of carbohydrate eaten in the form of a high-GI food (potatoes). The lower glycaemic and insulinaemic responses to the low-GI trial were reported to better maintain blood glucose and free fatty acid (FFA) concentrations during exercise, and reduce exercise respiratory exchange ratio (RER) values. A slower rate of glycogen synthesis was presumed, providing an explanation for the increased endurance seen in the low-GI trial (22).

Although the results of this study have quickly led to advice that low-GI, carbohydrate-rich foods are the preferred pre-event meal choice for the endurance athlete (24), the evidence for clear performance benefits is lacking. Other investigations comparing high- and low-GI foods eaten as pre-exercise meals have failed to find any differences in exercise performance between trials, even when disturbances to blood glucose and insulin profiles were attenuated with the low-GI meals (25–27). It is important to note that even small rises in insulin concentration cause a suppression of FFA concentrations and lipolysis; this appears to be an absolute rather than dose-dependent effect (27, 28). Therefore, although the insulin rise may be blunted with a low-GI carbohydrate food (27), or with the addition of fat to carbohydrate (e.g. as in chocolate) (28) compared with a high-GI food, this may not be sufficient to completely abolish other metabolic changes.

In practical terms, it appears that a small percentage of athletes suffer from an exaggerated and detrimental response to the insulin rise following the intake of carbohydrate. However, for the majority of the athletes the effects of any carbohydrate feeding on subsequent exercise metabolism are transient, of little significance to performance, or are offset by the increase in carbohydrate availability. Most importantly, the effects of pre-exercise feedings on subsequent metabolism and performance should not be taken in isolation.

As discussed in the next section, the ingestion of carbohydrate-rich drinks or foods *during* exercise is an effective and popular strategy used by endurance athletes to promote carbohydrate availability, and the interaction of the intake of carbohydrate before and during exercise requires further study. The authors' own work has indicated that ingestion of carbohydrate during prolonged cycling, according to guidelines summarized below, maintains fuel availability throughout the exercise and overrides any metabolic and performance effects arising from the choice of pre-event meal (29).

Carbohydrate intake during exercise

There is plentiful evidence that consuming carbohydrate during prolonged (>60–90 min), moderate-intensity exercise can improve work capacity. The majority of well-controlled trials employing either prolonged cycling or treadmill running have shown significant improvements in exercise 'performance' when carbohydrate is consumed during the exercise session (for a review, see (30)). Measurements of performance may include time to reach exhaustion, work completed in a given time, or time to complete a set amount of work. In some field studies, carbohydrate intake has been shown to improve the performance of sporting events such as orienteering (31), soccer (32) or a cycling road race (33). Even when there was no significant positive effect of carbohydrate ingestion on exercise capacity, neither was performance adversely affected by increasing the availability of carbohydrate (30). Although it was originally suggested that the intake of carbohydrate might allow 'sparing' of muscle glycogen stores, the majority of studies show no effect of carbohydrate feedings during moderate-intensity exercise on muscle glycogen utilization (for a review, see (34)). Rather, the major effect of carbohydrate feedings on prolonged exercise is to maintain plasma glucose concentration, sustain high rates of carbohydrate oxidation, and spare *liver* glycogen (35). Carbohydrate feedings have the potential to improve the performance of exercise in which body carbohydrate stores would otherwise become depleted, leading to reduced rates of carbohydrate oxidation and reduced work intensity (34).

Recent studies have shown that carbohydrate ingestion also enhances the performance of sustained high-intensity exercise of ~1 hour duration (36, 37). Since fuel availability is unlikely to be limiting in this type of exercise, another mechanism is needed to explain these findings. Perhaps athletes receive a 'central' boost or reduced perception of effort following carbohydrate ingestion. Further research is needed to support and explore these findings. Although it is more difficult to undertake performance studies in team and racquet sports, it is reasonable to assume that athletes in these highly skilled sports will also benefit from carbohydrate intake during training and competition sessions. Fuel stores may become depleted during prolonged games, particularly in the case of tournaments and daily training when there may be inadequate time to fully restore glycogen levels between sessions. The central effects of carbohydrate ingestion are also likely to benefit skill and cognitive function.

Tracer technology has allowed estimations to be made of the contribution of ingested carbohydrate to the fuel mix during prolonged exercise. Notwithstanding the methodological problems involved with these techniques, it appears that there are no physiologically significant differences in the maximal rates of oxidation of carbohydrate from a variety of types of carbohydrate of moderate and high GI. With the exception of fructose (a low-GI carbohydrate), it appears that all carbohydrate types ingested during exercise are ultimately oxidised at the same rate (for a review, see (38, 39)). Whereas only small amounts of the ingested carbohydrate are oxidised during the first hour of exercise, thereafter carbohydrate feedings may be oxidised at peak rates of ~1 g/min (38, 39).

Typically, studies show that carbohydrate intakes of 30–60 g/hour during exercise are needed to promote performance benefits, although higher rates of intake may be needed to support carbohydrate oxidation in the latter stages of a prolonged exercise bout. Carbohydrate intake is more effective when it is consumed in advance of fatigue, rather than when the athlete waits for the onset of symptoms of fuel depletion (40). More specifically, a recent study has shown that performance is enhanced when carbohydrate feedings begin early and continue throughout the exercise session, rather than as a large dose consumed late in the session (41). This confirms that factors in addition to improved fuel availability are responsible for the performance improvements.

Of course, practical issues may largely dictate the timing and frequency of feedings that athletes can follow. During endurance events, carbohydrate intake occurs while the athlete is literally 'on the run', and might be limited by consideration of the time lost in stopping or slowing down to consume food or fluid, or the impact of such ingestion on gastrointestinal discomfort (42). On the other hand, in many team sports, the opportunity to ingest fluid is governed by the official rules of the sport and is limited to formal breaks or informal stoppages in play (43).

Chocolate as a carbohydrate source during exercise

In studies and in real life, a variety of carbohydrate-rich foods and drinks have been consumed by athletes during exercise to provide additional fuel during the session. In several studies investigating the effect of carbohydrate intake during exercise on performance and metabolism, chocolate bars were chosen as the carbohydrate feeding. Compared with an artificially sweetened placebo, chocolate bars consumed during prolonged cycling were shown to maintain blood glucose concentrations and enhance performance in a time trial undertaken at the end of the exercise session (44, 45). This effect was seen whether the bars were fed hourly, or consumed in smaller portions at 30 min intervals (45).

Several studies have compared the effect of feeding equal amounts of carbohydrate in solid or liquid form during exercise; trials have compared a rice-based drink with a rice food bar (46) and a sports drink versus a sports bar (47) or meal replacement bar (48). The solid foods contained small amounts of protein and fat, in addition to their carbohydrate contribution. Exercise consisted of cycling at 65–70% of VO_{2max}, and fluid intakes were matched between trials. All studies reported maintenance of blood glucose profiles, and similar metabolic responses to exercise, and exercise performance between carbohydrate trials (46–48). Similarly, both a solid carbohydrate feeding (chocolate bar) or a carbohydrate drink fed 5 min prior to exercise of higher intensity (77% of VO_{2max}) improved time-trial performance compared with the intake of a placebo (10). Solid bananas were equally effective as a banana 'slurry', fed during a brief rest period between long cycling bouts, in prolonging the duration of exercise undertaken to the point of fatigue (49). From metabolic and gastrointestinal viewpoints, carbohydrate consumed immediately prior to exercise is considered similar to carbohydrate consumed during the session.

Thus, during moderate-intensity exercise in a thermoneutral environment, it appears that solid and liquid forms of carbohydrate are equally useful as a source of additional fuel, provided that fluid needs are also met. It is likely, however, that the intake of solid carbohydrate foods during exercise of higher intensity, particularly during running, incurs a higher risk of gastrointestinal problems. Some athletes may be at an increased risk of such problems and the presence of significant amounts of fat, protein or fibre in some carbohydrate-rich foods may exacerbate these problems. However, the range and boundaries of characteristics of carbohydrate sources that are suitable for various activities have not been systematically investigated. It is likely that considerable differences exist between individuals.

Apart from the gastrointestinal preferences of at least some athletes, the use of liquid forms of carbohydrate during exercise offers the advantage of simultaneous replacement of fluid losses. Dehydration has a major effect on performance and the perception of effort during exercise. The enhancement of exercise performance resulting from fluid replacement has been shown to be independent of and additive to the beneficial effects of carbohydrate intake (36). Commercial sports drinks (carbohydrate–electrolyte beverages of 4–8% carbohydrate concentration) have been tailor-made to promote efficient delivery of fluid and carbohydrate needs during exercise. At a typical concentration of ~6% carbohydrate, an intake of ~760 ml/hour will allow most athletes to achieve adequate fuel intake during prolonged exercise; this ratio of fluid and carbohydrate delivery can be altered according to the individual needs of the athlete and their exercise situation by changing the concentration of the drink (42). Given the widespread popularity of these drinks and the simplicity with which they allow nutrition guidelines to be achieved, it is not surprising that their use is supported, if not directly promoted, by the major groups involved in sports nutrition education (50).

Nevertheless, a range of other carbohydrate-rich drinks and foods, including chocolate bars, may be consumed successfully by athletes during exercise and may be chosen on the basis of practical issues such as taste, cost and availability. Solid foods such as chocolate bars are a portable carbohydrate supply for athletes who need to transport their own provisions (e.g. road cyclists, cross-country skiers and hikers). They may also offer some taste variety and satiety for athletes undertaking ultra-endurance events. Whether an athlete 'on the move' such as a runner has the opportunity to unwrap and consume a bar might be taken into consideration, as must the 'keeping' characteristics of chocolate bars in hot conditions.

Carbohydrate and post-exercise recovery

Rapid recovery of fuel stores is important when the athlete has to train or compete within 8–24 hours. This is a common issue in the training schedules of élite athletes, but may also occur in competition settings for athletes who compete in weekly or bi-weekly fixtures, tournaments or multi-stage events. In the absence of muscle damage, muscle glycogen is restored at an average rate of about 5% per hour, and requires about 24 hours before stores are normalised (4). Studies have shown that a high-carbohydrate diet consumed after exercise for 4 hours (51) or 24 hours (52) enhanced recovery such that the endurance capacity of the high-carbohydrate group was better than the control group consuming a placebo (51) or moderate carbohydrate intake (52) at a subsequent trial of running until exhaustion. While the benefits of enhanced fuel recovery on a single bout of exercise are well documented, it is interesting that longitudinal studies do not show clear advantages to training adaptations and performance when a high-carbohydrate diet is compared with a moderate carbohydrate intake during a period of daily training (for reviews, see (53, 54)). It is possible that athletes are able to adapt to chronic reductions in muscle glycogen levels; however, the most probable explanation for the results of the existing longitudinal studies is that methodological problems have masked the detection of real performance changes (54).

The most important determinant of muscle glycogen storage is the amount of carbohydrate consumed. Studies have shown that an optimal rate of refuelling occurs when ~1 g of carbohydrate/kg body mass is consumed immediately after exercise, towards a total intake of 7–10 g/kg over the next 24 hours (for a review, see (4)). There is evidence that the rate of glycogen storage is enhanced during the first couple of hours of post-exercise recovery (55). However, the main reason for recommending carbohydrate intake early in the recovery period is that substantial storage does not occur until carbohydrate substrate is provided. Thus, early feeding maximises the length of effective recovery between exercise sessions and may be important when the recovery time is less than 8 hours. When there are 8–24 hours or more between recovery sessions, it is likely that any small differences in the early rates of glycogen resynthesis are unimportant. As long as sufficient carbohydrate is consumed during the period, it appears that the timing of the first meal after exercise can be delayed for a couple of hours without significant penalty (56).

The question of whether carbohydrate intake during recovery is best consumed as large feedings or as a series of snacks has been addressed in several studies. The results show that as long as the total amount of carbohydrate ingested is sufficient, the repletion of muscle glycogen synthesis is unaffected by the frequency of food intake. Muscle glycogen storage after 24 hours was similar when 525 g of carbohydrate was fed as two or seven meals (7) or when 10 g/kg body mass of carbohydrate was consumed as either four large meals or 16 snacks (57).

Post-exercise recovery and chocolate bars

The effects of the type (low or high GI) and form (solid or liquid) of carbohydrate ingested during the post-exercise recovery period on the rates of muscle glycogen synthesis have been investigated. While glucose and sucrose have been shown to produce similar rates of muscle glycogen recovery, the ingestion of similar amounts of fructose promotes a much lower rate of glycogen storage (58). Studies of muscle glycogen resynthesis with different carbohydrate foods have produced conflicting results, principally because of the confusing classification of carbohydrate foods into 'simple' and 'complex' types, rather than true metabolic considerations such as the measured GI. For example, Costill *et al.* (7) reported that during the first 24 hours of recovery after exhaustive running, both 'simple' and 'complex' carbohydrate foods produced comparable rates of muscle glycogen synthesis. Over the subsequent 24-hour period, the consumption of 'complex' carbohydrate resulted in a significantly higher rate of glycogen storage. By contrast, others have shown that 'simple' and 'complex' carbohydrate diets were equally successful in producing muscle glycogen storage over 72 hours of recovery (59).

Now that the GI allows a more precise knowledge of metabolic characteristics of carbohydrate-rich foods, it has been proposed that those with a moderate-to-high GI should take priority in the post-exercise diet, and that foods with a low GI should not make up more than one-third of recovery meals (4). Results of a recent study support this notion. It was reported that a diet of high-GI, carbohydrate-rich foods promoted greater glycogen storage than an equal amount of carbohydrate eaten as low-GI foods in the 24-hour period after strenuous exercise (60). It is tempting to explain these results in terms of increased glucose and insulin responses; however, other factors may also be responsible for the lower rates of glycogen storage with low-GI foods (57, 60).

Consumption of carbohydrate in either solid or liquid form appears to be equally efficient in providing substrate for muscle glycogen synthesis (61, 62). The coingestion of protein with carbohydrate feedings does not confer additional advantages to glycogen storage when total energy content is controlled (63). This recent finding counters a previous investigation in which protein added to a carbohydrate feeding – thus increasing total energy intake – was shown to enhance glycogen storage during 4 hours of recovery from exercise (64). An investigation in which protein and fat were added to carbohydrate meals over 24 hours of recovery shows that muscle glycogen storage is unaffected by alterations in the metabolic response to meals, provided that total carbohydrate intake is adequate (65).

Taken together, it appears that the most important issue in post-exercise recovery is that athletes consume sufficient carbohydrate intake. In many real-life situations, practical issues such as poor appetite, reduced access to food and other time commitments may challenge the athlete's ability to consume food. Carbohydrate drinks or foods with a high liquid content may appeal to athletes who are tired and dehydrated (54). Foods that are easy to prepare, have good storage properties and are enjoyable to eat provide other practical advantages (54). In the case of limited appetite or restricted energy intake, foods with a lower fat and protein content are advised so that total carbohydrate intake is not displaced (65).

Summary

Strategies to match carbohydrate availability to the fuel needs of training and competition form an ongoing cycle for the athlete. Carbohydrate ingestion after exercise aids the resynthesis of muscle and liver glycogen in preparation for subsequent exercise sessions. The pre-exercise meal can assist to 'top up' body carbohydrate stores, while carbohydrate intake during exercise may be needed to provide additional fuel as body stores become depleted. In all cases, the amount of carbohydrate consumed by the athlete is important, and the type of carbohydrate-rich food or drink that might be consumed is dictated more by practical issues than the physiological characteristics of the carbohydrate source. Chocolate bars may be useful in many circumstances, since they provide a compact, portable and well-liked form of carbohydrate. Nevertheless, the athlete is encouraged to choose a variety of carbohydrate-rich foods to meet overall carbohydrate intake goals, and to let individual preference and experience dictate which particular choices are used in specific situations of pre-, during- and post-exercise refuelling.

References

1. Burke, L.M. and Read, R.S.D. (1989) Sports nutrition: approaching the nineties. *Sports Med.* **8**, 80–100.
2. Hawley, J.A. and Hopkins, W.G. (1995) Aerobic glycolytic and aerobic lipolytic power systems. A new paradigm with implications for endurance and ultra-endurance events. *Sports Med.* **19**, 240–250.
3. Hawley, J.A., Schabort, E.J., Noakes, T.D. and Dennis, S.C. (1997) Carbohydrate-loading and exercise performance: an update. *Sports Med.* **24**, 73–81.
4. Coyle, E.F. (1992) Timing and method of increased carbohydrate intake to cope with heavy training, competition and recovery. In *Food, Nutrition and Sports Performance* (Ed. by Williams, C. and Devlin, J.T.), pp. 35–62. E. & F. Spon, London.
5. Bergstrom, J., Hermansen, L., Hultman, E. and Saltin, B. (1967) Diet, muscle glycogen and physical performance. *Acta Physiol. Scand.* **71**, 140–150.
6. Sherman, W.M., Costill, D.L., Fink, W.J. and Miller, J.M. (1981) Effect of diet-exercise manipulation on muscle glycogen and its subsequent utilization during performance. *Int. J. Sports Med.* **2**, 114–118.
7. Costill, D.L., Sherman, W.M., Fink, W.J., Maresh, C., Witten, M. and Miller, J.M. (1981) The role of dietary carbohydrates in muscle glycogen re-synthesis after strenuous running. *Am. J. Clin. Nutr.* **34**, 1831–1836.
8. Burke, L.M. and Read, R.S.D. (1987) A study of carbohydrate loading techniques used by marathon runners. *Can. J. Sports Sci.* **12**, 6–10.

9. Hawley, J.A., Palmer, G.S. and Noakes, T.D. (1997) Effects of 3 days of carbohydrate supplementation on muscle glycogen content and utilization during a 1-h cycling performance. *Eur. J. Appl. Physiol.* **75**, 407–412.
10. Neufer, P.D., Costill, D.L., Flynn, M.G., Kirwan, J.P., Mitchell, J.B. and Houmard, J. (1987) Improvements in exercise performance: effects of carbohydrate feedings and diet. *J. Appl. Physiol.* **62**, 983–988.
11. Sherman, W.M., Brodowicz, G., Wright, D.A., Allen, W.K., Simonsen, J. and Dernbach, A. (1989) Effects of 4 h pre-exercise carbohydrate feedings on cycling performance. *Med. Sci. Sports Exerc.* **21**, 598–604.
12. Wright, D.A., Sherman, W.M. and Derbach, A.R. (1991) Carbohydrate feedings before, during, or in combination improve cycling endurance performance. *J. Appl. Physiol.* **71**, 1082–1088.
13. Coyle, E.F., Coggan, A.R., Hemmert, M.K., Lowe, R.C. and Walters, T.J. (1985) Substrate usage during prolonged exercise following a pre-exercise meal. *J. Appl. Physiol.* **59**, 429–433.
14. Robinson, T.A., Hawley, J.A., Palmer, G.S., *et al.* (1995) Water ingestion does not improve 1-h cycling performance in moderate ambient temperatures. *Eur. J. Appl. Physiol.* **71**, 153–160.
15. Rehrer, N.J., van Kemenade, M., Meester, W., Brouns, F. and Saris, W.H.M. (1992) Gastrointestinal complaints in relation to dietary intake in triathletes. *Int. J. Sport Nutr.* **2**, 48–59.
16. Gleeson, M., Maughan, R.J. and Greenhaff, P.L. (1986) Comparison of the effects of pre-exercise feeding of glucose, glycerol and placebo on endurance and fuel homeostasis in man. *Eur. J. Appl. Physiol.* **55**, 645–653.
17. Hargreaves, M., Costill, D.L., Katz, A. and Fink, W.J. (1985) Effects of fructose ingestion on muscle glycogen usage during exercise. *Med. Sci. Sports Exerc.* **17**, 360–363.
18. Foster, C., Costill, D.L. and Fink, W.J. (1979) Effects of pre-exercise feedings on endurance performance. *Med. Sci. Sports* **11**, 1–5.
19. Alberici, J.C., Farrell, P.A., Kris-Etherton, P.M. and Shively, C.A. (1993) Effects of pre-exercise candy bar ingestion on glycemic response, substrate utilization, and performance. *Int. J. Sport Nutr.* **3**, 323–333.
20. Decombaz, J., Sartori, D., Arnaud, M.J., Thelin, A.L., Schurch, P. and Howald, H. (1985) Oxidation and metabolic effects of fructose or glucose ingested before exercise. *Int. J. Sports Med.* **6**, 282–286.
21. Hargreaves, M., Costill, D.L., Fink, W.J., King, D.S. and Fielding, R.A. (1987) Effect of pre-exercise carbohydrate feedings on endurance cycling performance. *Med. Sci. Sports Exerc.* **19**, 33–36.
22. Thomas, D.E., Brotherhood, J.R. and Brand, J.C. (1991) Carbohydrate feeding before exercise: effect of glycemic index. *Int. J. Sports Med.* **12**, 180–186.
23. Hawley, J.A. and Burke, L.M. (1997) Effect of meal frequency and timing on physical performance. *Brit. J. Nutr.* **77** (Suppl.), S91–S103.
24. Brand Miller, J., Foster-Powell, K. and Colagiuri, S. (1996) *The G.I. Factor.* Hodder & Stoughton, Sydney.
25. Thomas, D.E., Brotherhood, J.R. and Brand Miller, J. (1994) Plasma glucose levels after prolonged strenuous exercise correlate inversely with glycemic response to food consumed before exercise. *Int. J. Sport Nutr.* **14**, 361–373.

26. Febbraio, M.A. and Stewart, K.L. (1996) CHO feeding before prolonged exercise: effect of glycemic index on muscle glycogenolysis and exercise performance. *J. Appl. Physiol.* **81**, 1115–1120.
27. Sparks, M.J., Selig, S.S. and Febbraio, M.A. (in press) Pre-exercise carbohydrate ingestion: effect of the glycemic index on endurance exercise performance. *Med. Sci. Sports Exerc.*
28. Horowitz, J.F. and Coyle, E.F. (1993) Metabolic responses to pre-exercise meals containing various carbohydrates and fat. *Am. J. Clin. Nutr.* **58**, 235–241.
29. Burke, L.M., Claassen, A., Hawley, J.A. and Noakes, T.D. (submitted) Glycemic index of pre-exercise meals does not affect metabolism or performance when carbohydrate is ingested during prolonged cycling.
30. Hawley, J.A., Dennis, S.C. and Noakes, T.D. (1995) Carbohydrate, fluid and electrolyte requirements during prolonged exercise. In *Sports Nutrition: Minerals and Electrolytes* (Ed. by Kies, C.V. and Driskell, J.A.), pp. 235–265. CRC Press, Boca Raton.
31. Kujala, U.M., Heinonen, O.J., Kvist, M., *et al.* (1989) Orienteering performance and ingestion of glucose and glucose polymers. *Br. J. Sports Med.* **23**, 105–108.
32. Leatt, P.B. and Jacobs, I. (1989) Effect of glucose ingestion on glycogen depletion during a soccer match. *Can. J. Sports Sci.* **14**, 112–116.
33. Edwards, T.L. and Santeusanio, D.M. (1984) Field test of the effects of carbohydrate solutions on endurance performance, selected blood chemistries, perceived exertion, and fatigue in world class cyclists (abst). *Med. Sci. Sports Exerc.* **16** (Suppl.), 190.
34. Coggan, A.R. and Coyle, E.F. (1991) Carbohydrate ingestion during prolonged exercise: effects on metabolism and performance. In *Exercise and Sports Science Reviews*, Vol. 19 (Ed. by Holloszy, J.O.), pp. 1–40. Williams and Wilkins, Baltimore.
35. Bosch, A.N., Dennis, S.C. and Noakes, T.D. (1993) Influence of carbohydrate loading on fuel substrate turnover and oxidation during prolonged exercise. *J. Appl. Physiol.* **74**, 1921–1927.
36. Below, P.R., Mora-Rodriguez, R., Gonzalez-Alonso, J. and Coyle, E.F. (1995) Fluid and carbohydrate ingestion independently improve performance during 1 h of intense cycling. *Med. Sci. Sports Exerc.* **27**, 200–210.
37. Jeukendrup, A.E., Brouns, F., Wagenmakers, A.J.M. and Saris, W.H.M. (1997) Carbohydrate-electrolyte feedings improve 1 h time trial cycling performance. *Int. J. Sports Med* **18**, 125–129.
38. Hawley, J.A., Dennis, S.C. and Noakes, T.D. (1992) Oxidation of carbohydrate ingested during prolonged exercise. *Sports Med.* **14**, 27–42.
39. Guezennec, C.Y. (1995) Oxidation rates, complex carbohydrates and exercise. *Sports Med.* **19**, 365–372.
40. Coggan, A.R. and Coyle, E.F. (1989) Metabolism and performance following carbohydrate ingestion late in exercise. *Med. Sci. Sports Exerc.* **21**, 59–65.
41. McConell, G., Kloot, K. and Hargreaves, M. (1996) Effect of timing of carbohydrate ingestion on endurance exercise performance. *Med. Sci. Sports Exerc.* **28**, 1300–1304.
42. Coyle, E.F. and Montain, S.J. (1992) Benefits of fluid replacement with carbohydrate during exercise. *Med. Sci. Sports Exerc.* **24** (Suppl.), S324–S330.
43. Burke, L.M. and Hawley, J.A. (1997) Fluid balance in team sports: guidelines for optimal practices. *Sports Med.* **24**, 38–54.

44. Hargreaves, M., Costill, D.L., Coggan, A., Fink, W.J. and Nishibata, I. (1984) Effect of carbohydrate feedings on muscle glycogen utilization and exercise performance. *Med. Sci. Sports Exerc.* **16**, 219–222.
45. Fielding, R.A., Costill, D.L., Fink, W.J., King, D.S., Hargreaves, M. and Kovaleski J.E. (1985) Effect of carbohydrate feeding frequencies and dosage on muscle glycogen use during exercise. *Med. Sci. Sports Exerc.* **17**, 472–476.
46. Mason, W.L., McConell, G. and Hargreaves, M. (1993) Carbohydrate ingestion during exercise: liquid vs solid feedings. *Med. Sci. Sports Exerc.* **25**, 966–969.
47. Lugo, M., Sherman, W.M., Wimer, G.S. and Garleb, K. (1993) Metabolic responses when different forms of carbohydrate energy are consumed during cycling. *Int. J. Sport Nutr.* **3**, 398–407.
48. Robergs, R.A., McMinn, S.B., Mermier, C., Leadbetter, G., Ruby, B. and Quinn, C. (1998) Blood glucose and glucoregulatory hormone responses to solid and liquid carbohydrate ingestion during exercise. *Int. J. Sport Nutr.* **8**, 70–83.
49. Murdoch, S.D., Bazzarre, T.L., Snider, I.P. and Goldfarb, A.H. (1993) Differences in the effects of carbohydrate food form on endurance performance to exhaustion. *Int. J. Sport Nutr.* **3**, 41–54.
50. American College of Sports Medicine. (1996) Position stand: exercise and fluid replacement. *Med. Sci. Sports Exerc.* **28**, i–vii.
51. Fallowfield, J.L., Williams, C. and Singh, R. (1995) The influence of ingesting a carbohydrate-electrolyte beverage during 4 hours of recovery on subsequent endurance capacity. *Int. J. Sport Nutr.* **5**, 285–299.
52. Fallowfield, J.L. and Williams, C. (1993) Carbohydrate intake and recovery from prolonged exercise. *Int. J. Sport Nutr.* **3**, 150–164.
53. Sherman, W.M. and Wimer, G.S. (1991) Insufficient dietary carbohydrate during training: does it impair performance? *Int. J. Sport Nutr.* **1**, 28–44.
54. Burke, L.M. (in press) Dietary carbohydrates. In *Encyclopaedia of Sports Medicine: Nutrition for Sport* (Ed. by Maughan, R.J.). Blackwell Science, Oxford.
55. Ivy, J.L., Katz, A.L., Cutler, C.L., Sherman, W.M. and Coyle, E.F. (1988) Muscle glycogen synthesis after exercise: effect of time of carbohydrate ingestion. *J. Appl. Physiol.* **65**, 1480–1485.
56. Parkin, J.A.M., Carey, M.F., Martin, I.K., Stojanovska, L. and Febbraio, M.A. (1996) Muscle glycogen storage following prolonged exercise: effect of timing of ingestion of high glycemic index food. *Med. Sci. Sports Exerc.* **29**, 220–224.
57. Burke, L.M., Collier, G.R., Davis, P.G., Fricker, P.A., Sanigorski, A.J. and Hargreaves, M. (1996) Muscle glycogen storage after prolonged exercise: effect of the frequency of carbohydrate feedings. *Am. J. Clin. Nutr.* **64**, 115–119.
58. Blom, P.C., Hostmark, A.T., Vaage, O., Vardal, K.R. and Maehlum, S. (1987) Effect of different post-exercise sugar diets on the rate of muscle glycogen synthesis. *Med. Sci. Sports Exerc.* **19**, 491–496.
59. Roberts, K.M., Noble, E.G., Hayden, D.B. and Taylor, A.W. (1988) Simple and complex carbohydrate-rich diets and muscle glycogen content of marathon runners. *Eur. J. Appl. Physiol.* **57**, 70–74.
60. Burke, L.M., Collier, G.R. and Hargreaves, M. (1993) Muscle glycogen storage after prolonged exercise: effect of glycemic index on carbohydrate feedings. *J. Appl. Physiol.* **75**, 1019–1023.

61. Keizer, H., Kuipers, H., van Kranenburg, G. and Geurten, P. (1986) Influence of liquid and solid meals on muscle glycogen resynthesis, plasma fuel hormone response, and maximal physical working capacity. *Int. J. Sports Med.* **8**, 99–104.
62. Reed, M.J., Brozinick, J.T., Lee, M.C. and Ivy, J.L. (1989) Muscle glycogen storage post-exercise: effect of mode of carbohydrate administration. *J. Appl. Physiol.* **66**, 720–726.
63. Zawadzki, K.M., Yaspelkis, B.B. and Ivy, J.L. (1992) Carbohydrate–protein complex increases the rate of muscle glycogen storage after exercise. *J. Appl. Physiol.* **72**, 1854–1859.
64. Tarnopolsky, M.A., Bosman, M., McDonald J.R., Vandeputte, D., Martin, J. and Roy, B.D. (1997) Post-exercise protein–carbohydrate and carbohydrate supplements increase muscle glycogen in men and women. *J. Appl. Physiol.* **83**, 1877–1883.
65. Burke, L.M., Collier, G.R., Beasley, S.K., *et al.* (1995) Effect of coingestion of fat and protein with carbohydrate feedings on muscle glycogen storage. *J. Appl. Physiol.* **78**, 2187–2192.

Chapter 17

Chocolate Craving: Biological or Psychological Phenomenon?

David Benton

Chocolate is a uniquely attractive substance that has an appeal unmatched by any other food item. Large sections of the population will readily admit to craving chocolate; some will even claim to be addicted (1), although this is lay self-diagnosis rather than an accurate or medically justifiable description. Articles in the media and popular books have speculated that chocolate's appeal can be explained in terms of the influence that it has on the brain's chemistry. For example, Debra Waterhouse, in her book *Why Women Need Chocolate* (2), stated that

> 'food cravings are Mother Nature's way of informing us that we need to eat a specific food in order to look and feel great ... Chocolate can cause a rush of both serotonin and endorphins into your brain cells ... it has been called the most effective non-drug anti-depressant ... the "Prozac of plants".'

Waterhouse claimed that chocolate has a calming influence for two major reasons. First, the sugar in chocolate was said to increase the synthesis of serotonin, the neurotransmitter whose activity is raised by Prozac and other anti-depressant drugs. Second, the fat in chocolate was said to release endorphins that induce a sense of well-being. In addition, the phenylethylamine, theobromine and magnesium supplied by chocolate add to the experience.

In reality, any certainty concerning the basis of chocolate's popularity is unjustified; it has been the subject of relatively little scientific attention. The existing evidence is described here and the plausibility of various possible mechanisms considered. A series of biological mechanisms is examined and the relative roles of psychological and physiological mechanisms compared.

Food craving

A survey of young Canadian adults found that 97% of women and 68% of men experienced food cravings (3). These cravings were, however, highly selective, chocolate being by far the most commonly and intensely craved item (4, 5, 6). As 85% of the Canadian sample reported that more often than not they gave in to their cravings, they are clearly powerful phenomena.

To date the term *food craving* has been used in a similar way to the lay definition; it is a strong desire or urge for a particular food. All research has been limited by the way that cravings have been measured. In the majority of cases, subjects have been asked simply to rate their desire to eat a particular food. A single-item scale is unreliable and there has been an implicit assumption that craving can be explained using a single dimension. To counter these problems, Benton *et al.* (7) asked 330 people to respond to 80 statements concerning chocolate and statistically established the dimensions that accounted for their reactions.

The first dimension that was found was labelled *craving*. It was associated with a considerable preoccupation with chocolate and acts of compulsion. In fact, the questions that defined this first dimension fell into two groups. Chocolate was a source of some distraction; it is 'overpowering', 'preys on my mind', you cannot 'take it or leave it' and 'can't get it out of my head'. Those scoring heavily on this factor liked the taste and mouth-feel of chocolate. The second type of questions on the first dimension reflected a weakness for chocolate when under emotional stress; it was eaten 'when I am bored', 'to cheer me up', 'when I am upset' and 'when I am down'. The coupling of these two groups of questions demonstrated a link between negative mood and an intense desire to consume chocolate. A colloquial way of describing those who eat chocolate for these reasons is that they are indulging in 'comfort eating'.

This association between negative mood and chocolate craving was found by Benton *et al.* (7) in a sample chosen to be representative of the population rather than having a history of psychiatric complaints. In contrast, others have described a group who reported 'self-medicating' with chocolate who were more likely to have personality traits associated with hysteroid dysphoria (8), a syndrome characterized by episodes of depression in response to feeling rejected. The experience of strong food cravings has been associated with being bored, anxious and having a dysphoric mood (6). Similarly, a desire for chocolate has been reported to be associated with depression, although not related to suicidal thoughts (9). Thus, there is considerable evidence that chocolate craving is associated with depression and other disturbances of mood, although it should be recognized that this is not causative; rather, it is seen by those affected as more 'curative'.

The second dimension found by Benton *et al.* (7) was labelled *guilt* and again it included two types of question. First, there were comments associating chocolate with negative experiences. I feel 'unattractive', 'sick', 'guilty', 'depressed', 'unhealthy' after eating chocolate. It is not surprising that after eating chocolate: 'I often wish I hadn't'. The second type of question related to weight and body image: 'I often diet', 'I look at the calorific value of a chocolate snack', 'if I ate less chocolate I think I would have a better figure'.

Benton *et al.* (7) asked subjects to press the space bar on a computer to earn chocolate buttons. The number of presses required to earn a chocolate button increased after each reinforcement according to a fixed ratio, 2, 4, 8, 16, 32, 64 and so on. The measure was the number of presses made to obtain chocolate. Those with a higher craving score were prepared to press the space bar more frequently to receive more chocolate buttons. When mild depression was induced, by playing miserable music, they pressed the space bar more often to receive more chocolate (10).

A third factor reflected a pragmatic approach to chocolate. It is eaten when it serves some useful purpose; 'to keep my energy levels up when doing physical exercise', 'in the winter when it is colder', 'only when I am hungry' and 'as a reward when everything is going well'. Unlike the guilt and craving factors, this third factor was not associated with mood.

In summary, a consistent picture has emerged. Chocolate craving is by far the most common food item that is craved, and this is particularly true in females. There is considerable evidence that chocolate craving and poor mood are related. The question that arises is the origin of this relationship. To what extent do the mood-enhancing properties of chocolate reflect biological or psychological mechanisms?

Carbohydrate intake and serotonin synthesis

It has been suggested that as the eating of chocolate is associated with a high carbohydrate intake, it may result in an increased level of brain serotonin. The behavioural consequences of increased serotonergic activity include changes in aggressiveness, mood and pain sensitivity. More specifically, it has been hypothesized that a high-carbohydrate, as opposed to a high-protein, meal will decrease alertness and be associated with a general decline in cognitive efficiency (11). There are, however, many problems with the idea that the high sugar content of chocolate leads to enhanced serotonin synthesis.

Wurtman and Wurtman (12) developed the hypothesis that carbohydrates can relieve depression, a speculation that depended on three disorders: carbohydrate craving obesity, the premenstrual syndrome (PMS, which is discussed below) and seasonal affective disorders. Given the widespread evidence that associates a deficiency in serotonergic functioning with depression, the Wurtmans proposed that an increase in carbohydrate intake reflected an attempt at self-medication, that carbohydrate intake enhanced serotonin synthesis. It was proposed that in normal individuals, increases in brain serotonin affect their food preferences and they no longer like carbohydrate to the same extent and eat other macro-nutrients. The Wurtmans proposed that some obese people suffer from a disturbance of this feedback mechanism, so after eating carbohydrate the mechanism that stops further carbohydrate intake fails to work.

When those suffering with carbohydrate craving obesity were offered snacks that differed in macro-nutrient composition, towards the late afternoon they snacked almost entirely on high-carbohydrate foods, although they did not eat more at mealtimes (12). When asked why they snacked the response was more likely to be that it made them calm or clear-headed rather than that they were hungry. Those who were carbohydrate cravers felt less depressed and more alert. Those who did not crave carbohydrate felt sleepy and fatigued.

Those suffering with seasonal affective disorder (SAD) typically eat more in the winter and put on weight. From the Wurtmans' perspective, this is an attempt to decrease depressive symptoms by eating carbohydrate-rich foods. It has been reported that the eating of carbohydrate-rich, protein-poor meals was associated with improved mood in those suffering with SAD (13). In this type of study, the prediction was that the consumption of a high-carbohydrate/low-protein meal should have a differential impact on those who were, and were not, depressed. However, as is discussed later, there is evidence that carbohydrate does not selectively influence the depressed. The extent to which such a response reflects carbohydrate intake selectively is unclear. Often a high intake of carbohydrate is taken as part of a high-fat or high-protein meal.

De Castro (14) asked normal subjects to keep dietary diaries from which he calculated the proportion of calories that came in the form of carbohydrates. Over a period of 9 days, he found a significant negative relationship between the proportion of energy consumed as carbohydrate and self-reported depression. A higher intake of carbohydrate was also associated with feeling more energetic. Depression was positively correlated with the proportion of meals consumed as protein. An interesting aspect of these data is that the associations were not apparent when particular meals were related to mood around the time of eating. When experimental diets containing low, medium or high levels of carbohydrate were eaten by students for a week, the consumption of the low-carbohydrate diet was associated with increased anger, depression and tension (15). These findings confirmed the previous report that the feeding of a low-carbohydrate/high protein breakfast for 3 weeks resulted in increased levels of anger (16).

The consistency of the finding that the consumption of a diet high in carbohydrate is associated with better mood suggested that it was a robust phenomenon. Such a view does, however, conflict with the theory of the Wurtmans (17). The results were obtained using non-depressed individuals while, according to the Wurtmans, they should only be observed in the depressed.

The Wurtman hypothesis linked carbohydrate and protein intake to brain serotonin levels (12). Based on the study of rats, a sequence of events was proposed. After a meal, the increase of blood glucose stimulates the release of insulin from the pancreas. Insulin in turn causes the uptake of most amino acids, but not tryptophan, by peripheral tissues such as muscle. In contrast, tryptophan is bound to blood albumin, and insulin increases the affinity of albumin for tryptophan. The result of this sequence of events is that the ratio of tryptophan to the other amino acids in the blood increases. It was thus suggested that a high- as opposed to a low-carbohydrate meal increases the ratio of tryptophan to 'large neutral amino acids' (tyrosine, phenylalanine, leucine, isoleucine and valine) in the blood.

Tryptophan and the other large neutral amino acids compete with each other for a transporter molecule that allows entry into the brain. Thus, when a high-carbohydrate meal increases the ratio of tryptophan to other large amino acids, relatively more tryptophan is transported into the brain. Tryptophan is the precursor of the neurotransmitter serotonin, into which it is transformed by the enzyme tryptophan hydroxylase. Normally tryptophan hydroxylase is not fully saturated, and any increased transportation of tryptophan into the brain results in increased serotonin synthesis (17).

A test of the hypothesis that meals differing in protein and carbohydrate content will have different physiological consequences is offered by the measurement of the ratio of tryptophan to long-chain neutral amino acids (LNAA) in plasma. A series of studies have found, as predicted, differences in this ratio after high-carbohydrate as opposed to high-protein meals (18–21). Similar effects have been observed in the obese (22) and those suffering with SAD (23). Although superficially supportive of the Wurtmans' theory, the practical significance of these findings is unclear.

Table 17.1 The influence of different amounts of protein and carbohydrate on the ratio between tryptophan and LNAAs.

% of calories as protein	Number of subjects	Tryptophan/LNAA (% of baseline)	Studies producing significant results
<2	140	123	10/14
4	10	109	0/1
5–10	34	101	0/4
12–19	18	94	0/2
20–49	24	75	2/3
>50	67	67	4/6

Table 17.1 summarizes the results of 30 human studies that have looked at the influence of meals that differed in the percentage of calories coming from protein rather than carbohydrate. There is clear support for the Wurtmans' theory that the ratio of carbohydrate to protein in a meal influences the ratio between tryptophan and LNAAs. However, these data do not support more than the first step of the theory. Clearly, when protein offers less than 2% of the calories, the

make-up of amino acids in plasma markedly favours tryptophan. However, as little as 5% of the calories in the form of protein are enough to ensure that this does not happen. Table 17.1 summarizes the data examining the influence of the percentage of calories in a meal as protein or carbohydrate, on the ratio of tryptophan to LNAAs in the blood. The data are reported as a percentage of baseline values. Only when protein is less than 2% of the calories is the ratio of tryptophan to LNAAs significantly increased. It also shows that a high-protein meal markedly changes this ratio to favour the LNAAs. It appears that it is easier to decrease the availability of tryptophan to the blood–brain transport molecule, by consuming a large amount of protein, than it is to increase it by consuming a large amount of carbohydrate.

The data in Table 17.1 cause serious problems for the Wurtmans' hypothesis. An increased availability of tryptophan is only going to occur when protein offers less than 5% of the calories, perhaps only when it is under 2%. Wurtman and Wurtman used their hypothesis to explain the choice of food rich in carbohydrate by those wishing to enhance their mood. However, it is difficult to find meals that contain so little protein that the uptake of tryptophan is likely to be increased. In potatoes, 10% of the calories come in the form of protein; in bread, it is 15%; and in milk chocolate, it is 13%. Looking at Table 17.1, it appears that no increase in the availability of tryptophan can be expected with bread, potato or chocolate. If anything, less tryptophan will be available to enter the brain and the synthesis of serotonin would decrease rather than increase.

A second problem is that the time-scale is wrong. Any increase in the level of brain tryptophan will occur only after insulin has been released in response to the rise in blood glucose. As much of the digestion of protein takes place in the intestine, the release of amino acids into the blood stream will not be immediate. The response to mood after eating chocolate takes place in a few minutes rather than after an hour or longer. In summary, there is no reason to suggest that the attraction of chocolate results from an increased availability of tryptophan in the blood.

Endorphins

The endorphins are a family of peptides that are released by the brain and act at the same site as morphine. The term *endorphin* is a contraction of the term *endogenous opiates*. There is increasing evidence that the response to high-fat and sweet foods is endorphin mediated. In animals, the preference for a sweet taste, and the intake of sweet solutions, is increased by an opiate agonist and decreased by an opiate antagonist such as naloxone or naltrexone (23). In rats, the consumption of chocolate has been associated with an increased release of β-endorphin (24). The palatability of the food appears to be important, as in the rat, naloxone decreased the consumption of chocolate chip cookies more than the intake of standard rat food (25). The rate that rats will press a lever to stimulate electrodes placed in the lateral hypothalamus – one of the brain's pleasure centres – is enhanced by food deprivation (26).

As this food deprivation-induced increase in self-stimulation was blocked by naloxone, it was proposed that endogenous opiate activity promoted eating by enhancing the reward value of the food. The amount of β-endorphin occupying receptors in the rat hypothalamus has been reported to increase when chocolate milk and candy are eaten (24). Opioid antagonists are more effective in reducing food intake in obese animals than in those of a normal weight (27). Thus, there is increasing evidence that in rodents, endogenous opiates regulate food intake by modulating the extent to which pleasure is induced by palatable foods.

In humans, opioid antagonists decreased thinking about food, feelings of hunger (28, 29) and food intake (30). Naltrexone reduced the preference for sucrose (31). Spontaneous eating has been associated with an increased release of β-endorphin (32). There have been limited attempts to relate the effects of nutrients on the level of plasma β-endorphins. The eating of meat soup resulted in a rapid rise in β-endorphin in 15 min, that then fell and rose again after 60 min (33). In a similar study, 100 g of glucose resulted in a rise in β-endorphin after 150–180 min (34).

The increase in β-endorphin may reflect either a pleasant sensory experience or alternatively a reaction to the metabolic consequences of eating. In rats, the anticipation of eating palatable food is associated with β-endorphin release (24), suggesting that psychological factors play a role. When the reactions to two chocolate drinks were compared in humans (35), plasma endorphins were more elevated with drinks sweetened with aspartame rather than sucrose. The authors suggested that β-endorphin release is more associated with glucose homeostasis than a pleasant taste.

Mandenoff *et al.* (36) proposed that with a monotonous diet, in a predictable environment, the endogenous opiate system is not necessary for the control of eating. However, with stress, fasting and the consumption of highly palatable foods it plays a role. In the rat, a stressor, such as pinching the tail, will induce a naloxone-reversible increase in eating (37). Mandenoff *et al.* (36) suggested that if a stress-induced release of endorphin is not enough to protect the animal, it is adaptive to eat and increase blood glucose levels. In this way further endorphin release can be stimulated. The authors pointed to the parallels between the stress-induced increase in rodent eating and the stressed humans who snack on palatable foods. As discussed above, there is a close association between negative mood and chocolate craving. The possibility that the binge eating of palatable foods may be modulated by the endogenous opiate systems has also been considered. Abnormally high levels of β-endorphin have been reported in both bulimic and obese women (38).

The suggestion that opiate mechanisms selectively influence the pleasure associated with palatable food was supported by a study in which healthy human males were given nalmefene, a long-lasting opioid antagonist (39). Treatment with nalmefene decreased caloric intake by 22%, although the subjective ratings of hunger did not alter. The consumption of fat and protein, but not carbohydrate, decreased. Nalmefene did not influence the intake of particular macronutrients, but rather it influenced the intake of palatable foods, for example high-fat cheese such as Brie. The choice was between various savoury food items; chocolate and sweet foods were not on offer. In a similar study, naloxone differentially decreased the intake of high-fat/high-sugar foods (40).

Drewnowski *et al.* (41) examined the hypothesis that the influence of opiate antagonists on taste preferences and food consumption would be greater amongst binge eaters. Bingeing is typically associated with food cravings. They compared patients with a diagnosis of bulimia nervosa with a group without this history. Snack foods were presented and divided into four categories depending on whether they contained high or low levels of sugar or fat. The high-sugar/high-fat category contained chocolate bars and chocolate-containing cookies and candies. The infusion of naloxone, rather than saline, significantly reduced the total energy intake of binge eaters. The reduction in intake was most marked for the high-sugar/high-fat foods such as those containing chocolate. The obvious explanation is that such foods are highly palatable and that the pleasure associated with eating such foods is mediated via opioid mechanisms. Thus drugs such as naloxone, that block the action of endogenous opioids, reduce the pleasantness of high-sugar/high-fat foods such as chocolate.

In summary, a major theory is that the eating of palatable foods is associated with the release of endorphins. The blocking of the action of endorphins, with drugs such as naloxone or naltrexone, decreases the intake of chocolate.

Phenylethylamine

Phenylethylamine is found in chocolate at levels that vary from 0.4–6.6 μg/g of chocolate (42). Thus, a 50 g bar at the most contains about a third of a milligram. Naturally phenylethylamine is present in low concentrations in the brain; it has distinct binding sites, although it acts as a neuromodulator rather than as a neurotransmitter. If injected into the brain of animals it causes stereotyped behaviour, acting in a similar way to amphetamine. As phenylethylamine does not bind directly to dopamine sites, but its effects can be blocked by dopamine antagonists, it is assumed to release dopamine (43). At physiological doses, it potentiates the neurotransmitters dopamine and norepinephrine (44). Monoamine oxidase B preferentially oxidizes phenylethylamine, although this enzyme also oxidizes dopamine. Phenylethylamine is present in low concentrations in the brain (<10 ng/g) and has a rapid turnover (half-life is 5–10 min).

Two New York psychoanalysts associated passionate love with chocolate. They found that a group of 'love-addicted' women whom they were treating produced a large amount of phenylethylamine. When the women's infatuation stopped, so did the production of phenylethylamine (45). The implication sometimes drawn in popular writing is that chocolate is a substitute for love.

Given the biological role played by phenylethylamine, it is obvious to suggest that it may be responsible for the attraction of chocolate. The drug will increase the rate of lever pressing to receive stimulation of 'pleasure centres' in the brain (46), something true of all drugs of abuse. All drugs to which humans become addicted influence the activity of brain dopamine. It is likely that if chocolate supplied sufficient phenylethylamine then addiction would occur. The question is whether this is a likely scenario. Is the level of the compound in chocolate sufficient to cause addiction and account for craving?

Although the level of phenylethylamine in chocolate is high compared with most food (47), the levels are exceeded in some cheeses and sausage, foods that are rarely craved. It is instructive to consider the dose that influences behaviour. The behaviour of rats, trained to press a lever to obtain electrical stimulation of the hypothalamus, was influenced by doses of 25 and 50 mg/kg phenylethylamine (48). Barr *et al.* (49) reported that doses of 16 and 32 mg/kg inhibited mouse killing by rats. Goudie and Buckland (50) reported that 20–60 mg/kg influenced food-rewarded responding. These studies used doses typical of the animal literature. If the doses of phenylethylamine that were effective in rats were administered to humans at the same level, they would be taking 2 or 3 g. In fact, there is a report that a dose of 2–6 g/day enhanced the mood of depressed patients (51). Clearly, the most extensive chocolate binge could not offer anything approaching this dose. The rapid rate at which phenylethylamine is broken down by monoamine oxidase makes it largely ineffective in animals, unless they are treated with a drug that inhibits monoamine oxidase. Sabelli and Javaid (51) reviewed the role played by phenylethylamine in the modulation of effect. Phenylacetic acid, the major metabolite of phenylethylamine, is decreased in those suffering with depression. The administration of its precursor L-phenylalanine, or phenylethylamine itself, improved the mood of some depressed patients treated with a selective monoamine oxidase B inhibitor. Although phenylethylamine can influence mood, the levels in chocolate are clearly far too low to influence central nervous system activity.

Methylxanthines (see also Chapter 10)

Whether cocoa-containing products contain methylxanthines in doses sufficient to influence psychological functioning has been little considered. Theobromine (3,7-dimethylxanthine) is a naturally occurring alkaloid found in cocoa products. Although the stimulant action of caffeine (1,3,7-trimethylxanthine) is well established, theobromine has been rarely considered. The level of methylxanthines in chocolate differs even within a brand. A 40–50 g Hershey chocolate bar supplies 86–240 mg of theobromine and 9–31 mg of caffeine (52). Shively and Tarka (53) reported that milk chocolate products average about 2 mg/g theobromine and 0.2 mg/g caffeine (about 88 mg of theobromine and 9 mg of caffeine per 44 g bar).

The question that arises with caffeine is whether the small dose in a chocolate bar is high enough to influence neural functioning. The level in chocolate should be compared with brewed coffee that has 85 mg of caffeine/150 ml and tea with 50 mg/150 ml. With theobromine, we need in addition to ask whether there is evidence that it influences psychological functioning at all.

Caffeine

The majority of the literature concerning the impact of caffeine on psychological measures has used doses of at least 200 mg. At this dose, there is clear evidence of stimulant activity. For example, there is consistent evidence that doses of 200 mg of caffeine improve reaction times, although the findings with lower doses are less consistent (54). There is, however, a report that a dose of 32 mg of caffeine improved reaction times (55), although Kuznicki and Turner (56) were unable to find a significant finding with 20 mg.

Similarly, doses over 100 mg improve the ability to sustain attention in vigilance tasks, a finding also reported with doses of 32 and 64 mg of caffeine (57). However, there is a consistent finding that caffeine, at any dose, does not influence learning and memory (58–60).

Mood is the most commonly examined psychological parameter. Studies consistently report that 100 mg of caffeine increases measures of alertness and vigour, and decreases boredom and fatigue (61–64). There have been very few studies of low doses. Lieberman *et al.* (55) found that 64 mg increased feelings of alertness and vigour. The finding that low doses of caffeine improve mood is supported by the reports that low doses of caffeine are reinforcing. When, under double-blind conditions, subjects are given the choice of two coffees, one without caffeine and the other with varying doses, there was evidence that some individuals chose a dose of 25 mg but were unable to distinguish 12.5 mg (65).

Thus, the evidence is that a reliable psychological response to caffeine is most readily observed with doses in excess of 100 mg. Although there have been a few reports that lower doses are active, in no case has a response to a dose of 9 mg of caffeine been reported. It is therefore unlikely that the eating of a typical chocolate bar would offer a dose of caffeine sufficient to influence functioning. The possibility cannot be excluded that a chocolate product offering caffeine towards the top of the dose range (31 mg) might influence weakly some psychological measures. Alternatively, bingeing could result in an intake sufficient to offer a pharmacologically active dose. Such possibilities would not, however, account for the widespread reporting of craving throughout the population.

Theobromine

The response to theobromine, as measured by behavioural measures, is less than that to caffeine. In fact, based on studies with animals, some have concluded that it is behaviourally inert (66–68). However, others have reported a modest impact on the behaviour of mice (69), rats (70) and cats (71), although earlier reviews found no evidence that theobromine has a behavioural influence in humans (72, 73).

Studies of drug discrimination are amongst the most sensitive methods for establishing subtle drug effects. In rats trained to discriminate 32 mg/kg caffeine from saline, theobromine at doses up to 75 mg/kg did not cause a caffeine-like response (74). In a second study, rats were trained to discriminate either 10 or 30 mg/kg caffeine from saline (75). Doses of theobromine, up to 300 mg/kg, produced at best a 50% caffeine-like response. There is little reason to suggest that theobromine accounts for craving.

Other possible biological mechanisms

Chocolate is a chemically complex substance, rich with many pharmacologically active compounds including histamine, tryptophan, serotonin, phenylethylamine and octopamine. As these are all found in higher levels in other food items, lacking the appeal of chocolate, it is improbable that they play a major part in chocolate craving. Chocolate is a major source of certain minerals including copper, magnesium and iron. A 50 g bar of plain chocolate offers 1.2 mg of iron and milk chocolate 0.8 mg. These levels compare with the US recommended daily amount of 15 mg/day for an adult female and 10 mg/day for an adult male. In a sample of young British adults, Fordy and Benton (76) found that 52% of females and 11% of males had levels of ferritin, the storage protein for iron, below the recommended level. Given the widespread instance of iron-deficiency anaemia, both in industrialized and developing countries, any source of iron is likely to be valuable. Although there is no reason to believe that an enhanced intake of iron will improve your mood if you have sufficient of the mineral, without doubt iron-deficiency anaemia is associated with feelings of lethargy and lowered mood. As one iron-containing food, amongst others, chocolate could offer a valuable source of the mineral.

Schifano and Magni (77) described a series of case studies where taking the drug ecstasy (MDMA) was associated with a craving for chocolate. These individuals reported that bingeing on chocolate could result in the consumption of 1000–2000 calories per episode. They speculated that phenylethylamine, with its amphetamine-like properties, and theobromine could be useful in counteracting MDMA withdrawal. It should be remembered that as ecstasy is known to damage serotonergic mechanisms, we are likely to be dealing with a pathological population. In addition, the above discussion concerning the low doses of phenylethylamine and theobromine obtained from eating chocolate makes it extremely unlikely that this is the correct explanation.

The significance of the ecstasy finding for the rest of the population is unclear. Although some have suggested that the finding with ecstasy users gives credibility to the suggestion that chocolate craving reflects abnormal neural functioning, it should be remembered that these individuals could well be suffering from brain damage. In contrast, in a sample of students who reported taking ecstasy at least 20 times, there was no greater report of chocolate craving than in those who had not taken the drug (M. Morgan, personal communication). This sample needs to be compared with the patients of Schifano and Magni (77) who reported taking ecstasy as many as 2000 times, although some on as few as 20 occasions. Ecstasy-induced chocolate craving appears to be unusual, or alternatively only results when ecstasy has been taken on numerous occasions.

Anandamide is a brain lipid that binds with high affinity to cannabinoid receptors and therefore has an influence similar to the active ingredients of cannabis (78). As anandamide is released from neurones, and is rapidly broken down by enzymic activity, it may be an endogenous cannabinoid neurotransmitter or neuromodulator. It is interesting that anandamide has been found as a constituent of chocolate (79); it contains three unsaturated *N*-acylethanolamines that could act either directly at cannabinoid sites, or indirectly by increasing the levels of anandamide. No active ingredients were found in white chocolate. Some have speculated that the endogenous cannabinoid system influenced the subjective feelings associated with eating chocolate and with chocolate craving. Perhaps elevated levels of anandamide intensify the sensory properties of chocolate. These findings must be treated as very preliminary as they are based on *in vitro* studies. It remains to be shown that the levels of these cannabinoids in chocolate are high enough to produce similar actions *in vivo*. Even if the anandamide in chocolate is present in high levels, it remains to be shown that it can survive digestion and absorption. To be active it needs to cross the blood–brain barrier and arrive in sufficient concentrations to influence neuronal activity in those areas of the brain with cannabinoid receptors.

Flavonoids are potent antioxidants whose consumption has been inversely related to coronary heart disease. Waterhouse *et al.* (80) assayed the level of flavonoid polyphenols in chocolate and suggested that they can contribute a significant proportion of dietary antioxidants. Although there have been no studies of their potential behavioural influences, such a response is not inconceivable following the chronic consumption of chocolate. Since a major theory of human ageing is that it reflects the activity of free radicals, the possible benefit of antioxidants in the diet is of current interest. If future research supports the view that antioxidants are beneficial, any food offering antioxidant will counter age-related changes. Any beneficial effect would, however, be expected following chronic rather than acute consumption and such a mechanism could not account for craving.

Premenstrual syndrome and chocolate craving

It has been reported repeatedly that the incidence of chocolate craving increases in the premenstrual stage. As early as the 1950s, surveys began to find that the premenstrual stage was associated with cravings for sweet items, particularly chocolate, although there was also a general increase in appetite (81). As one example, a survey found that 58% of women reported that appetite increased and 61% reported an increased desire for sweet foods in the premenstrual stage (81). The consistency of these findings has led to the view that food craving is a symptom of PMS.

There is a large body of evidence that energy intake increases significantly during the luteal phase of the cycle. Vlitos and Davies (81), when they reviewed the topic, listed 13 studies that reported this finding. The mean difference in energy intake between these phases ranged from 87–674 kcal/day, that is a 4–35% rise from the follicular phase of the cycle. There is substantial evidence that the basal metabolic rate changes over the menstrual cycle in both animals and humans (82). Webb (83) found that 80% of women showed a rise of 8–16% in basal metabolic rate between the follicular and luteal phase of the cycle

Thus, there is a parallel between increased premenstrual appetite, energy intake and increased metabolic rate. The question arises as to whether the anecdotal observation that there is an increased craving for sugary items results in changes in the type of macro-nutrient consumed? Vlitos and Davies (81) found that the majority of studies did not support such a view.

If there is no clear evidence that the intake of carbohydrate increases specifically during the premenstrual stage, is there support for the anecdotal reports that there is an increased craving for sweet foods? Sucrose intake has been found to be higher in the luteal phase (84). The intake of dietary fibre is lower (85), a reflection of an increased choice of sweets, cake and chocolate. During menstruation a preference for chocolate foods, rather than similar non-chocolate alternatives, has been found (86). Fong and Kretsch (87) found that carbohydrate intake was higher when bleeding compared with the time around ovulation, a reflection of an increase in sweet consumption, mainly chocolate. There is little evidence that carbohydrate is the subject of the craving, rather than pleasant-tasting, high-fat/high-carbohydrate foods.

In summary, there is no consistent evidence that the intake of carbohydrate increases in the premenstrual stage. It is possible that the common view that carbohydrate craving occurs is a reflection of a generally increased appetite, rather than a specific increase in carbohydrate intake. There is, however, some evidence that the premenstrual and menstrual periods are associated with a higher sugar intake and a corresponding decrease in the intake of fibre. In particular, the consumption of chocolate increases, although not exclusively.

A possibility that has not been systematically considered is that the attraction of chocolate in the premenstrual stage reflects an attempt to increase the intake of magnesium. It has been found that the level of plasma (88) and erythrocyte (89, 90) magnesium is lower in those suffering with PMS.

In a double-blind trial, women who suffered with premenstrual problems took either magnesium or a placebo (91). In the second, but not the first, cycle the taking of magnesium was associated with fewer symptoms. Given such evidence, some have suggested that the craving for chocolate is an attempt to self-medicate. Although chocolate is a good source of magnesium, the explanation is not convincing. A 50 g bar of milk chocolate supplies 26 mg of magnesium whereas plain chocolate would offer 50 mg. Those on the trial (91), in contrast, took 360 mg of magnesium, three times a day, for the second half of the menstrual cycle. Even then, the improvement did not occur until the procedure had been followed for two cycles. The much smaller amount of magnesium offered by chocolate, and the time-scale involved, make it improbable that the eating of chocolate reduces craving by supplying magnesium.

It might be argued that chocolate can contribute to the US recommended daily allowance of 280 mg of magnesium for an adult female; indeed, if eaten regularly chocolate may help to prevent a deficiency. However, the evidence of Michener and Rozin (92) was that a capsule of cocoa powder that supplied the same amount of magnesium as milk chocolate lacked the ability to satisfy chocolate craving (see below). There is no evidence that an acute intake of magnesium relieves chocolate craving.

A psychological or physiological reaction?

Only one study has attempted to compare the relative contributions of the psychological and physiological mechanisms that underlie chocolate craving (92). The approach taken was to see which of the various constituents of chocolate satisfy craving. Cocoa butter is the fat that, when removed from chocolate liquor, leaves cocoa powder. The known pharmacological ingredients are all in the cocoa powder. Therefore, if you eat white chocolate – made from the cocoa butter – you have the fat and sugar intake of chocolate but not the pharmacological constituents. If you consume cocoa powder you take the pharmacological ingredients but not the fat and sugar.

Michener and Rozin (92) studied subjects who reported cravings for chocolate at least once a week. They were given boxes containing one of six treatments:

(1) A chocolate bar
(2) Capsules containing the same amount of cocoa powder as in the chocolate bar

(3) Placebo capsules containing flour that offered the same calories as the cocoa capsules
(4) A bar of white chocolate
(5) White chocolate plus capsules containing cocoa powder
(6) Nothing.

When the subjects experienced chocolate craving they opened the box, consumed what was offered and rated the extent to which the craving for chocolate was satisfied.

Table 17.2 lists various theories of the origin of chocolate craving tested by Michener and Rozin (92). When subjects experienced chocolate craving they consumed one of the items in the left-hand column. The ability to satisfy the craving is presented in the right-hand column: ++, full effect; +, partial effect; 0, no effect. The predictions of the various theories of the origin of chocolate craving are outlined in columns two to four. If it is the sensory experience that is important then chocolate itself, and to a lesser extent white chocolate, should be satisfying. If it is the increase in blood glucose that is important then brown and white chocolate should help cravings. If the pharmacological ingredients or magnesium are important, then both cocoa powder and chocolate should satisfy cravings.

Table 17.2 Theories of chocolate craving – the ability of different constituents of chocolate to satisfy craving.

Item	Sensory	Calories	Magnesium	Drug	Result
Milk chocolate	++	++	++	++	++
White chocolate	+	++	0	0	+
Cocoa capsules	0	0	++	++	0
Placebo capsules	0	0	0	0	0
White chocolate plus cocoa	+	++	++	++	+
Nothing	0	0	0	0	0

In this study, only chocolate itself, and to a lesser extent white chocolate, had the ability to satisfy chocolate craving. Capsules containing the possible pharmacological ingredients had a similar effect to taking nothing. The addition of cocoa capsules to white chocolate did not increase the less than optimal effect of white chocolate. The obvious conclusion was that it was the sensory experience associated with eating chocolate, rather than pharmacological constituents, that was important.

Discussion

This review has produced little, if any, evidence to support the suggestion that chocolate craving reflects a biological need that is satisfied, in a drug-like manner, by some constituent of chocolate. The evidence is that theobromine, caffeine, phenylethylamine and magnesium are likely to be provided in insufficient amounts to have a psychotropic influence. Similarly, the level of protein in chocolate is sufficient to ensure that there is no increase in the levels of blood tryptophan. In fact, the proportion of tryptophan in the blood is likely to fall. It follows that there is no reason to expect that the synthesis of serotonin in the brain will increase after eating chocolate (Table 17.1).

What does seem to be important is that chocolate tastes good (Table 17.2). Animals, including humans, prefer foods that are both sweet and high in fats. Drewnowski and Greenwood (93) considered the palatability of combinations of fat and sugar and found the optimal combination was 7.6% sugar with cream containing 24.7% fat. The fat content of chocolate is close to this hedonic ideal, although the sugar content of chocolate is greater. As the profile was derived from a combination of cream and sugar, an obvious explanation is that more sugar is needed to counteract the bitterness of chocolate. When we eat something that tastes pleasant, endorphin mechanisms in the brain are stimulated.

Opioid antagonists such as naloxone influence the eating of pleasant-tasting food such as chocolate, in both animals (23–27) and humans (28–30, 39, 40). It is widely believed that drugs of addiction influence mechanisms in the brain that have as their normal function the control of some rewarding activity, such as eating or drinking (94). Opioids play an important role in the initiation and maintenance of drug dependence (95); in the limbic forebrain they are associated with drug craving and relapse following withdrawal. Alcohol craving in both animals and humans is reduced after taking naltrexone (95). Heroin (96), alcohol (97) and nicotine (98) addictions are all associated with the perception of a sweet taste as more pleasant.

The context of the present discussion should be remembered. There is relatively little data dealing directly with the topic of chocolate craving. Rather the present review has looked at related data to consider the possibility that various constituents may or may not play a part in craving. To a large extent the absence of studies that have directly used chocolate must make any conclusions tentative. A parsimonious explanation that might be further considered is that the attractiveness of chocolate reflects its taste and mouth-feel. For many, chocolate offers a near optimally pleasant taste that potently stimulates endorphin release in the brain. There is no convincing evidence that there are substances in chocolate that act directly on the brain in a pharmacological manner.

Summary

- Chocolate is by far the most commonly reported food item that is craved, particularly by females during the premenstrual stage.

- Those who crave chocolate tend to do so when they are emotionally distressed. A second consideration is whether feelings of guilt are associated with the consumption of chocolate.
- There are two major explanations of chocolate craving. First, it results from a pleasant taste. Second, it reflects physiological mechanisms. It has been suggested that these may include increased serotonin production, the release of endorphins, the actions of methylxanthines, phenylethylamine and the supply of magnesium.
- There are repeated suggestions that a high intake of carbohydrate will increase the ratio of tryptophan to LNAAs in the blood. An increased intake of tryptophan by the brain is suggested to increase the synthesis of serotonin and hence improve mood. Although the phenomenon can be demonstrated in the laboratory, a food item that contains any more than a minimal level of protein does not stimulate this mechanism. Chocolate contains protein in levels too high for this mechanism to function. In fact, if this mechanism is active chocolate will decrease rather than increase serotonin synthesis.
- Drugs that block the action of endorphins decrease the intake of palatable foods such as chocolate. The most plausible biological explanation is that chocolate, in a similar manner to other palatable foods, induces the release of brain endorphins.
- Chocolate contains the methylxanthines theobromine and caffeine. However, the levels are so small that it is unlikely that they influence mood.
- Phenylethylamine has powerful pharmacological properties that are similar to amphetamine. However, the active dose is so large, and the rate at which it is broken down so rapid, that the amount supplied by chocolate cannot influence mood.
- Neither the administration of the pharmacological constituents of chocolate nor white chocolate are able to satisfy chocolate craving. It seems, therefore, that the major factor that underlies chocolate craving is the hedonic experience.
- The attraction of chocolate lies in its taste. Sweet, fatty food items have a pleasant taste; chocolate approaches the ideal combination of sweetness and fat content. Pleasant-tasting foods induce the release of endorphins in the brain.

References

1. Hetherington, M.M. and MacDiarmid, J.I. (1993) Chocolate addiction: a preliminary study of its description and its relationship to problem eating. *Appetite* **21**, 233–246.

2. Waterhouse, D. (1995) *Why Women Need Chocolate*. Vermilion, London.
3. Weingarten, H.P. and Elston, D. (1991) Food cravings in a college population. *Appetite* **17**, 167–175.
4. Rozin, P., Leveine, E. and Stoess, C. (1991) Chocolate craving and liking. *Appetite* **17**, 199–212.
5. Rodin, J., Mancuso, J., Granger, J. and Nelbach, E. (1991) Food craving in relation to body mass index, restraint and estradiol levels: a repeated measures study in healthy women. *Appetite* **17**, 177–185.
6. Hill, A.J., Weaver, C.F.L. and Blundell, J.E. (1991) Food craving, dietary restraint and mood. *Appetite* **17**, 187–197.
7. Benton, D., Greenfield, K. and Morgan, M. (1998) The development of the attitudes to chocolate questionnaire. *Per. Ind. Diff.* **24**, 513–520.
8. Schuman, M., Gitlin, M.J. and Fairbanks, L. (1987) Sweets, chocolate and atypical depressive traits. *J. Nerv. Ment. Disord.* **175**, 491–495.
9. Lester, D. and Bernard, D. (1991) Liking for chocolate, depression and suicidal preoccupation. *Psychol. Rep.* **69**, 570.
10. Willner, P., Benton, D., Brown, E., *et al.* (1998) Depression increases craving for sweet rewards in animal and human models of depression and craving. *Psychopharmacology* **136**, 272–283.
11. Spring, B., Chiodo, J. and Bowen, D.J. (1987) Carbohydrates, tryptophan and behavior: a methodological review. *Psychol. Bull.* **102**, 234–256.
12. Wurtman, R.J. and Wurtman, J.J. (1989) Carbohydrates and depression. *Sci. Am.* **260**, 50–57.
13. Rosenthal, N., Genhart, M., Caballero, B., *et al.* (1989) Psychological effects of carbohydrate- and protein-rich meals in patients with seasonal affective disorder. *Biol. Psychiatry* **25**, 1029–1040.
14. de Castro, J.M. (1987) Macro-nutrient relationships with meal patterns and mood in spontaneous feeding behavior of humans. *Physiol. Behav.* **39**, 561–569.
15. Keith, R.E., O'Keefe, K.A., Blessing, D.L. and Wilson, D.G. (1991) Alternations in dietary carbohydrate, protein and fat intake and mood state in trained female cyclists. *Med. Sci. Sports Exerc.* **23**, 212–216.
16. Deijen, J.B., Heemstra, M.L. and Orlebeke, J.F. (1989) Dietary effects on mood and performance. *J. Psychiatr. Res.* **23**, 275–283.
17. Wurtman, R.J., Hefti, F. and Melamed, E. (1981) Precursor control of neurotransmitter synthesis. *Pharmacol. Rev.* **32**, 315–335.
18. Ashley, D.V.M., Liardom, R. and Leathwood, P.D. (1985) Breakfast meal composition influences plasma tryptophan to large neutral amino acids ratios of healthy lean young men. *J. Neural Transm.* **63**, 271–283.
19. Lieberman, H.R., Caballero, B. and Finer, N. (1986) The composition of lunch determines afternoon plasma tryptophan ratios in humans. *J. Neural Transm.* **65**, 211–217.
20. Teff, K.L., Young, S.N. and Blundell, J.E. (1989) The effect of protein or carbohydrate breakfasts on subsequent plasma amino acid levels, satiety and nutrient selection in normal males. *Pharmacol. Biochem. Behav.* **34**, 829–837.
21. Christensen, L. and Redig, C. (1993) Effect of meal composition on mood. *Behav. Neurosci.* **107**, 346–353.

22. Pijl, H., Koppeschaar, H.P.F., Cohen, A.F., *et al.* (1993) Evidence for brain serotonin-mediated control of carbohydrate consumption in normal weight and obese humans. *Int. J. Obesity* **17**, 513–520.
23. Reid, L.D. (1985) Endogenous opioid peptides and regulation of drinking and feeding. *Am. J. Clin. Nutr.* **42**, 1099–1132.
24. Dum, J., Gramsch, C.H. and Herz, A. (1983) Activation of hypothalamic β-endorphin pools by reward induced by highly palatable food. *Pharmacol. Biochem. Behav.* **18**, 443–447.
25. Giraudo, S.Q., Grace, M.K., Welch, C.C., Billington, C.J. and Levine, A.S. (1993) Naloxone's anoretic effect is dependent upon the relative palatability of food. *Pharmacol. Biochem. Behav.* **46**, 917–921.
26. Carr, K.D. and Simon, E.J. (1983) The role of opioids in feeding and reward elicited by lateral hypothalamic electrical stimulation. *Life Sci.* **33** (Suppl. 1), 563–566.
27. Kanarek, R.B. and Marks-Kaufman, R. (1991) Opioid peptides: food intake nutrient selection and food preferences. In *Neuropharmacology of Appetite* (Ed. by Cooper, S.J. and Lieberman, J.M.). Oxford University Press, London.
28. Spiegel, T.A., Stunkard, A.J., Shrager, E.E., O'Brien, C.P. and Morrison, M.F. (1987) Effect of naltrexone on food intake, hunger and satiety in obese men. *Physiol. Behav.* **40**, 135–141.
29. Wolkowitz, O.M., Doran, M.R., Cohen, R.M., Cohen, T.N., Wise, T.N. and Picckar, D. (1988) Single-dose naloxone acutely reduces eating in obese humans: behavioral and biochemical effects. *Biol. Psychiatry* **24**, 483–487.
30. Trenchard, E. and Silverstone, T. (1982) Naloxone reduces the food intake of normal human volunteers. *Appetite* **4**, 249–257.
31. Fantino, M., Hosotte, J. and Apfelbaum, M. (1986) An opioid antagonist naltrexone reduces preference for sucrose in humans. *Am. J. Physiol.* **251**, R91–R96.
32. Davis, J.M., Lowy, M.T., Yim, G.K.W., Lam, D.R. and Malven, P.V. (1983) Relationships between plasma concentrations of immunoreactive beta-endorphin and food intake in rats. *Peptides* **4**, 79–83.
33. Matsumura, M., Fukuda, N., Saito, S. and Mori, H. (1982) Effect of a test meal, duodenal acidification and tetragastrin on the plasma β-endorphin-like immunoreactivity in man. *Regul. Pept.* **4**, 173–181.
34. Getto, C.J., Fullerton, D.T. and Carlson, I.H. (1984) Plasma immunoreactive β-endorphin response to glucose ingestion. *Appetite* **5**, 329–335.
35. Mechinor, J.C., Rigaud, D., Colas-Linhart, N., Petiet, A., Giraard, A. and Apfelbaum, M. (1991) Immunoreactive β-endorphin increases after an aspartame chocolate drink in healthy human subjects. *Physiol. Behav.* **50**, 941–944.
36. Mandenoff, A.F., Fumerton, M., Apfelbaum, M. and Margules, D.L. (1982) Endogenous opiates and energy balance. *Science* **215**, 1536–1537.
37. Koch, J.E. and Bodnar, R.J. (1993) Involvement of mu1 and mu2 opioid receptor subtypes in tail-pinch feeding in rats. *Physiol. Behav.* **53**, 603–605.
38. Waller, D.A., Kiser, R.S., Hardy, B.W., Fuchs, I., Feigenbaum, L.P. and Uauy, R. (1986) Eating behavior and plasma β-endorphin in bulimia. *Am. J. Clin. Nutr.* **44**, 20–23.
39. Yeomans, M.R., Wright, P., Macleod, H.A. and Critchley, J.A.J.H. (1990) Effects of nalmefene on feeding in humans. *Psychopharmacology* **100**, 426–432.

40. Drewnowski, A., Gosnell, B., Krahn, D.D. and Canum K. (1989) Sensory preferences for sugar and fat: evidence for opioid involvement. *Appetite* **12**, 206.
41. Drewnowski, A., Krahn, D.D., Demitrack, M.A., Nairn, K. and Gosnell, B.A. (1995) Naloxone and opiate blocker reduces the consumption of sweet high-fat foods in obese and lean female binge eaters. *Am. J. Clin. Nutr.* **61**, 1206–1212.
42. Hurst, W.J. and Toomey, P.B. (1981) High-performance liquid chromatographic determination of four biogenic amines in chocolate. *Analyst* **106**, 394–402.
43. Webster, R.A. and Jordan, C.C. (1989) *Neurotransmitters Drug and Disease*. Blackwell Science, Oxford.
44. Paterson, I.A., Juorio, A.V. and Boulton, A.A. (1990) 2-Phenylethylamine: a modulator of catecholamine transmission in the mammalian central nervous system? *J. Neurochem.* **55**, 1827–1837.
45. Weil, A. (1990) *Natural Health Natural Medicine*. Houghton Mifflin, Boston.
46. Greenshaw, A.J. (1984) β-Phenylethylamine and reinforcement. *Prog. Neuro-psychopharmacol. Biol. Psychiatry* **8**, 615–620.
47. Hirst, W.J., Martin, R.A., Zoumas, B.L. and Tarka, S.M. (1982) Biogenic amines in chocolate – a review. *Nutr. Rep. Int.* **26**, 1081–1087.
48. Greenshaw, A.J., Sanger, D.J. and Blackman, D.E. (1985) Effects of D-amphetamine and β-phenylethylamine on fixed-interval responding maintained by self-regulated lateral hypothalamus stimulation in rats. *Pharmacol. Biochem. Behav.* **23**, 519–523.
49. Barr, G.A., Gibbons, J.L. and Bridger, W.H. (1979) A comparison of the effects of acute and subacute administration of β-phenylethylamine and D-amphetamine on mouse-killing behaviour of rats. *Pharmacol. Biochem. Behav.* **11**, 419–422.
50. Goudie, A.J. and Buckland, C. (1982) Serotonin receptor blockade potentiates the behavioural effects of β-phenylethylamine. *Neuropharmacology* **21**, 1267–1272.
51. Sabelli, H.C. and Javaid, J.I. (1995) Phenylethylamine modulation of affect: therapeutic and diagnostic implications. *J. Neuropsychiatry Clin. Neurosci.* **7**, 6–14.
52. Mumford, G.K., Evans, S.M., Kaminski, B.J., *et al.* (1994) Discriminative stimulus and subjective effects of theobromine and caffeine in humans. *Psychopharmacology* **115**, 1–8.
53. Shively, C.A. and Tarka, S.M. (1984) Methylxanthine composition and consumption patterns of cocoa and chocolate products. In *The Methylxanthine Beverages and Foods: Chemistry Consumption and Health Effects* (Ed. by Spiller, G.), pp. 149–178. Alan R. Liss, New York.
54. Lieberman, H.R. (1992) Caffeine. In *Handbook of Human Performance*, Vol. 2 (Ed. by Smith, A.P. and Jones, D.M.), pp. 49–72. Academic Press, London.
55. Lieberman, H.R., Wurtman, R.J., Embe, G.G. and Coviella, I.L.G. (1987) The effects of caffeine and aspirin on mood and performance. *J. Clin. Psychopharmacol.* **7**, 315–320.
56. Kuznicki, J.T. and Turner, L.S. (1986) The effects of caffeine on caffeine users and non-users. *Physiol. Behav.* **37**, 397–408.
57. Lieberman, H.R., Wurtman, R.J., Embe, G.G., Roberts, C. and Coviella, I.L. (1987) The effects of low doses of caffeine on human performance and mood. *Psychopharmacology* **92**, 308–312.
58. Battig, J.J., Buzzi, R., Martin, J.R. and Feierabend, J.M. (1984) The effects of caffeine on physiological functions and mental performance. *Experientia* **40**, 1218–1223.

59. Lieberman, H.R. (1988) Beneficial effects of caffeine. In *Twelfth International Scientific Colloquium on Coffee*. ASIC, Paris.
60. Loke, W.H. (1988) Effects of caffeine on mood and memory. *Physiol. Behav.* **44**, 367–372.
61. Goldstein, A., Kaizer, S. and Warren, R. (1965) Psychotropic effects of caffeine in man II. Alertness, psychomotor co-ordination and mood. *J. Pharmacol. Exp. Ther.* **150**, 146–151.
62. Leathwood, P. and Pollit, P. (1982) Diet-induced mood changes in normal populations. *J. Psychiatr. Res.* **17**, 147–154.
63. Roache, J.D. and Griffiths, R.R. (1987) Interactions of diazepam and caffeine: behavioral and subjective dose effects in humans. *Pharmacol. Biochem. Behav.* **26**, 801–812.
64. Fagan, D., Swift, C.G. and Tiplady, B. (1988) Effects of caffeine on vigilance and other performance tests in normal subjects. *J. Psychopharmacol.* **2**, 19–25.
65. Oliveto, A.H., Hughes, J.R., Pepper, S.L., Bickel, W.K. and Higgins, S.T. (1990) Low doses of caffeine can serve as reinforcers in humans. In *Problems of Drug Dependence 1990. US Department of Health and Human Services* (Ed. by Harris, L.). National Institute on Drug Abuse, Rockville, MD.
66. Spugel, W., Mitznegg, P. and Heim, F. (1977) The influence of caffeine and theobromine on locomotor activity and the brain of cGMP/cAMP ratio in white mice. *Biochem. Pharmacol.* **26**, 1723–1724.
67. Snyder, S.H., Katims, J.J., Annau, Z., Bruns, R.F. and Daly, J.W. (1981) Adenosine receptors and behavioural actions of methylxanthines. *Proc. Natl. Acad. Sci. USA* **78**, 3260–3264.
68. Carney, J.M., Cao, W., Logan, L., Rennert, O.M. and Seale, T.W. (1986) Differential antagonism of the behavioural depressant and hypothermic effects of 5′-(N-ethylcarboxamide) adenosine by theobromine. *Pharmacol. Biochem. Behav.* **25**, 769–773.
69. Katims, J.J., Annau, Z. and Snyder, S.H. (1983) Interactions in the behavioural effects of methylxanthines and adenosine derivatives. *J. Pharmacol. Exp. Ther.* **227**, 167–173.
70. Beer, B., Chasin, M., Clody, D.E., Vogel, J.R. and Horovitz, Z.P. (1972) Cyclic adenosine monophosphate phosphodiesterase in brain: effect on anxiety. *Science* **176**, 428–430.
71. Herz, A., Neteler, B. and Teschemacher, H.J. (1968) Vergleichende Untersuchungen über zentrale Wirkungen von Xanthindeerivaen in Hinblick auf deren Stoffweechsel und Verteilung im Organismus. *Naunyn Schmiedebergs Arch. Pharmakol. Exp. Pathol.* **261**, 1123–1132.
72. Tarka, S.M. (1982) The toxicology of cocoa and methylxanthines: a review of the literature. *Crit. Rev. Toxicol.* **9**, 275–312.
73. Stavric, B. (1988) Methylxanthines: toxicity to humans. 3. Theobromines, paraxanthine and the combined effects of methylxanthines. *Food Chem. Toxicol.* **26**, 725–733.
74. Carney, J.M., Holloway, F.A. and Modrow, H.E. (1985) Discriminative stimulus properties of methylxanthines and their metabolites in rats. *Life Sci.* **36**, 913–920.
75. Holtzman, S.G. (1986) Discriminative stimulus properties of caffeine in the rat: noradrenergic mediation. *J. Pharmacol. Exp. Ther.* **239**, 706–714.

76. Fordy, J. and Benton, D. (1994) Does low-iron status influence psychological functioning? *J. Hum. Nutr. Dietet.* **7**, 127–133.
77. Schifano, F. and Magni, G. (1994) MDMA (ecstasy) abuse: psychopathological features and craving for chocolate: a case series. *Biol. Psychiatry* **36**, 763–767.
78. DiMarzo, F.A., Caadas, H., Schinelli, S., Cimino, G., Schwaartz, J.C. and Piomelli, D. (1994) Formation and inactivation of endogenous cannabinoid anandamide in central neurons. *Nature* **372**, 686–691.
79. DiTomaso, E., Beltramo, M. and Piomelli, D. (1996) Brain cannabinoids in chocolate. *Nature* **382**, 677–678.
80. Waterhouse, A.L., Shirley, J.R. and Donovan, J.L. (1996) Antioxidants in chocolate. *Lancet* **348**, 834.
81. Vlitos, A.L.P. and Davies, G.J. (1996) Bowel function, food intake and the menstrual cycle. *Nutr. Res. Rev.* **9**, 111–134.
82. Buffenstein, R., Poppitt, S.D., McDevitt, R.M. and Prentice, A.M. (1995) Food intake and the menstrual cycle: a retrospective analysis with implications for appetite research. *Physiol. Behav.* **58**, 1067–1077.
83. Webb, P. (1986) Twenty-four hour energy expenditure and the menstrual cycle. *Am. J. Clin. Nutr.* **44**, 614–619.
84. Gong, E.J., Garrel, D. and Calloway, D.H. (1989) Menstrual cycle and voluntary food intake. *Am. J. Clin. Nutr.* **49**, 252–258.
85. Davies, G.J., Collins, A.L.P. and Mead, J.J. (1993) Bowel habit and dietary fibre intake before and during menstruation. *J. Royal Soc. Health* **113**, 64–67.
86. Tomelleri, R. and Grunewald, K.K. (1987) Menstrual cycle and food cravings in young college women. *J. Am. Diet. Assoc.* **87**, 311–315.
87. Fong, A.K.H. and Kretsch, M.J. (1993) Changes in dietary intake, urinary nitrogen and urinary volume across the menstrual cycle. *Am. J. Clin. Nutr.* **234**, E243–E247.
88. Posaci, C., Erten, O., Uren, A. and Acar, B. (1994) Plasma copper, zinc and magnesium levels in patients with premenstrual tension syndrome. *Acta Obstet. Gynecol. Scand.* **73**, 452–455.
89. Sherwood, R.A., Rocks, B.F., Stewart, A. and Saxton, R.S. (1986) Magnesium and the premenstrual syndrome. *Ann. Clin. Biochem.* **23**, 667–670.
90. Rosenstein, D.L., Elin, R.J., Hosseini, J.M., Grover, G. and Rubinow, D.R. (1994) Magnesium measures across the menstrual cycle in the premenstrual syndrome. *Biol. Psychiatry* **15**, 557–561.
91. Fachinetti, F., Borella, P., Sances, G., Fioroni, L., Nappi, R.E. and Genazani, A.R. (1991) Oral magnesium successfully relieves premenstrual mood changes. *Obstet. Gynecol.* **78**, 177–181.
92. Michener, W. and Rozin, P. (1994) Pharmacological versus sensory factors in the satiation of chocolate craving. *Physiol. Behav.* **56**, 419–422.
93. Drewnowski, A. and Greenwood, M.R.C. (1983) Cream and sugar: human preferences for high-fat foods. *Physiol. Behav.* **30**, 629–633.
94. Di Chiara, G. and North, R.A. (1992) Neurobiology of opiate abuse. *Trends Pharmacol. Sci.* **13**, 185–193.
95. Van Ree, J.M. (1996) Endorphins and experimental addiction. *Alcohol* **13**, 25–30.
96. Shufman, P.E., Vas, A., Luger, S. and Steiner, J.E. (1997) Taste and odor reactivity in heroin addicts. *Isr. J. Psychiatry Relat. Sci.* **34**, 290–299.

97. Kampov-Polevoy, A., Garbutt, J.C. and Janowsky, D. (1997) Evidence of preference for a high-concentration sucrose solution in alcoholic men. *Am. J. Psychiatry* **154**, 269–270.
98. Kos, J., Hassenfratz, M. and Battig, K. (1997) Effects of a 2-day abstinence from smoking on dietary, cognitive, subjective and physiologic parameters among younger and older female smokers. *Physiol. Behav.* **61**, 671–678.

Miscellaneous Considerations

Chapter 18

Cocoa, Chocolate and Acne

Ian Knight

Cocoa and its more familiar relative chocolate date back to pre-Columbian times in the Americas. Chocolate was known as 'the food of the gods' through mythological stories of the Mexican Indians describing the origins of cacao as being delivered to lessen mortal man's exile on earth (1). One of its first written references comes from a drawing on the frontispiece of a Latin book published in 1639.

Despite these historic origins, and its demonstrably popular taste, or perhaps because of it, chocolate has been blamed for a great many complaints, one of which is acne. Acne is defined as an inflammatory disease of the skin with the formation of an eruption of papules or pustules, more particularly acne vulgaris (2). Acne can vary across a wide range from mild, superficial and transient types through more serious clinical acne all the way to severe acne.

Acne vulgaris is a disease of the sebaceous glands and pilosebaceous ducts (3). The sebaceous glands are activated by androgens, and increased androgen production during puberty is most likely an important factor in the pathogenesis of acne. It leads to an increased sebum excretion and seems to influence the keratinization and bacterial flora of the pilosebaceous duct. Acne lesions are inflammatory reactions mainly in or around this duct. Clearly bacterial action also plays a part in the etiology of acne. Treatments with antibiotics are frequently employed and concern is surfacing in the literature about bacterial resistance in acne (4).

The prevalence of acne is about three out of four teenagers in the USA (5). In Sweden, it is estimated that of those teenagers having acne, approximately 10% have acne of a severe type (3).

There are few, if any, well-controlled studies on the effects of various dietary restrictions in acne (3), and there are unfortunately no suitable animal models for the study of acne. The experimental data listed in the literature generally are either animal studies to determine the effects of diet on free fatty acids and sebum excretion rate, or human studies to investigate the relationship of diet to free fatty acids, sebum excretion rate and clinical acne.

Acne and diet

Several foods such as fats (fatty meat, fried foods, nuts), carbohydrates (all sweets, cola drinks, chocolate) and iodides (seafood, iodized salt) have been commonly implicated as causing acne (6,7). Tomatoes, citrus fruits and spices have also been blamed by some patients.

It has been shown that human sebum composition and its rate of excretion can be changed by diet. Diets high in fats resulted in increased free fatty acids in all subjects and increased sebum excretion rate in the majority (7). Conversely, fasted subjects demonstrated a 40% reduction in sebum excretion rate (7). Similarly Macdonald reported that varying proportions of chocolate, sucrose and skim milk can significantly change surface lipid levels over a minimum 5-day time interval (8, 9). Unfortunately, these studies involve gross dietary maneuvers and experience has shown that in reality, compliance with even simple treatments for short-term disease is low.

Since serum lipid levels do not correspond to serum excretion rate and acne patients generally show no reproducible tendency to elevated cholesterol or triglycerides (7), it seems unlikely that such dietary factors alone could be responsible for acne. Acne patients as a group do not have abnormalities in glucose tolerance (7), nor do they consume abnormal quantities of sweets. Limiting sugar in the diet has not been shown to significantly improve acne. In fact, acne patients treated with intravenous carbohydrates noted an improvement (3). Studies designed to evaluate the effects of iodides on acne have had mixed conclusions.

Two dietary factors that seem to be linked to acne are zinc and retinol-binding protein (3), a close relative of vitamin A. Baer *et al.* (10) suggest that low zinc diets may worsen or aggravate acne and this was confirmed by Michaëlsson (3), who observed these relationships in a number of patients. There are as yet no explanations for these connections, although they were reported as presently under study.

Acne and chocolate

It is a widely held belief by the public that chocolate ingestion by those with acne is contraindicated. A questionnaire was completed by some 1023 students at Boston suburban high schools, of whom 85% reported that they suffered from acne (11). Of these, 365 (41.9%) were restricting their diet to minimize chocolate and fats; this was without medical advice to do so. Of these, 43% and 36.2% reported that their dietary avoidance had respectively improved their acne substantially or slightly. The investigators were unable to ascertain whether the perceived improvement reported by 43% was due to physical or psychological influence. In short, the authors concluded that they had not resolved the controversy, yet it is discouraging that this study is widely reported as indicating a positive relationship between chocolate and acne (12).

Rasmussen (7) criticized the work of Fulton *et al.* (13) as not employing sufficiently specific means to evaluate a shift in severity from equal numbers of comedones to pustules. Nevertheless, the work of Fulton *et al.* is generally recognized as disproving any link between chocolate and acne and is cited positively in a letter to the editor of the *Journal of the American Medical Association* (14). This letter also pointed out that many erroneous assumptions have been made in relation to chocolate and allergy. It also speaks to the discrepancy between patients' perceptions and clinical symptoms. A study is described in which 500 allergic individuals, of whom 33% have been told to avoid chocolate, were tested. Of the 500, 16% thought that chocolate caused allergic symptoms. In fact, only 10 of the 500 manifested specific allergic symptoms within a predictable time after consuming chocolate. Eight of these ten were further tested using the double-blind technique, finding that only three of the eight reproduced any effects with the sample, but not with the placebo. Interestingly, only one of these three tested positively in a skin test.

Anderson (15) conducted a study over 6 years in what is described as 'the unusually favorable situation of the (Missouri) University Health Service'. Student patients with acne, most of whom believed certain foods substantially exacerbated their acne (within 36 hours), were initially questioned about these foods. Of these patients, 8–10% believed chocolate was a cause, and 3–4% blamed nuts, cola or milk. These subjects were provided with and consumed large daily amounts of chocolate, milk, nuts or cola under supervision. The author concluded, 'to the constant amazement of both the patients and medical students, absolutely no major flares of acne were produced by the foods'.

There are two other papers seeking to draw a correlation between chocolate consumption and acne (as well as, in one case, dental caries). The first describes Eskimos leaving their native habitat to thereafter enjoy chocolate consumption and at the same time suffer acne vulgaris (6, 12); the second attempts to connect the availability of dietary fats in certain regions to acne prevalence (12). Both are papers relying upon inductive reasoning without providing evidence and without adding new data to the presumed controversy.

There are many examples of authors eloquently describing their beliefs along with anecdotal support of the connection between chocolate consumption and acne. Hard evaluation of the data and their sources provide little, if any, real correlation. Three previous major reviews on the subject (3, 7, 12) concluded no effects of chocolate on the production of acne. The American Dietetic Association in its consumer publication *Complete Food and Nutrition Guide*, described the view that chocolate causes acne as a myth: 'That misconception has captured the attention of teens for years. However hormonal changes during adolescence are the usual causes of acne, not chocolate' (16).

Summary

The supposed connection between chocolate and acne probably originates in folklore. Conscientious attempts to demonstrate the connection have been unable to do so, but that has not prevented those with strongly held views from expressing them persuasively, sometimes disguised as scientific facts. The actual causes of acne appear not to be understood even today. However, factors such as hormonal changes, both in adolescents and in women during menstruation, and bacterial status can adversely affect acne eruptions. There also appear to be connections to deficiencies of zinc and retinol-binding protein, but these are as yet unexplainable. Certainly, expert bodies such as the American Dietetic Association (16), the American Academy of Dermatology (5) and the Mayo Clinic (17) all state that there is no link between chocolate and acne, as do previous reviewers. To paraphrase Maslansky and Wein (14), 'is it possible that we have been guilty of taking candy away from babies?'

References

1. Cook, L.R. and Muersing, E.H. (1982) *The Apocrypha of Cacao in Chocolate Production and Use*. Harcourt Brace Jovanovich, New York.
2. Friel, J.P. (ed.) (1977) *Dorland's Pocket Medical Dictionary*. W.B. Saunders, Philadelphia.
3. Michaëlsson, G. (1981) Diet and acne. *Nutr. Rev.* **39** (2), 104–106.
4. Eady, E.A. (1998) Bacterial resistance in acne. *Dermatology* **196** (1), 59–66.
5. Anon (1990) Clearing up beliefs about blemishes. *Tufts University Diet Nutr. Lett.* **8** (8), 3–6.
6. Glickman, F.S. and Silvers, S.H. (1972) Dietary factors in acne vulgaris. *Arch. Dermatol.* **106**, 129.
7. Rasmussen, J.E. (1977) Diet and acne. *Int. J. Dermatol.* **16** (6), 488–492.
8. Macdonald, I. (1968) Effects of a skim milk and chocolate diet on serum and skin lipids. *J. Sci. Food Agric.* **19**, 270.
9. Macdonald, I. (1967) Dietary carbohydrates and skin lipids. *Br. J. Dermatol.* **79**, 119.
10. Baer, M.T., King, J.C., Tamura, T. and Margen, S. (1978) Acne in zinc deficiency. *Arch. Dermatol.* **114**, 1093.
11. Emerson, G.W. and Strauss, J.S. (1972) Acne and acne care. *Arch. Dermatol.* **105**, 407–411.
12. Fries, J.H. (1978) Chocolate: a review of published reports of allergic and other deleterious effects, real or presumed. *Ann. Allergy* **41** (4), 195–207.
13. Fulton, J.E., Jr, Plewig, G. and Kligman, A.M. (1969) Effect of chocolate on acne vulgaris. *JAMA* **210** (11), 2071.
14. Maslansky, L. and Wein, G. (1970) Effect of chocolate on acne vulgaris. Letter to the editor. *JAMA* **211** (11), 1856.
15. Anderson, P.C. (1971) Foods as the cause of acne. *Am. Fam. Physician* **3** (3), 102–103.

16. Duyff, R.L. (1996) Melt away myths about chocolate. In *The American Dietetic Association's Complete Food and Nutrition Guide* (Ed. by Braun, J. and Hornick, B.), p. 131. Chronimed Publishing, Minneapolis.
17. Gilliland, S.C. and Palumbo, P.J. (Eds) (1990) Chocolate: the truth behind the temptations. *Mayo Clinic Newsletter* **3** (2), 1–2.

Chapter 19

Liver, Kidney and Gastrointestinal Effects

Ian Knight

With very few exceptions, chocolate and by association cocoa, from which it is made, have arguably been subjected to more criticism than any other food. This condemnation has not been limited to those in the medical community, but has extended to, as one reviewer observed, 'purveyors of folklore'. This review is intended to examine the available scientific literature in an attempt to ascertain the reality behind some of the accusations, specifically in the areas of liver, kidney and gastrointestinal effects. Evidence is scant, yet in some cases the views are surprisingly strongly held.

Liver

It was not possible to find any references in the literature to liver problems associated with cocoa or chocolate administration or consumption. Indeed, what little information exists appears to be confined to lipid and cholesterol levels present in the livers of test animals, which is addressed in Chapters 5 and 6.

Kidney

Again, this is not a subject that appears often in the literature. There are references to incidental observations of renal pelvic mineralization in one generation of a three-generation rat study, designed to evaluate any reproductive effects of methylxanthine consumption via cocoa powder administration (1). This observation is accompanied by a recognition that such exposure over three generations 'caused no remarkable effects on the thymus, spleen, kidney or reproductive organs'. Investigators from the same laboratory, studying toxicity/carcinogenicity potential in rats consuming cocoa powder (2), noted increases in both renal pelvic dilatation and pelvic microcalculi, although found these observations to be lacking in temporal and dose-related qualities. The development of renal pelvic dilatation has been shown to be a polygenic heritable trait (3) and so the investigators concluded this to be a factor. Additionally, dietary protein is also implicated as a causative factor in the development of pelvic dilatation (4), so this observation was not considered persuasive.

Cocoa and chocolate contain intrinsic methylxanthines, predominantly theobromine with small amounts of caffeine. Methylxanthines as a group are well-known diuretics and reportedly contribute to urinary mineral excretion when consumed at high levels (see Chapter 10 (5)). However, no other impact upon the kidney has been reported from the consumption of cocoa or chocolate.

The subject of microcalculi formation and urinary excretion of minerals is covered in more detail in Chapter 9.

Gastrointestinal effects

Some patients have tended to report heartburn symptoms following chocolate consumption and reportedly with some frequency. This led Babka and Castell (6) to further investigate this claim. They demonstrated that immediate and sustained lowering of the pressure of the lower sphincter of the esophagus could be induced experimentally with ingestion of chocolate as well as some other foods.

The esophagus is a tube-like structure connecting the mouth to the stomach (Fig. 19.1). Since the stomach usually has a higher pressure than that of the esophagus, a specialized muscle closes off the esophagus from the stomach at the point of joining. This is called the lower esophageal sphincter (LES). During normal eating, the LES relaxes after swallowing to allow food to pass into the stomach, and then quickly closes again. In most people, this system functions adequately, but in approximately 10% of the US population who suffer daily and 33% who suffer occasionally, this system is inadequate. In these individuals, the LES is either weak, or more commonly, relaxes inappropriately, allowing a backwash of acidic stomach contents into the esophagus, thereby irritating the esophageal lining (7).

The original studies by Babka and Castell (6) measured pressure of the LES in normal subjects following administration of either water (control), whole milk, non-fat milk, orange juice, water/tomato paste mixture and dilutions of chocolate syrup. Whole milk was found to lower LES pressure, to a significant ($P < 0.05$) but lesser extent than the chocolate syrup ($P < 0.005$). Orange juice and the tomato preparation both showed transient decreases in pressure followed by a gradual return to baseline and 'considerable pressure variation and secondary contractions'. It was first thought that the high fat content alone was responsible for this effect; however, a further study (8) was conducted using chocolate syrup

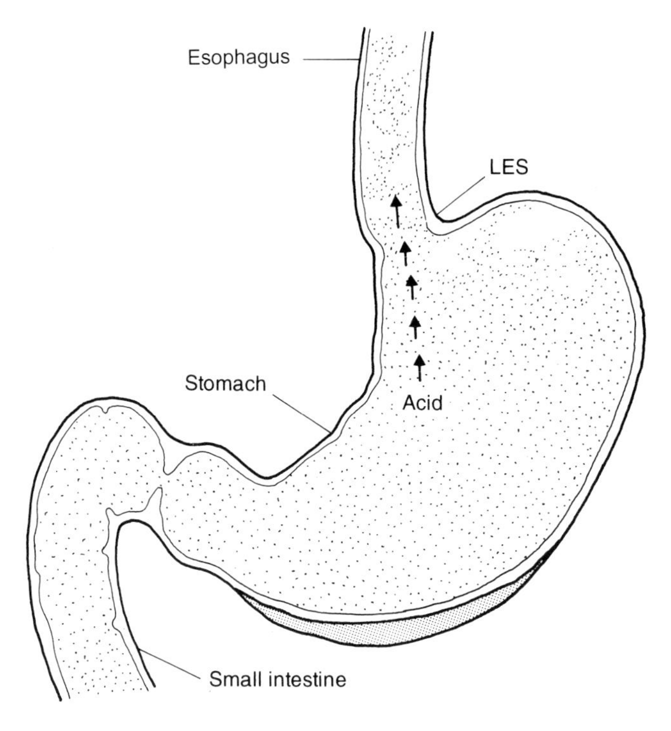

Fig. 19.1 Mechanism of esophageal reflux resulting from weak lower esophageal sphincter (LES) pressure valve, or inappropriate relaxation of the LES.

having a low fat content (1%). From this, it was hypothesized that the methylxanthines may be partially responsible for this effect. Subsequent studies (9, 10) established that the reduced LES pressure did indeed contribute to esophageal reflux, supporting recommendations that patients suffering from reflux esophagitis should avoid chocolate, particularly solid, dark chocolate, as well as caffeine and fats.

Summary

Although cocoa and chocolate are habitually blamed for many complaints, there is little scientific research on which to base such generalizations. There appears to be no work to support liver problems. In the kidney, there exist some scattered observations in animals, relating to renal pelvic dilatation and the formation of microcalculi; however, their causes were by no means defined. The only real effect for which there appears to be solid evidence is that of chocolate on patients who are susceptible to reflux esophagitis, leading to serious heartburn. Such people, when identified, are recommended to avoid chocolate, particularly solid, dark chocolate, but equally smoking, alcohol, caffeine and high-fat products generally.

References

1. Hostetler, K.A., Morrissey, S.M., Tarka, S.M., Jr, Apgar, J.L. and Shively, C.A. (1990) Three-generation reproductive study of cocoa powder in rats. *Food Chem. Toxicol.* **28**, (7), 483–490.
2. Tarka, S.M., Jr, Morrissey, S.M., Apgar, J.L., Hostetler, K.A. and Shively, C.A. (1991) Chronic toxicity/carcinogenicity studies of cocoa powder in rats. *Food Chem. Toxicol.* **29** (1), 7–19.
3. Van Winkle, T.J., Womack, J.E., Barbo, W.D. and Davis, T.W. (1988) Incidence of hydronephrosis among several production colonies of outbred Sprague–Dawley rats. *Lab. Anim. Sci.* **38**, 402–406.
4. Greaves, P. and Faccini, J. (1984) Urinary tract. In *Rat Histopathology*, pp. 145–146. Elsevier, Amsterdam.
5. Tarka, S.M., Jr and Shively, C.A. (1987) Methylxanthines. In *Toxicological Aspects of Food* (Ed. by Miller, K.) Elsevier, London.
6. Babka, J.C. and Castell, D.O. (1973) On the genesis of heartburn: the effects of specific foods on the lower esophageal sphincter. *Am. J. Dig. Dis.* **18** (5), 391–397.
7. Castell, D.O. and Vernalis, M.N. undated. *Heartburn – Fact Sheet.* National Digestive Diseases Education and Information Clearinghouse. US Dept of Health and Human Services, Public Health Services, NIH. Inter America Research Associates, Rosslyn, VA.
8. Castell, D.O. (1975) Diet and the lower esophageal sphincter. *Am. J. Clin. Nutr.* **28**, 1296.
9. Wright, L.E. and Castell, D.O. (1975) The adverse effect of chocolate on lower esophageal sphincter pressure. *Am. J. Dig. Dis.* **20**, 703–707.
10. Murphy, D.W. and Castell, D.O. (1988) Chocolate and heartburn: evidence of increased esophageal acid exposure after chocolate ingestion. *Am. J. Gastroenterol.* **83** (6), 633–636.

Section V

Consumption of Cocoa and Chocolate

Chapter 20

Chocolate Consumption Patterns

Judith S. Douglass and Mary M. Amann

Consumption of cocoa and chocolate has been documented from the time of the ancient Aztecs and Mayas, who worshipped the cacao bean. The first Latin name for the cacao plant, *Amygdalae pecuniariae*, or 'money almond', reflects its use as currency in ancient cultures (1).

In 1519, the Spanish adventurer Hernando Cortés observed the Aztecs using cacao beans to make *xocoatl*, a beverage thought to cure diarrhea and dysentery (2). Montezuma is said to have consumed more than 50 cups per day of the thick chocolate beverage (1).

By the early 17th century, chocolate houses became alternatives to pubs as gathering places for 'gambling, intrigue making, and conspirations' (1). Several European governments tried to charge taxes on chocolate products in efforts to restrict consumption, but chocolate houses flourished in spite of these efforts.

Today, chocolate beverages, chocolate confectionery and other chocolate products are used on every continent. The quantities of chocolate available for use by various world populations vary widely, but time-trend data on chocolate availability indicate general increases in use worldwide.

Unfortunately, little is known about patterns of chocolate consumption within populations. In many countries, periodic food consumption surveys are conducted to provide information on food availability and nutrition status for population subgroups. Information on chocolate intake is collected in most of these surveys, but because chocolate is considered to be a minor contributor to nutrient intake, data on chocolate consumption generally are not compiled for publication or for public access. This chapter reviews data on chocolate availability and use by various world populations, and introduces newly compiled data on patterns of chocolate use within selected populations.

Sources of data on chocolate use and consumption

Food consumption survey data are collected for a variety of purposes, and the survey methods used depend to a great extent on the intended purposes. Data collected for one purpose may, however, be useful for answering questions outside the scope of the original survey if limitations of the data, particularly the population studied and the methods used, are taken into account.

Food supply (market disappearance) surveys, household budget surveys, and surveys of individuals represent the major types of food consumption surveys conducted.

Food supply surveys

Food supply surveys are conducted by national government institutions and by food industry groups to provide information on food availability for a specific population during a specified time period. These data are used in setting priorities, analyzing trends, developing policy, and formulating food programs.

The Food and Agriculture Organization (FAO) of the United Nations compiles national supply data on raw agricultural commodities into 'food balance sheets' (3). These data may be used to compare availability of food commodities in different countries or regions.

Similarly, the International Office of Cocoa, Chocolate, and Sugar Confectionery (IOCCC) generates data useful for comparing supplies of confectionery products, including chocolate confectionery, in different countries (4). However, the IOCCC data cannot be used to characterize confectionery consumption by individuals within the countries. Appropriate use of these data is limited primarily because waste is not considered. It must be assumed that a portion of the chocolate available for consumption is discarded by industry after shelf-life expiration or thrown away uneaten by individuals. In addition, a certain proportion of chocolate confectionery counted as available to the consumer may be used by industry or food service facilities for non-confectionery applications.

The US Department of Commerce (DOC) conducts an annual survey to produce detailed information on US confectionery supplies (5). The data compilation methods are similar to IOCCC methods, and as with the IOCCC data, neither waste nor the factors described above are considered.

Products included in 'chocolate confectionery' categories differ somewhat in the DOC and IOCCC surveys. Actually, even within the IOCCC system, definitions of chocolate confectionery and other products vary somewhat from country to country (6). The IOCCC chocolate confectionery category includes unfilled chocolate; filled tablets and bars; 'bonbons', pralines and 'other chocolate'; sugar confectionery containing cocoa; and white chocolate. The chocolate confectionery category used in DOC reporting includes enrobed/molded, solid, solid with inclusions, panned and 'other types of chocolate products'.

Household budget surveys

Household budget surveys are conducted by many countries to provide food supply data at the household rather than national level. Summary data on household food use of chocolate have been included in budget survey reports (7), but as with chocolate supply data, waste by household members is not considered. In addition, chocolate purchased and consumed outside the home is not considered. Users of chocolate within a household cannot be distinguished, and individual variation cannot be assessed. Chocolate intakes by subpopulations based on age, sex, health status and other variables in most cases can be estimated only by using standard proportions or equivalents for age/sex categories.

Food consumption surveys of individuals

Methods used for collecting data on food consumption by individuals are categorized as *retrospective* or *prospective* methods.

Retrospective

These methods focus on food consumed during a time period that has already passed. Commonly used retrospective methods include 24-hour or other short-term recalls, food frequencies and diet histories.

Recall methods require that survey respondents identify and quantify foods and beverages consumed during a specific period, usually the preceding day. Pictures, household measures or two- or three-dimensional food models may be used to help respondents quantify the food consumed. The interviewer may probe for certain foods or beverages that are frequently forgotten. However, such probing has also been shown to introduce bias by encouraging respondents to report items not actually consumed.

Recall interviews are relatively easy to conduct, require a minimum of time (about 20 min or less for a 24-hour recall) for completion, and can provide high-quality food consumption data for populations with low literacy (8, 9).

Many nationwide food consumption surveys have been conducted using short-term recall methods. The *Third National Health and Nutrition Examination Survey* (NHANES 3) (10) and the 1994–96 US Department of Agriculture (USDA) *Continuing Survey of Food Intakes by Individuals* (CSFII) (11) are examples of surveys conducted using recall methods (Figs 20.1 and 20.2).

Use of a *food frequency questionnaire* (FFQ), or checklist, allows determination of the frequency of consumption of a limited number of foods. FFQ food lists are in many cases developed to allow the collection of data relevant for a very specific nutrition-related issue. Respondents are asked how many times a day, week or month each food on the list is usually consumed. Semi-quantitative FFQs

NHANES 3, conducted from 1988–94 by the National Center for Health Statistics, US Department of Health and Human Services, included five major components: a household questionnaire, medical history questionnaire, dietary questionnaire, physical examination and clinical tests. The survey sample was designed and recruited using multi-stage area probability sampling procedures. Parts of the survey were conducted in the home, the rest in specially designed mobile examination centers (MECs), where the dietary and clinical procedures were conducted.

NHANES 3 dietary components consisted of a 24-hour recall, a food frequency questionnaire, and questions about special diets, medications and nutritional supplements. Recall data were collected for over 25 000 people. These data may be analyzed to characterize consumption of chocolate confectionery for various population groups. The major limitation of NHANES 3 data is that only one 24-hour recall is obtained for each person, so intra-individual variation cannot be estimated.

Fig. 20.1 Third National Health and Nutrition Examination Survey (NHANES 3).

In the 1994–96 CSFII, respondents provided recall data on each of 2 days, approximately 1 week apart. These data were obtained for about 15 000 individuals of all ages. The survey sample was designed and recruited using multi-stage area probability sampling procedures.

CSFII data on age, sex, race, region, season and other variables allow estimation of consumption of foods, including chocolate confectionery, for a wide variety of population subgroups. Results can be reported based on consumption per day or on 2-day averages.

Fig. 20.2 1994–96 US Department of Agriculture Continuing Survey of Intakes by Individuals (CSFII).

allow an estimation of amounts consumed by asking subjects if their usual portion size is small, medium or large compared to a stated 'medium' portion. The size of the medium portion is usually based on mean intakes of large populations but may be standardized for various age/sex groups.

The *diet history* is used to obtain information from individuals about the usual pattern of eating over an extended period of time (12, 13). Used primarily in epidemiological research, the diet history is a more in-depth and time-consuming procedure than the recall, record and FFQ methods. A recall or FFQ may be included as a diet history component.

Prospective

In *prospective* food consumption studies, survey participants are asked to record information on foods and beverages as the foods are consumed during a specific period. Quantities of foods and beverages consumed are entered in the record usually after weighing or measuring the amounts served and subtracting leftover amounts.

The *Dietary and Nutritional Survey of British Adults* (NDS) (14) and the

German *Nationale Verzehrsstudie* (NVS) (15) are examples of surveys conducted using diary/record methods (Figs 20.3 and 20.4). British national surveys using methodology similar to that used in the NDS for adults have been conducted to assess dietary intakes of infants (16), toddlers (17) and school-aged children (18).

The Dietary and Nutritional Survey of British Adults was carried out between October 1986 and August 1987 by the Ministry of Agriculture, Fisheries and Food (MAFF). The survey sample was recruited using a multi-stage random probability design, with the goal of obtaining a nationally representative sample of non-institutionalized adults ages 16 to 64.

Approximately 2200 survey respondents completed 7-day dietary records. The results can be analyzed to estimate chocolate confectionery intakes by population subgroups per day or per week; percentages of populations consuming chocolate confectionery on a particular day; or percentages of populations consuming chocolate confectionery on at least 1 day of the survey period.

Fig. 20.3 The Dietary and Nutritional Survey of British Adults.

The German NVS was conducted from October 1985 to January 1989. Over 20 000 individuals completed 7-day weighed food records and 7-day activity records. Basic socio-demographic data were collected on all individuals; one 'target individual' 18 years or older in each household also provided information on food frequency, health history, lifestyle and nutrition knowledge and attitudes.

As with the British survey data, NVS data can be analyzed to estimate chocolate confectionery intakes by population subgroups per day or per week; percentages of populations consuming chocolate confectionery on a particular day; or percentages of populations consuming chocolate confectionery on at least 1 day of the survey period.

Fig. 20.4 The *Nationale Verzehrsstudie*.

In general, data from short-term recalls and from food diaries are the most accurate and flexible data to use in estimating food consumption by individuals. Data from these surveys can be used to calculate distributions of intake, and estimates can be calculated for subpopulations based on age, sex, ethnic background, socio-economic status and other demographic variables, provided that such information is collected for each individual.

It must be recognized that, although data on foods consumed by individuals in general provide a better picture of intake patterns than do food supply data or household budget data, the validity and reliability of the methods and the potential for error in data collection must be considered.

Methods for assessing food consumption by individuals have been validated by surreptitious observation and by using biological markers, but the validity of one survey method for use in obtaining accurate food consumption data generally is tested using another common survey method. Method reliability – the extent to which a method yields reproducible results – depends somewhat on the number of days of dietary intake data collected for each individual in the population. The number of days of food consumption data required for reliable estimation of population intakes is related to each subject's day-to-day variation in diet (intra-individual variation) and the degree to which subjects differ from each other in their diets (inter-individual variation) (19, 20). When intra-individual variation is small relative to inter-individual variation, population intakes can be reliably estimated with consumption data from a smaller number of days than should be obtained if both types of variation are large. The number of intake days required for reliable estimation of consumption is smaller for a frequently consumed food (e.g. milk) than for an infrequently consumed food (e.g. organ meats).

Potential sources of error in individual food consumption surveys include chance and measurement factors. The size of error due to chance depends somewhat on the extent to which the sample population reflects the actual population. Error due to chance may also arise from data collection at different times of the day, on different days of the week, or at different seasons of the year. The survey instrument may introduce measurement error if questions are not clear, if standard probes lead the subject to a desired answer, if questions are culture specific, or if questions do not follow a logical sequence. Interviewer bias may contribute to measurement error if the interviewer is judgmental or uses nonstandard probes to obtain data.

Respondents may contribute to measurement error if they omit reporting foods actually consumed, misrepresent the quantity of foods consumed, or report consumption of foods which were not actually consumed. Foods that are not staples in the diet or that might be viewed as luxury items – such as chocolate – are more likely to be under-reported by some populations than are staple foods such as wheat, rice or potatoes. Dietary under-reporting by obese individuals has been documented in a number of studies (21–23). Heitmann and Lissner (22) identified snack-type foods as the category most likely to be omitted from dietary reports.

Consumption patterns for chocolate confectionery

Availability of chocolate confectionery in the USA

Chocolate confectionery shipment data compiled by the DOC (5) indicate that from 1992–96, per capita availability of these products increased by 13%, from 12.8 g/day to 14.5 g/day (Table 20.1). DOC data show that enrobed and molded chocolate confectionery with candy, fruit, nut or granola centers were most popular and accounted for nearly half of all types of chocolate confectionery available to Americans. Data compiled by the IOCCC (4) show similar increases in per capita availability of chocolate confectionery in recent years, from 12.6 g/day in 1991 to 14.5 g/day in 1996 (Fig. 20.5).

Table 20.1 Availability of chocolate confectionery in the USA.

Chocolate and chocolate-type confectionery	Availability (g/day)		Types of chocolate confectionery (%)	
	1992	1996*	1992	1996*
All chocolate and chocolate types	12.8	14.5	—	—
Solid	1.7	2.0	13.4	13.6
Solid with inclusions	1.3	1.3	10.2	8.7
Enrobed or molded: with candy, fruit, nut or granola center	6.1	6.3	47.5	43.3
Enrobed or molded: with bakery product center	0.9	1.0	6.6	6.9
Panned	1.8	2.6	14.4	17.6
Assortments and other	1.0	1.4	7.9	9.8

Source: Bureau of Census, US DOC (5).
* Figures have not been rounded, hence sums of individual rows do not necessarily equal totals.

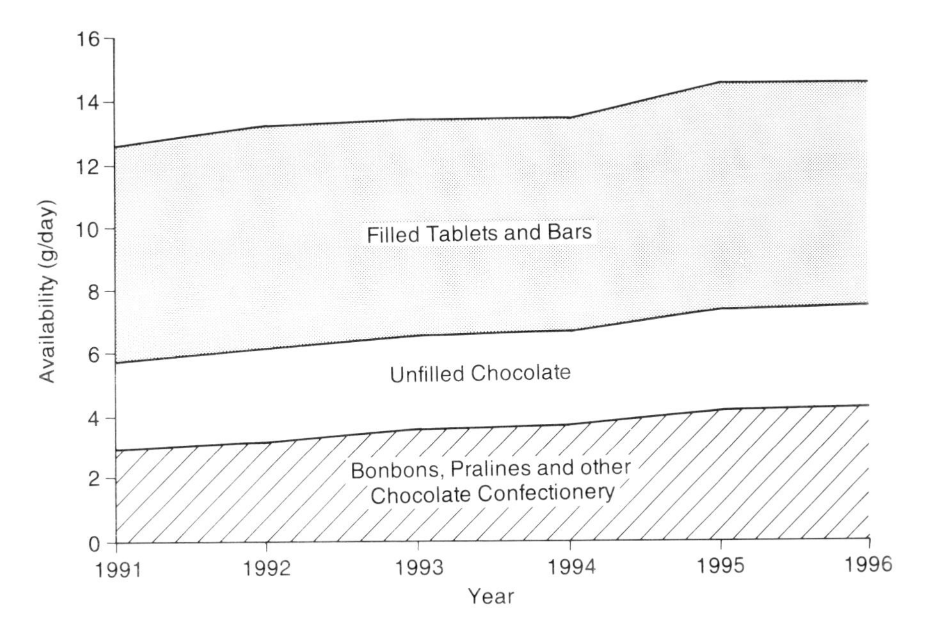

Fig. 20.5 Trends in chocolate confectionery availability in the USA. *Source:* IOCCC (4).

Availability of chocolate confectionery internationally

Chocolate confectionery availability varies widely from country to country (Fig. 20.6). European countries, where production of chocolate and chocolate confectionery as we know it today originated, tend to have the highest availability, while countries in Asia and tropical climates tend to have very low availability.

IOCCC data on international chocolate confectionery supplies reveal that overall availability generally increased throughout the world from 1986–96, though there were slight changes in rank among the countries (Fig. 20.6).

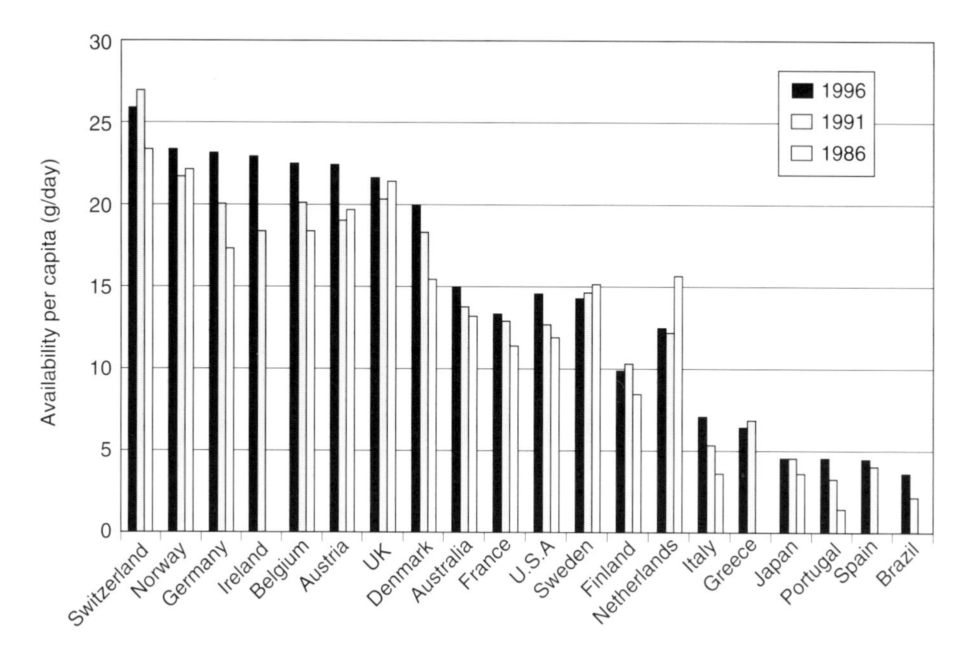

Fig. 20.6 Trends in international availability of chocolate confectionery. *Source:* IOCCC (4).

The age ranges of the CSFII, NDS and NVS survey populations differed, and comparisons of data on mean food intakes by the total survey populations therefore are inappropriate. However, intake data may be compared for males and females ages 19–50, as individuals in this age range are included in all three surveys.

Specific chocolate confectionery items coded in the CSFII, NDS and NVS surveys vary, as would be expected in three different consumer markets. Numbers of chocolate confectionery food codes ranged from 12 in the UK NDS for adults to about 70 in the 1994–96 CSFII.

The NDS for adults and NVS surveys are 7-day diary/record surveys. The NDS for toddlers was a 4-day survey, with weighting factors calculated to project 7-day average food and nutrient intakes. The 1994–96 CSFII data were collected using dietary recalls on each of 2 days, approximately but not exactly 2 weeks apart. It would be appropriate to compare overall survey average food consumption intake data from the NDS for adults and the NVS, as both are 7-day surveys. It would not be appropriate to compare these averages with unadjusted 4-day average consumption by UK toddlers or 2-day average consumption of foods reported in the CSFII because of differences in probability of consumption of rarely consumed foods over the survey periods.

CSFII, NDS and NVS survey results may, however, be appropriately compared when each person-day of intake is considered separately. In the following sections, we examine CSFII, NDS and NVS data on chocolate confectionery consumption, including:

- Mean intakes per user (on survey days on which reported)
- Mean per capita daily intakes
- Percent of survey days on which chocolate confectionery consumption was reported.

USA

Results of the 1994–96 CSFII survey show a wide range in mean amounts of chocolate confectionery consumed by different US subpopulations (Table 20.2). Intake reported by males in each age range exceeded intake reported by females in the same age range. The highest mean daily intakes of chocolate confectionery by males were reported by teenagers and young adults. Highest mean daily intakes by females were reported by teenagers. In each country, both the quantity of chocolate confectionery consumed and the frequency of consumption declined with age.

Table 20.2 Consumption of chocolate confectionery in the USA.

Population group (sex/age (years))	Percent of survey days on which reported	Mean daily intake (g)	
		Per user	per capita
MF 9–13	14	45.7	6.2
M 14–18	12	59.2	6.9
M 19–30	9	56.1	4.9
M 31–50	9	60.3	5.3
M 51–70	8	51.0	3.9
M >70	5	44.2	2.4
F 14–18	10	52.9	5.1
F 19–30	10	49.9	4.8
F 31–50	8	46.7	3.8
F 51–70	8	41.6	3.2
F >70	5	32.3	1.6
MF (19–50)	9	53.4	4.7
Total (ages 4+)	9	47.9	4.2

Source: 1994–96 USDA CSFII (11).

Mean intake of chocolate confectionery on days when these products were consumed ranged from 32.3 g for females over 70 to 60.3 g for males 31–50, with 53.4 g as the mean consumption on days consumed by males and females 19–50.

Consumption of chocolate confectionery was reported in 6% of all 1994–96 CSFII daily food intake recalls. The mean percentage of CSFII respondents consuming chocolate confectionery on a particular day ranged from 5% for females over 70 years of age to 14% for males and females ages 9–13. Per capita intake of chocolate confectionery in each 1994–96 CSFII subpopulation reflects the low percentages of consumers; overall mean per capita intake of chocolate confectionery for ages 19–50 was 4.7 g and for all ages 4+ was 4.2 g.

Chocolate confectionery consumption reported in the 1994–96 CSFII follows age–gender consumption patterns similar to those seen in previous US surveys. However, amounts reported by consumers appear to be lower in the 1994–96 survey than in previous surveys. This could be due to changing patterns of eating behavior for chocolate confectionery and does not necessarily reflect an actual reduction in consumption. In the 1987–88 USDA Nationwide Food Consumption Survey, teenage males reported the largest mean daily intake of chocolate confectionery (90 g per day), followed closely by females aged 30–39 years old (87 g per day) (24). Males 12–19 years of age had the highest consumption of chocolate confectionery (over 60 g per day) in the 1989–91 CSFII, although men aged 20–49 and women 40–49 years old also reported mean daily intakes of approximately 55 g (8). Results of the 1987–88 USDA Nationwide Food Consumption Survey, the 1989–91 CSFII and 1994–96 CSFII indicate that mean daily intake of chocolate confectionery, particularly among women of all ages, has declined in the past decade.

UK

Individuals in the 1986–87 NDS reported an overall mean intake on days consumed of 68.3 g chocolate confectionery per day (Table 20.3). Among adults aged 19–50, mean daily intake of chocolate confectionery was 69.7 g on days when consumed.

Table 20.3 Consumption of chocolate confectionery in the UK.

Population group (sex/age (years))	Percent of survey days on which reported	Mean daily intake (g)	
		Consumers only	Per capita
M 16–18	29	87.4	25.0
M 19–30	21	90.9	19.0
M 31–50	16	64.4	10.6
M 51–64	12	73.0	9.0
F 16–18	27	86.2	22.9
F 19–30	24	64.8	15.4
F 31–50	18	64.2	11.8
F 51–64	17	50.1	8.2
MF 19–50	19	69.7	13.4
Total (16–64)	18	68.3	12.4

Source: UK National Diet and Nutrition Surveys (7, 14, 17, 18).

Males between the ages of 19 and 30 consumed the largest quantity of chocolate confectionery at a given eating occasion; women of the same age reported smaller quantities of intake although in general a lower percentage of women reported chocolate confectionery consumption. Overall, chocolate confectionery consumption was reported on 18% of adult survey person-days.

Germany

NVS survey data on consumption of chocolate confectionery in the former West Germany (1985–89) are presented in Table 20.4. Consumption of chocolate confectionery on days consumed ranged from more than 40 g per day for both men and women aged 70 and greater, to 76.2 g per day among 14–18-year-old males. Among individuals under the age of 70, men consumed larger quantities than did women. Overall, consumption of chocolate confectionery was reported on 15% of the total survey days and the mean daily intake among all chocolate confectionery consumers was 56.2 g.

Table 20.4 Consumption of chocolate confectionery in former West Germany.

Population group (sex/age (years))	Percent of survey days on which reported	Mean daily intake (g)	
		Consumers only	Per capita
MF 9–13	24	55.6	13.6
M 14–18	22	76.2	16.6
M 19–30	15	70.3	10.8
M 31–50	11	61.1	6.8
M 51–70	8	55.6	4.6
M >70	8	45.6	3.6
F 14–18	22	60.8	13.1
F 19–30	18	55.7	9.9
F 31–50	13	49.5	6.2
F 51–70	10	45.9	4.5
F >70	8	42.7	3.3
MF 19–50	14	58.7	8.2
Total (4+)	15	56.2	8.3

Source: NVS (15).

Consumption of chocolate confectionery as snacks

The CSFII, NDS and NVS surveys each allow the characterization of consumption of specific foods per eating occasion, but differences in reporting conventions make comparison of results of analyses difficult. In the NDS, food consumption was recorded in consecutively numbered eating occasions, but respondents did not identify these occasions by name. Foods consumed by NVS respondents, on the other hand, were recorded as being consumed at one of six specific eating occasions, each essentially treated as a meal.

Food consumption by CSFII respondents was recorded by occasions self-described as breakfast, brunch, lunch, supper, dinner or one of any number of snack periods. While common use of descriptors for specific meals varies regionally throughout the USA, distinctions between meal and snacking occasions are more uniform.

US CSFII respondents reported consumption of chocolate confectionery at self-described snacks more often than at meals, with snacks accounting for 79% of eating occasions at which chocolate confectionery was consumed and 82% of total chocolate confectionery intake (Table 20.5).

Table 20.5 Consumption of chocolate confectionery at meals and snacks in the USA by the total population.

Meal	Mean consumption per occasion (g)	Percent of eating occasions	Percent of chocolate confectionery consumption
Breakfast	49.9	2	3
Brunch	52.8	<1	<1
Lunch	35.5	10	8
Dinner	37.3	5	4
Supper	33.9	3	2
Snacks	47.7	79	82

Source: 1994–96 USDA CSFII (11).

These results indicate a possible increase in the proportion of chocolate consumed at snacking periods, reported at only 66% of total intake in 1990 (1). Results from the 1994–96 survey indicate that lunch was the second most popular eating occasion for chocolate confectionery, accounting for over 10% of eating occasions at which chocolate confectionery was consumed and 8% of total chocolate confectionery intake. The mean amount of chocolate confectionery consumed at lunch was somewhat lower than that consumed as snacks (35.5 versus 47.7 g).

CSFII data indicate that most chocolate confectionery is consumed as a snack (79% of eating occasions) by individuals with relatively low body mass index (Table 20.6). It must be noted, however, that body mass indices for CSFII respondents were calculated by the USDA based upon weights and heights reported by respondents rather than upon measured values. It must also be noted that, as discussed previously, dietary under-reporting by obese individuals has been documented in a number of studies (21–23).

In a British study (25) of the relative contributions of meals and snacks to total energy intake, adolescents were seen to obtain a greater proportion of total energy intake from snacks than were older age groups, and the contribution of chocolate confectionery to total energy intake at snacking periods was sig-

Table 20.6 Chocolate confectionery consumed as snacks: percentage of total consumption by respondents within different body mass index ranges.

Adult consumers (M, F, 19–70 years)	Body mass index (kg/m²)				
	<25	≥25 and <30	≥30 and <35	≥35 and <40	≥40
	(% of consumption by subpopulation)*				
M 19–30	61	27	9	2	0
F 19–30	63	25	5	3	3
M 31–50	30	34	28	6	2
F 31–50	52	22	10	5	12

Source: 1994–96 USDA CSFII (11).
* Figures have not been rounded, hence sums of some rows do not equal 100%.

nificantly greater for adolescents than for the older age groups. A previous British study (26) attributes the high proportion of snack-based energy intake by children and adolescents to use of pocket money to purchase snacks on the way to, at, and on the way home from, school.

In a study conducted to evaluate the belief, commonly held by Australians, that consumption of snack foods is characteristic of people of low socio-economic status, 1984 and 1988/9 Household Expenditure Survey data on chocolate confectionery and other snack foods were examined (27). Findings indicated that chocolate confectionery expenditures by households of low and average/high socio-economic status were about equal for households consisting of one or two individuals. For households including children, chocolate confectionery expenditures by those with average/high socio-economic status were over 50% higher than expenditures by those with low socio-economic status.

Discussion and conclusions

Availability data from the IOCCC and the US DOC are useful for identifying international patterns and trends related to chocolate confectionery. However, as noted previously, these data must be regarded as overestimates of consumption.

The ratio of chocolate confectionery availability to consumption at comparable time periods varies from country to country (Table 20.7).

In the 1980s when survey data was collected in the UK, over 20 g of chocolate confectionery were available per day per capita based upon balance sheet data. Estimates of consumption show that on a per capita basis, individuals 16–64 years of age consumed an average of 8.3 g per day, suggesting a greater than twofold overestimate of consumption. Similarly, in Germany in the late 1980s, slightly less than 20 g of chocolate confectionery was available daily on a per capita basis.

Table 20.7 Chocolate confectionery consumption (g per capita/day) as estimated using availability data and individual food consumption survey data.

Country	Time period	Chocolate confectionery consumption (g per capita/day)	
		Based on availability data	Based on individual food consumption survey data
USA	1994–96	14.5	4.2
UK	1986–87	20.0	8.3
Germany	1985–89	19.0	12.4

Sources: Apgar and Seligson (6), 1994–96 USDA CSFII (11), UK National Diet and Nutrition Surveys (7, 14, 17, 18), NVS (15).

Analyses of food intake data indicate that individuals 4 years of age and above consumed 12.4 g per capita per day.

The greatest difference between chocolate confectionery availability and intake estimates, however, is seen for the USA. At the time of the most recent CSFII data collection in the USA, 14.5 g of chocolate confectionery were estimated to be available for each individual, while survey data indicate that on a per capita basis, Americans consumed 4.2 g per day.

Country-to-country and time-related differences in the relationship between chocolate confectionery availability and consumption are to be expected. Economic factors and cultural differences in food production, wastage, industrial reuse and shelf-life affect the extent to which products counted as 'available' for consumption actually are consumed.

Data from the IOCCC indicate that the quantities of chocolate confectionery available in Germany, Switzerland, and Belgium are greater than those available to American, Asian, African, Oceanic and other European populations. This evidence for high per capita chocolate confectionery consumption by European populations relative to other populations is supported by results of food consumption surveys of individuals, which indicate that daily consumption in the USA is lower than that in the UK and Germany on both the per user and per capita basis.

Although the mean amounts of chocolate confectionery consumed by survey respondents in the USA, UK and former West Germany differed, patterns of consumption by age–gender groups within the populations are similar. In each survey, intake reported by males in each age range exceeded intake reported by females in the same age range. The highest mean daily intakes of chocolate confectionery by males were reported by teenagers and young adults. Highest mean daily intakes by females were reported by teenagers. In each country, both the quantity of chocolate confectionery consumed and the frequency of consumption declined with age.

National survey data on chocolate confectionery consumption by school-aged children in the UK are available only for children aged 10–11 and 14–15, making comparisons with US and German data somewhat difficult. However, data obtained from a survey of British and German primary school children in 1994 (28) indicate that chocolate confectionery consumption may be lower in Germany than in England.

Chocolate confectionery intake per consumer appears to be lower in the 1994–96 CSFII survey than in previous US surveys. This may reflect increased volume sales of smaller 'snack-size' bars, as noted by Apgar and Seligson in 1994 (6). However, decreases in intake per consumer do not necessarily reflect an actual reduction in consumption on a population basis.

US 1994–96 CSFII data indicate that Americans view chocolate confectionery as 'snack food'. The American Dietetic Association recommends frequent snacking for people with diabetes and for the population at large, as this pattern of food consumption is thought to provide more constant blood glucose levels and to be more healthful in general than the three-meals-a-day pattern traditional in the USA. Chocolate confectionery contains substantial amounts of sugar and saturated fat, and previously was not thought of as a healthful snack. However, opinion on this matter is changing. Cedermark *et al.* (29) found that isocaloric substitution of a milk chocolate bar for the regular afternoon snack consumed by teenagers with diabetes had no effect on the postprandial blood glucose curve. Most of the fat in chocolate is in the form of stearic, oleic and palmitic acids. Stearic and oleic acids are saturated fatty acids thought to have neutral effects on serum cholesterol. Kris-Etherton *et al.* (30–31) demonstrated that milk chocolate consumed as a snack actually may have hypocholesterolemic effects, and theorized that stearic acid in the cocoa butter is responsible for these effects.

References

1. Rössner, S. (1997) Chocolate – divine food, fattening junk or nutritious supplementation? *Eur. J. Clin. Nutr.* **51**, 341–345.
2. Trager, J. (1995) *A Food Chronology*. Henry Holt, New York.
3. Food and Agriculture Organization (1994) *AGROSTAT; Food Balance Sheets 1961–93* (computer version). FAO, Rome.
4. IOCCC (1997) *International Statistical Review of the Biscuit, Chocolate and Sugar Confectionery Industries*. International Office of Cocoa, Chocolate and Sugar Confectionery, Brussels.
5. Bureau of the Census, US Department of Commerce, Economics and Statistics Administration (1997) *Current Industrial Reports, Confectionery 1996*. GPO, Washington, DC.
6. Apgar, J.L. and Seligson, F.H. (1995) Consumer consumption patterns of chocolate and confectionery. *The Manufacturing Confectioner* **31**, 31–36.
7. Ministry of Agriculture, Fisheries and Foods (1997) *National Food Survey 1996:*

Annual Report on Food Expenditure, Consumption and Nutrient Intakes. The Stationery Office, London.

8. Block, G. (1989) Human dietary assessment: methods and issues. *Preventive Medicine* **18**, 653–660.
9. Dwyer, J.T. (1988) Assessment of dietary intake. In *Modern Nutrition in Health and Disease*, 7th edn. (Ed. by Shils, M.E. and Young, V.R.). Lea and Febiger, Philadelphia.
10. US Department of Health and Human Services, National Center for Health Statistics (1997) *National Health and Nutrition Examination Survey, III 1988–94* (CDROM series). National Technical Information Service, Springfield, VA.
11. US Department of Agriculture, Agricultural Research Service (1998) *1994–96 Continuing Survey of Food Intakes by Individuals* (CDROM series). National Technical Information Service, Springfield, VA.
12. Burke, B.S. (1947) The dietary history as a tool in research. *J. Am. Diet. Assoc.* **23**, 1041–1046.
13. Hankin, J.H. (1989) Development of a diet history questionnaire for studies of older persons. *Am. J. Clin. Nutr.* **50**, 1121–1127.
14. Gregory, J., Foster, K., Tyler, H. and Wiseman, M. (1990) *The Dietary and Nutritional Survey of British Adults*. HMSO, London.
15. Speitling, A., Hüppe, R., Kohlmeier, M., *et al.* (1992) *VERA Publication Series Volume 1A (English). Methodological Handbook, Nutrition Survey and Risk Factors Analysis*. Wissenschaftlicher Fachverlag Dr. Fleck, Niederkleen.
16. Mills, A. and Tyler, H. (1992) *Food and Nutrient Intakes of British Infants Aged 6–12 Months*. HMSO, London.
17. Gregory, J.R., Collins, D.L., Davies, P.S.W., Hughes, J.M. and Clarke, P.C. (1995) *National Diet and Nutrition Survey: children aged $1\frac{1}{2}$ to $4\frac{1}{2}$ years. Volume 1: Report of the Diet and Nutrition Survey*. HMSO, London.
18. Subcommittee on Nutritional Surveillance, Committee on Medical Aspects of Food Policy (1989) *The Diets of British Schoolchildren*. HMSO, London.
19. Basiotis, P.P., Welsh, S.O., Cronin, J. *et al.* (1987) Number of days of food intake records required to estimate individual and group intakes with defined confidence. *J. Nutr.* **117**, 1638–1641.
20. Nelson, M., Black, A.E., Morris, J.A. and Cole, T.J. (1989) Between- and within-subject variation in nutrient intake from infancy to old age: estimating the number of days required to rank dietary intakes with desired precision. *Am. J. Clin. Nutr.* **50**, 155–167.
21. Braam, L.A., Ocke, M.C., Bueno-de-Mesquita, H.B. and Seidell, J.C. (1998) Determinants of obesity-related under-reporting of energy intake. *Am. J. Epidemiol.* **147**, 1081–1086.
22. Heitmann, B.L. and Lissner, L. (1995) Dietary under-reporting by obese individuals – is it specific or non-specific? *BMJ* **311**, 986–989.
23. Klesges, R.C., Eck., L.H. and Ray, J.W. (1995) Who under-reports dietary intake in a dietary recall? Evidence from the Second National Health and Nutrition Examination Survey. *J. Consult. Clin. Psychol.* **63**, 438–444.
24. Seligson, F.H., Krummel, D.A. and Apgar, J.L. (1994) Patterns of chocolate consumption. *Am. J. Clin. Nutr.* **60**, 1060S–1064S.
25. Summerbell, C.D., Moody, R.C., Shanks, J., Stock, M.J. and Geissler, C. (1995)

Sources of energy from meals versus snacks in 220 people in four age groups. *Eur. J. Clin. Nutr.* **49**, 33–41.
26. British Nutrition Foundation (1985) *Eating in the Early 1980s*. BNF, London.
27. Santich, B.J. (1995) Socio-economic status and consumption of snack and take-away foods. *Food Aust.* **47**, 121–126.
28. Körtzinger, I., Neale, R.J. and Tilston, C.H. (1994) Children's snack food consumption patterns in Germany and England. *Br. Food J.* **96**, 10–15.
29. Cedermark, G., Selenius, M. and Tullus, K. (1993) Glycaemic effect and satiating capacity of potato chips and milk chocolate bar as snacks in teenagers with diabetes. *Eur. J. Pediatr.* **152**, 635–639.
30. Kris-Etherton, P.M., Derr, J., Mitchell, D.C., *et al.* (1993) The role of fatty acid saturation on plasma lipids, lipoproteins, and apolipoproteins: I. Effects of whole food diets high in cocoa butter, olive oil, soybean oil, dairy butter, and milk chocolate on the plasma lipids of young men. *Metabolism* **42**, 121–129.
31. Kris-Etherton, P.M., Derr, J., Mustad, V.A., Seligson, F.H. and Pearson, T.A. (1994) Effects of milk chocolate bar per day substituted for a high-carbohydrate snack in young men on an NCEP/AHA Step 1 Diet. *Am. J. Clin. Nutr.* **60** (Suppl. 6), 1037S–1042S.

Chapter 21

Sensory and Taste Preferences of Chocolate

Marcia L. Pelchat and Gary K. Beauchamp

What makes chocolate the 'flavor of the gods'? Chocolate is the most frequently mentioned food in surveys of cravings (usually defined as intense desires or longings for particular foods) (1, 2). This is especially true of young women (1, 3). Why is chocolate so frequently craved? Although many accounts focus on the presumed pharmacological properties of chocolate, there is little evidence that they influence the liking or ingestion of chocolate. Indeed, one recent study (4) indicates just the opposite – that chocolate craving is satisfied by the sensory properties of chocolate. Pharmacology plays no role. What are the sensory properties that make chocolate so valued? First, we briefly define flavor and then discuss the flavor attributes of chocolate as it is normally consumed.

Flavor

Flavor is defined as the overall perception arising from sensory stimulation of the systems of taste, smell and oral texture. Though these sensory systems are anatomically distinct, when we consume a food or beverage, the sensations often meld into a single experience, which seems to arise from the tongue. Only by concentrating (and even then it may be difficult) can we separate these sensations. It must be in the way the sensations interact that makes chocolate so profoundly appreciated.

Taste

It is common to hear people say 'I love the taste of chocolate'. However, what they should really be saying is 'I love the flavor of chocolate'. In fact, taste is only a small (though important) part of flavor.

The number of taste qualities is small. Most people agree that they include sweet, sour, bitter and salty. Another candidate is umami, which is a savory or meaty taste exemplified by the taste of monosodium glutamate (MSG) (see (5)). The taste of chocolate as it is normally consumed in the form of candy, cake, ice cream, etc., is primarily bitter (the Aztec term *chocolatl* means bitter water (6)), and sweet (due to the sugar added in processing). The bitterness of chocolate comes from the xanthines, theobromine and caffeine, and also from substances produced during the roasting process (7). The latter is necessary to produce the characteristic aroma of chocolate.

Bitter tastes are generally disliked. This dislike can be seen soon after birth in the facial expressions of newborn infants (8–11). Although a dislike for bitter seems to be innate, it is also modifiable by experience. For example, Moskowitz *et al.* (12) reported on a population of Indian laborers who showed a high preference for sour and bitter tastes. The unusual level of liking was attributed to familiarity with such flavors in their cuisine. Bitterness can also come to be appreciated in the context of many foods and beverages including chocolate, coffee and beer.

Bitterness in chocolate is reduced by taste–taste interactions. Sweetness and bitterness tend to inhibit each other (13). Salt also inhibits bitterness, so that addition of salt, for example, to hot cocoa not only adds some saltiness, but reduces bitterness and thereby reveals sweetness (14). Indeed, a pinch of salt is a recommended ingredient in the hot cocoa recipe on the Hershey's can.

In contrast, sweetness seems to be universally liked (15). Liking for sweet tastes can be seen in the facial expressions of and intake by newborn infants (8–11, 16) and even in premature infants, sweetness stimulates sucking responses (17, 18). Although liking for sweets is likely to be innate, the amount of sweetness that is preferred in a particular food context is probably also influenced by learning.

There is evidence for many individual differences in responses to sweet and bitter tastes and some of these differences may have a genetic basis. Consider first bitter taste. It is thought that there are several classes of bitterness (13, 19). Thus, extreme sensitivity to one bitter substance does not mean that an individual is very sensitive to all bitter compounds. In humans, the most widely studied genetically based individual difference in taste sensitivity is for the bitter compound propylthiouracil (PROP).

Recent work has suggested that there are three levels of sensitivity to PROP in human populations perhaps related to a single locus with two alleles and incomplete dominance (20). Extremely sensitive individuals are referred to as super tasters, moderately sensitive persons are referred to as tasters and less-sensitive individuals are referred to as non-tasters. About 70% of white people in the USA are tasters or super tasters. However, the proportion is higher in most other populations: for Brazilian aborigines, the proportion may be 98%. There is some evidence that PROP sensitivity is correlated with sensitivity to other bitter compounds (including caffeine; (21)) and to certain sweet compounds including saccharine and sucrose (22, 23). Although surrounded by controversy, there have been numerous reports that PROP taster status is related to food preferences: sensitive individuals tend to dislike certain bitter tastes more than do others (24, 25).

There are individual differences in responses to other bitter compounds such as quinine and urea (19). Unfortunately, only relatively few bitter compounds have been studied extensively and some of the major ones contributing to the bitterness of chocolate (xanthines, theobromine and others) have not been studied at all. Thus, the range of sensitivities to these substances and the possible role of genetic differences underlying individual differences in sensitivity and preference remain to be investigated.

There is conflicting evidence that individual differences in human sweet perception and preference may also be influenced by genes. Ritchey and Olson (26) reported that pre-school children's degree of liking for sweet foods was not correlated with the degree of parental liking for sweet foods. Logue *et al.* (27) found weak but statistically significant evidence for family resemblance in liking for a category that they called 'junk food' which included carbonated beverages. Of course in family studies, the mechanism for resemblance could be genetic and/or environmental.

Gent and Bartoshuk (23) have shown that suprathreshold sweetness intensity judgments for sucrose, saccharin and neohesperidin dihydrochalcone (an intensely sweet derivative of the bitter flavonoid, neohesperidin) are greater in tasters of PROP compared with non-tasters. Other evidence consistent with genetic effects on sweet taste responsiveness includes work of Looy *et al.* (28, 29) demonstrating a relationship between sweet-liking and PROP sensitivity. Their claim is that for non-tasters of PROP, liking increases monotonically with increasing sugar concentration, but that for PROP-sensitive individuals, very sweet solutions are disliked. However, Drewnowski *et al.* (30) have failed to find such a relationship.

Green *et al.* (31) studied the hedonic evaluation of sucrose and lactose solutions of young monozygotic and same-sex dizygotic twin pairs. They reported no heritable component for sucrose preference, although they did report a strong racial difference in the liking for sweet tastes, which may be caused by differences in genes across *populations*. However, the methods used in this study were very crude, so results must be treated with even more caution than usual. Krondl *et al.* (32) measured the sensitivity to sweet tastes for monozygotic (mz) and dizygotic (dz) twin pairs and reported that heritability for sweet sensitivity was 0.52 which approached, but did not achieve, statistical significance, due to the low number of twin pairs. They asked about liking for and use of a number of sweet foods, including honey, jam, ice cream and doughnuts (but, unfortunately, not chocolate).

There were no significant mz/dz differences for any of these foods for either use or liking. In another twin study (33), heritabilities for preference were reported for 3 of the 17 foods tested. Interestingly, these three were sweetened (but *not* unsweetened) cereal, orange juice and cottage cheese. The cereal and orange juice may be considered sweet. However, although most of the remaining foods tested were not sweet (e.g. beans, broccoli, chicken, corn), snack cake and cola, two clearly sweet foods, did not exhibit significant heritabilities. Rozin and Millman (34) also found no mz/dz difference in liking for plain sugar or for a peppermint lifesaver. None of these twin studies examined a range of carbohydrate and non-carbohydrate sweeteners and most had a very small number of twin pairs. Moreover, most unfortunately, none of these twin studies provided any data specifically on chocolate.

The existence of sex differences could also be taken as suggestive evidence for a genetic influence on perception of and preference for sweetness. There are many reports documenting greater preference for sweets by females than males (35, 36) and greater frequency of cravings for sweets by females (3).

Finally, there are several studies of group differences that may involve a genetic influence. Beauchamp and Moran (37) reported greater preferences for sucrose in short-term intake tests by black compared to white 6-month-old infants. Similarly, Desor *et al.* (38) reported that adolescent black subjects preferred higher concentrations of sucrose and lactose than did adolescent white subjects; a similar but non-significant trend was evident for adults. One study (39) comparing Taiwanese and US students found differences in optimal preferences for sweets that depended upon the food context. The Chinese-descended students tended to prefer lower sucrose concentrations in a solid food (cookies) than did the US group; for liquids, the converse tendency was evident. This difference was suggested to be a consequence of differential experiences with sweetened solid and liquid foods. As a whole, most of these ethnic/racial differences can be explained equally well by genetic and/or experiential factors.

Another possible source of individual differences in taste preferences is age. Many recent investigators have reported that old age-related changes in gustation are modest when compared to changes in olfaction and are quality (i.e. sweet, sour, bitter and salty) specific for both threshold and suprathreshold measures (40–42). Usually, larger decreases in sensitivity are reported for bitter and sour stimuli than for sweet and salty stimuli. However, changes in sensitivity with age have been reported for all of the basic tastes (40, 43). The difference across taste qualities is not as surprising as it might initially seem, because different biochemical mechanisms are involved in the detection of different taste qualities (44). There may even be differential loss within a taste quality: a recent study has demonstrated a pronounced age-related loss of sensitivity to quinine but not to urea which are both bitter (19).

Although changes in taste sensitivity with age tend to be modest, there have been a number of studies showing a relationship between aging and shifts in preference for basic tastes. One of the earliest studies (45) suggested that there is an increased preference for tartness over sweetness as we age. Several other authors report an increase in the most preferred concentration of tastants with age (from young adulthood to old age (43)). The study by Murphy and Withee (43) is notable because it was based on a sample size of 100 subjects in each of three age groups. An increase in peak preference concentration can be explained by a decrease in sensitivity: a flavored solution that seems appropriate to a young adult subject might seem too weak to an elderly subject, while a flavored solution that seems too strong to a young adult subject might seem appropriate to an older person.

Preference for sweet taste changes with age early in life as well. Children and adolescents prefer higher concentrations of sweet water than do young adults, as has been shown by both cross-sectional (38) and longitudinal (46) studies. The underlying reason for this is not known although it has been speculated that a decrease in energy needs once growth is complete is responsible.

Given these individual differences in liking for sweetness and bitterness, it is notable that chocolate candy is available in a variety of forms (bittersweet to sweet) which allow individuals to choose the balance between sweetness and bitterness that they like the best. For example, the age-related changes in perception and preference for sweet and bitter tastes may contribute to a shift in preference from very sweet milk chocolate in youth to more bitter, less sweet chocolates in adulthood and old age.

Olfaction

The nose and the mouth are connected. This concept is critical to understanding the role of smell (olfaction) in flavor. As evidence for the connection, most people are familiar with what happens when someone laughs while drinking milk. A neater way to demonstrate the anatomy is to inhale through the nose with the mouth closed. There will be a sensation of cool air at the back of the mouth. Therefore, during eating, food odors reach the olfactory epithelium (the site of the olfactory receptors) by both the orthonasal pathway (from the front of the nose) and the retronasal pathway (from the back of the nose). Most people experience flavor as emanating from the food on the tongue and are not even aware of the contribution of olfaction.

The misidentification of odor as taste is known as taste–smell confusion (47–49). One could easily argue that odor is the primary component of flavor (50). There are only a few basic tastes. So if taste were synonymous with flavor, the number of flavor experiences would be limited as well. In terms of taste, raspberry, mango, grape and peach would all be sweet and tart. Chocolate and coffee ice creams would both be bittersweet. Taste alone cannot account for the subtle nuances of flavor experience. It is the odor component that makes flavors unique, that gives a seemingly endless variety of flavor experiences. This is why people think that they cannot taste when they have colds – the real problem is that they are unable to smell anything.

The aroma of chocolate is extremely complex and difficult to duplicate. Indeed, no single compound has been found that mimics this odor (51). Many of the components of chocolate flavor develop during two crucial stages of processing of the bean: fermentation and roasting (7, 51). Outside of the context of chocolate candy, aroma is the sensory characteristic that conveys chocolate flavor in products such as chocolate ice cream, or cookies or cake. So an individual with a bad cold would not be able to distinguish chocolate ice cream from coffee ice cream with eyes closed (both would have a bitter-sweet taste and the aroma cue would be blocked by the congestion).

As with taste, there are age-related changes in sensitivity to odors and individual differences in odor sensitivity that can be due to genetic and/or environmental factors. Age-related olfactory loss is an important source of individual differences. In general, compared to young adults, elderly individuals are less able to detect weak odors (52, 53), rate detectable odors as being less intense (52, 54) and show deficits in identification of blended foods and food odors (55–57).

There is evidence that olfactory loss is associated with poor dietary selection in community-dwelling elderly women (58) and that elderly subjects are unable to discriminate soup containing a standard amount of marjoram from soup without the herb (59). It has also been reported that elderly people with poor olfaction prefer foods with enhanced odors to plain foods (60). This provides indirect evidence that even in subjects who insist that they enjoy their food, the lack of aroma has had a negative impact. A recently published study indicates that odor enhancement of foods eaten by nursing home residents can modestly improve their nutritional status (61). So as the population ages, more and more individuals should have difficulty detecting the aroma of chocolate.

Parallel to studies on PROP taste, there is one genetic effect on olfaction that has been extensively studied: the ability to smell the steroid androstenone. This substance, found in some foods (e.g. truffles, but not the chocolate kind), human sweat and male pig saliva, smells to some, who are very sensitive, of urine whereas others find even high concentrations slightly pleasant or even undetectable (62). Genetic differences between individuals play a major role in these individual differences although age and individual experience are also considerably involved (62, 63). It is presumed that there are many other such polymorphisms of the many genes underlying the variety of olfactory receptors (64) that make it possible for humans to detect so many different and distinctive odors and hence flavors. Unfortunately, nothing is known about potential individual differences in the major (or minor) volatile characteristics of chocolate or how such differences, if they exist, might contribute to individual differences (including, perhaps, sex differences) in chocolate liking.

Texture

Taste and smell are only two of many sensory qualities that make up flavor. The temperature, texture and the mild, pleasant irritation produced by carbonated beverages and 'hot' spices are others. Such sensations arise from the skin senses, modalities that are distinct from gustation and olfaction. In contrast to taste and smell, there is no specific organ for such sensations. Throughout most of the oral cavity, this information is carried by the trigeminal nerve, which also innervates the surface of the face. Additionally, the distribution of nerve endings sensitive to irritation, texture and temperature in the mouth is broader than the distribution of receptors for taste. So all oral surfaces can experience these sensations. Unfortunately, food texture has been difficult to study and little is known in this area. One source of difficulty is that all of the relevant textural properties of food are usually not present until the food is actively manipulated and deformed in the mouth. It is undoubtedly the case that there are individual differences in how food is processed in the mouth, but evidence for this is scant. Another source of difficulty in the study of texture is that this characteristic, like flavor, requires integration of inputs from several sources. In this case, these include both tactile and kinesthetic receptors (65, 66).

The mouth-feel of chocolate is perhaps its most unique and important characteristic. Cocoa butter has a very narrow melting point, which happens to be close to body temperature (6, 51). So when chocolate candy is eaten, it changes from a solid with some snap to a luxuriously smooth, mouth-coating liquid. Also important to the mouth-feel of chocolate are processes that reduce the particle size of ingredients.

Preference for high-fat foods may be universal (67). High-fat foods comprise the bulk of craved foods (3) and sweet–fat combinations are among the most highly preferred (68). Chocolate candy is an excellent example of a highly palatable sugar–fat mixture.

In contrast to sweet and bitter tastes, we know very little about the early development of fat preference. However, there is some evidence that experience can modify hedonic responses to high-fat foods. Mattes (69) reported that individuals who were deprived of the sensory experience of fat for 12 weeks showed declines in pleasantness ratings for high-fat foods and in preferred fat content of some foods. Participants who achieved a similar reduction in fat intake, but were permitted to use fat substitutes to obtain a fat-like sensory experience, did not show these changes in hedonic response to fats.

Summary

Flavor is a complex combination of several sensory inputs including the senses of taste, smell, chemical irritation, touch and temperature. Chocolate is the flavor of the gods (a take-off on its Latin name *Theobroma cacao*, which means 'food of the gods' (6)) because it combines sweet taste tempered by a touch of bitterness, because of its wonderful and complex aroma and its sensuous mouth-feel.

References

1. Rozin, P., Levine, E. and Stoess, C. (1991) Chocolate craving and liking. *Appetite* **17**, 199–212.
2. Weingarten, H. P. and Elston, D. (1991) Food cravings in a college population. *Appetite*, **17**, 167–175.
3. Pelchat, M.L. (1997) Food cravings in young and elderly adults. *Appetite* **28**, 103–113.
4. Michener, W. and Rozin, P. (1994) Pharmacological versus sensory factors in the satiation of chocolate craving. *Physiol. Behav.* **56**, 419–422.
5. (1991) *Physiol. Behav.* **49** (5) (whole issue).
6. McGee, H. (1984) *On Food and Cooking*. Charles Scribner's Sons, New York.
7. Hoskin, J.C. (1994) Sensory properties of chocolate and their development. *Am. J. Clin. Nutr.* **60** (Suppl.), 1068S–1070S.
8. Bergamasco, N.H., and Beraldo, K.E. (1990) Facial expressions of neonate infants in response to gustatory stimuli. *Braz. J. Med. Biol. Res.* **23** (3/4), 245–249.
9. Rosenstein, D. and Oster, H. (1988) Differential facial responses to four basic tastes in newborns. *Child Devel.* **59** (6), 1555–1568.
10. Steiner, J.E. (1974) Discussion paper: innate, discriminative human facial expressions to taste and smell stimulation. *Ann N. Y. Acad. Sci.* **237**, 229–233.
11. Steiner, J.E. (1973) The gustofacial response: observation on normal and anencephalic newborn infants. *Symp. Oral Sensation Perception* **4**, 254–278.
12. Moskowitz, H., Kumaraiah, V., Sharma, K.N., Jacobs, H.L. and Sharma, S.D. (1975) Cross-cultural differences in simple taste preferences. *Science* **190** (4220), 1217–1218.
13. Breslin, P.A. and Beauchamp, G.K. (1995) Suppression of bitterness by sodium: variation among bitter taste stimuli. *Chem. Senses* **20** (6), 609–623.
14. Breslin, P.A. and Beauchamp, G.K. (1997) Salt enhances flavour by suppressing bitterness. *Nature* **387** (6633), 563.
15. Rozin, P. (1976) The selection of food by rats, humans and other animals. In *Advances in the Study of Behavior*, Vol. 6 (Ed. by Rosenblatt, J., Hinde, R.A., Beer, C. and Shaw, E.), pp. 21–76. Academic Press, New York.
16. Desor, J.A., Maller, O. and Andrews, K. (1975) Ingestive responses of human newborns to salty, sour, and bitter stimuli. *J. Comp. Physiol. Psychology* **89** (8), 966–970.
17. Maone, T.R., Mattes, R.D., Bernbaum, J.C. and Beauchamp, G.K. (1990) A new method for delivering taste without fluids to preterm and term infants. *Dev. Psychobiol.* **23** (2), 179–191.
18. Mattes, R.D., Maone, T., Wager-Page, S., *et al.* (1996) Effects of sweet taste stimulation on growth and sucking in preterm infants. *J. Obstet. Gynecol. Neonatal Nurs.* **25** (5), 407–414.
19. Yokomukai, Y., Cowart, B.J. and Beauchamp, G.K. (1993) Individual differences in sensitivity to bitter-tasting substances. *Chem. Senses* **18** (6), 669–681.
20. Reed, D.R., Bartoshuk, L.M., Duffy, V., Marino, S. and Price, R.A. (1995) PROP

tasting: determination of underlying threshold distributions using maximum likelihood. *Chem. Senses* **20**, 529–533.
21. Hall, M.J., Bartoshuk, L.M., Cain, W.S. and Stevens, J.C. (1975) PTC taste blindness and the taste of caffeine. *Nature* **253** (5941), 442–443.
22. Bartoshuk, L.M. (1979) Bitter taste of saccharin related to the genetic ability to taste the bitter substance 6-*N*-propylthiouracil. *Science* **205** (4409), 934–935.
23. Gent, J.F. and Bartoshuk, L.M. (1983) Sweetness of sucrose, neohesperidin dihydrochalcone and saccharin is related to genetic ability to taste the bitter substance 6-*N*-propylthiouracil. *Chem. Senses* **7**, 265–272.
24. Drewnowski, A., Henderson, S.A. and Shore, A.B. (1997) Taste responses to naringin, a flavinoid, and the acceptance of grapefruit juice are related to genetic sensitivity to 6-*N*-propylthiouracil. *Am. J. Clin. Nutr.* **66** (2), 391–397.
25. Fischer, R., Griffin, F., England, S. and Garn, S.M. (1961) Taste thresholds and food dislikes. *Nature* **191**, 1328.
26. Ritchey, N. and Olson, C. (1983) Relationships between family variables and children's preference for and consumption of sweet foods. *Ecol. Food Nutr.* **13**, 257–266.
27. Logue, A.W., Logue, C.M., Uzzo, R.G., McCarty, M.J. and Smith, M.E. (1988) Food preferences in families. *Appetite* **10**, 169–180.
28. Looy, H., Callaghan, S. and Weingarten, H.P. (1992) Hedonic responses of sucrose likers and dislikers to other gustatory stimuli. *Physiol. Behav.* **52**, 219–225.
29. Looy, H. and Weingarten, H. (1991) Facial expressions and genetic sensitivity to 6-*N*-propylthiouracil predict hedonic response to sweet. *Physiol. Behav.* **52**, 75–82.
30. Drewnowski, A., Henderson, S.A., Shore, A.B. and Barratt-Fornell, A. (1997) Nontasters, tasters and supertasters of 6-*N*-propylthiouracil (PROP) and hedonic response to sweet. *Physiol. Behav.* **62** (3), 649–655.
31. Green, L.S., Desor, J.A. and Maller, O. (1975) Heredity and experience: Their relative importance in the development of taste preference in man. *J. Comp. Physiol. Psychology* **89**, 279–284.
32. Krondl, M., Coleman, P., Wade, J. and Milner, J. (1998) A twin study examining the genetic influence on food selection. *Hum. Nutr. Appl. Nutr.* **37A**, 189–198.
33. Falciglia, G.A. and Norton, P.A. (1994) Evidence for a genetic influence on preference for some foods. *J. Am. Diet. Assoc.* **94**, 154–158.
34. Rozin, P. and Millman, L. (1987) Family environment, not heredity, accounts for family resemblances in food preferences and attitudes: a twin study. *Appetite* **8**, 125–134.
35. Drewnowski, A. (1991) Obesity and eating disorders: cognitive aspects of food preference and food aversion. *Bull Psychonomic Soc.* **29**, 261–264.
36. Logue, A.W. and Smith, M.E. (1986) Predictors of food preferences in adult humans. *Appetite* **7**, 109–125.
37. Beauchamp, G.K. and Moran, M. (1982) Dietary experience and sweet taste preference in human infants. *Appetite* **3**, 139–152.
38. Desor, J.A., Green, L.S. and Maller, O. (1975) Preferences for sweet and salty in 9- to 15-year-olds and adult humans. *Science* **190**, 686–687.
39. Bertino, M., Beauchamp, G.K. and Jen, K.C. (1989) Rated taste perception in two cultural groups. *Chem. Senses* **8**, 3–15.
40. Cowart, B. (1989) Relationships of taste and smell across the adult life span. In

Murphy, C., Cain, W.S. and Hegsted D.M. (Eds), Nutrition and the chemical senses in aging. *Ann. N. Y. Acad. Sci.* **561**.

41. Murphy, C. and Gilmore, M.M. (1989) Quality-specific effects of aging on the human taste system. *Perception Psychophys.* **45**, 121–128.
42. Weiffenbach, J.M., Baum, B.J. and Burghauser, R. (1982) Taste threshold: quality-specific variation with aging. *J. Gerontol.* **37**, 372–377.
43. Murphy, C. and Withee, J. (1986) Age-related differences in the pleasantness of chemosensory stimuli. *Psychol. Aging* **1**, 312–318.
44. Teeter, J.H. and Brand, J.G. (1987) Peripheral mechanisms of gustation: physiology and biochemistry. In *Neurobiology of Taste and Smell* (Ed. by Finger, T.E. and Silver, W.L.), pp. 299–329. Wiley, New York.
45. Laird, D.A. and Breen, W.J. (1939) Sex and age alterations in taste preferences. *J. Am. Diet. Assoc.* **15**, 549–550.
46. Desor, J.A. and Beauchamp, G.K. (1987) Longitudinal changes in sweet preferences in humans. *Physiol. Behav.* **39** (5), 639–641.
47. Murphy, C. and Cain, W.S. (1980) Taste and olfaction: independence vs. interaction. *Physiol. Behav.* **24**, 601–606.
48. Murphy, C., Cain, W.S. and Bartoshuk, L.M. (1977) Mutual action of taste and olfaction [Abstract]. *Sensory Processes* **1**, 204–211.
49. Rozin, P. (1982) 'Taste–smell confusions' and the duality of the olfactory sense. *Perception Psychophys.* **31**, 397–401.
50. Mozell, M., Smith, B.P., Smith, P.E., Sullivan, R.L. and Swender, P. (1969) Nasal chemoreception in flavor identification. *Arch. Otolaryngol.* **90**, 131–137.
51. Morgan, J. (1994) Chocolate: a flavor and texture like no other. *Am. J. Clin. Nutr.* **60** (Suppl.), 1065S–1067S.
52. Cain, W.S. and Stevens, J.C. (1989) Uniformity of olfactory loss in aging. In Murphy, C., Cain, W.S. and Hegsted, D.M. (Eds) Nutrition and the chemical senses in aging. *Ann. N. Y. Acad. Sci.* **561**, 29–38.
53. Chalke, H.D. and Dewhurst, J.R. (1957) Accidental coal-gas poisoning. *BMJ* October 19, 915–917.
54. Enns, M.P. and Hornung, D.E. (1988) Comparisons of the estimates of smell, taste and overall intensity in young and elderly people. *Chem. Senses* **13**, 131–139.
55. Murphy, C. (1985) Cognitive and chemosensory influences on age-related changes in the ability to identify blended foods. *J. Gerontol.* **40**, 47–52.
56. Schiffman, S. (1977) Food recognition by the elderly. *J Gerontol.* **32**, 586–592.
57. Schiffman, S. and Pasternak, M. (1979) Decreased discrimination of food odors by the elderly. *J Gerontol.* **34**, 73–79.
58. Duffy, V.B., Backstrand, J.R. and Ferris, A. M. (1995) Olfactory dysfunction and related nutritional risk in free-living, elderly women. *J. Am. Diet. Assoc.* **95** (8), 879–884.
59. Cain, W.S., Reid, F. and Stevens, J.C. (1990) Missing ingredients: aging and the discrimination of flavor. *J. Nutr. Elderly* **9**, 3–15.
60. Schiffman, S.S. and Warwick, Z.S. (1988) Flavor enhancement of foods for the elderly can reverse anorexia. *Neurobiol. Aging* **9**, 24–26.
61. Schiffman, S.S. and Warwick, Z.S. (1993) Effect of flavor enhancement of foods for the elderly on nutritional status: food intake, biochemical indices, and anthropometric measures. *Physiol. Behav.* **53**, 395–402.

62. Dorries, K.M., Schmidt, H.J., Beauchamp, G.K. and Wysocki, C.J. (1989) Changes in sensitivity to the odor of androstenone during adolescence. *Dev. Psychobiol.* **22** (5), 423–435.
63. Wysocki, C.J., Dorries, K.M. and Beauchamp, G.K. (1989) Ability to perceive androstenone can be acquired by ostensibly anosmic people. *Proc. Nat. Acad. Sci. USA* **86** (20), 7976–7978.
64. Buck, L. and Axel, R. (1991) A novel multigene family may encode odorant receptors: a molecular basis for odor recognition. *Cell* **65** (1), 175–187.
65. Christensen, C.M. (1984) Food texture perception. *Adv. Food Res.* **29**, 159–198.
66. Szczesniak, A.S. (1990) Texture: is it still an overlooked food attribute? *Food Technol.* **44**, 86–95.
67. Drewnowski, A. (1997). Why do we like fat? *J. Am. Diet. Assoc.* **97** (Suppl. 7), S58–S62.
68. Drewnowski, A. and Greenwood, M.R. (1983) Cream and sugar: human preferences for high-fat foods. *Physiol. Behav.* **30** (4), 629–633.
69. Mattes, R.D. (1993) Fat preference and adherence to a reduced-fat diet. *Am. J. Clin. Nutr.* **57** (3), 373–381.

Chapter 22

Cultural and Psychological Approaches to the Consumption of Chocolate

Matty Chiva

Chocolate is a very widespread product which can be found nearly anywhere in the world nowadays. It is available in many forms and presentations, from the well-known bars of chocolate to speciality tarts and cakes, not forgetting chocolate-coated sweets and chocolate-coated bars. It is easy to eat, the price is affordable, it is easy to store and it has many good qualities.

We will see later that these 'qualities' have been attributed to it since its consumption was learnt of. They were important enough in any event for the tree to be called *Theobroma cacao*, when the cocoa tree was introduced to the Western botanical world, *Theobroma* meaning, in Greek, drink of the gods or food of the gods (1). The scientists in this case were only following the first accounts by the Spanish explorers who described the place and role of the cocoa tree in Amerindian cultures.

Anyhow, and even if it has become a product for everyday consumption, chocolate has the advantage of an excellent brand image. It is associated both with hedonism, festivity and tenderness, and with the image of a food, a little of which gives a lot of energy. And despite its abundance and availability, a gift of chocolates always gives pleasure.

At the same time, it seems that despite its omnipresence chocolate is not the same everywhere. There are great differences in presentation, in usage and especially in taste between different cultures. Similarly its consumption does not obey the same rules everywhere.

To try to understand more fully the origin and determinants of consumption behaviour, therefore, this chapter distinguishes between three different facets in this approach to consumer behaviour. These are: historical aspects; socio-cultural considerations in the very construction of the image of a food; and, finally, psychological approaches concerning the determinants of our behaviour in this sphere.

Chocolate: a food that has been a success

In studying human food one has to take several facts into consideration:

(1) The fact of eating, essential to the maintenance of life and health, is determined in the first place by biological considerations. These vary between species, and therefore largely determine their adaptation and ways of life. In the case of the human species, one of the major determinants is the fact that we are omnivorous. Unlike specialist eaters (herbivores, carnivores, etc.), in fact, humans need a varied diet, since they are incapable of synthesizing certain components which are essential for life. Being omnivorous is therefore both a constraint – that of finding the components of a diversified diet – and an extraordinary property, enabling our species to adapt to the most varied media and living conditions (2, 3). Being omnivorous also has other consequences, which are dealt with later, concerning the attitude to food.

(2) The search for food, sharing and storing it among members of human groups has played a major role in the very process of construction of human societies. This process has not only enabled survival, and then an increasingly better life, but has also had a powerful role in the structuring of human societies and the working out of the rules which govern them (4, 5). This has been expressed in the establishment of structured groups, the organization of hunting and distribution, the invention of animal husbandry and cultivation, and by the introduction of what amount to rules for consumption, table manners and timetables for meals (6).

(3) Another fact is obvious and important in the context of what is of interest to the subject matter of this chapter: humans have always travelled. These travels were determined as much by the need to find new living room, by curiosity, by a taste for knowledge, as by pleasure. To travel involved and still involves meeting other people, other ways and other products. The travellers' return home was not only a source of tales but also the opportunity to bring back new things, unknown in the civilization to which they belonged. From this perspective, foodstuffs, whether fauna, flora or mineral products, have occupied a major place.

A source of knowledge, but often a source of profit as well, the transportation of food has played a major role in the very history of humanity. Trade in spices or, although not a food, in tulips created enormous sources of wealth (1, 7). The planting of sugar cane in the New World and the resulting sugar trade resulted in major social changes, such as the slave trade (8): the remote consequences of this practice can still be felt. This is confirmed again even now, in a general way, with the economic and political importance of trade in agro-food products.

But, more modestly, this movement of food has led throughout the world to trade in and substitutions of one food for another, more appropriate or easier to store or consume. Finally, in certain cases, new products have arrived which did not exist in the culture of origin and which have become incorporated into it. These movements, for which there is evidence since earliest times, have had peaks of dietary innovation with the great discoveries (1, 7, 9). But the arrival of new products did not guarantee their immediate acceptance into the host culture.

Although the Middle East, for example, was the source of the spread more particularly of cereals or of the many varieties of beans, the discovery of the New World led to the discovery of a whole series of other products, plants in particular. Among the best known are maize, potatoes, tomatoes, peppers and, of course, cocoa. All of them are now 'successful products'. But one ought to remember that their acceptance by the Old World was never immediate and took longer than one may now think.

Although the New World was discovered by Christopher Columbus in 1492, cocoa, and with it chocolate, only arrived in Spain about 1520 (or, according to other sources, 1527 when Cortés is said to have given beans of cocoa to Charles V). It reached the Netherlands (which were then Spanish provinces) about 1609, spread to England in 1657, and reached France only in 1659. Furthermore, it was only in 1660 that Benjamin d'Acosta inaugurated the first plantation of cocoa trees in the Windward Islands, now known as the Lesser Antilles (9, 10).

As previously mentioned, chocolate as a drink played a large role in the religious ceremonies of the Aztec Court. Similarly, cocoa beans played the role of currency; anything one wanted could be bought with them: gold, clothing, slaves or prostitutes (10). But it was not until Diaz del Castillo, a companion of Cortés, recounted for the first time the place of this beverage at the court of Montezuma, Emperor of the Aztecs, in Mexico, that people learnt what this beverage was like: cocoa, finely milled, was mixed with pimento, boiled, and stirred vigorously with special tools to make it foam before it was drunk.

Other preparations based on this fundamental recipe existed; apart from the pimento, which was always included, achiote (roucou), ground maize, fruits or even hallucinogenic mushrooms could be added to it. These various preparations bestowed on it the status of a food, but also of a drug, or even of an aphrodisiac (10).

This way of preparing 'chocolate' is quite different from our present tastes. But the use of cocoa or of bitter black chocolate to make a pimento and spiced sauce is still common today in Mexico: this is the famous chocolate turkey (*mole poblano do guajolote*).

Nonetheless, in its original form 'chocolate' was not very popular from the start; it began to be accepted later, only towards the end of the 16th century, thanks to the conjunction of two factors:

- The first is what can be described as its meeting with sugar. According to certain Mexican traditions it was the nuns in the convent of Oaxaca who wanted to sweeten the beverage to make it acceptable to them. They therefore replaced the pimento with sugar. That was at the time of the boom in sugar-cane plantations. There was increased availability of sugar as such, and also it arrived in Europe in bulk thanks to Portuguese trade[1].

Most noteworthy was that the addition of sugar, an accepted and well-known taste, was required to temper the bitterness of the cocoa and replace the pimento, thus making chocolate closer to accepted tastes. On the other hand, for a long time after that, spices, such as vanilla, cinnamon, anise, orange flowers, etc. were readily added in Spain. Vestiges of this preparation exist to this day.

- The second factor comes from what may be called the civilization of the court. According to this model, devised by Elias (11), royal society served as a reference. This model was all the more valued because it was difficult to reach, reserved for a few, even prohibited for ordinary mortals. The habits, fashions and values of the court were first copied by those near to it: nobles, courtiers, etc. The latter imitated and at the same time, deliberately or not, transformed the initial model; they themselves served as models for those beneath them.

  Thus, through what one may term a virtual apprenticeship through imitation, the manners and fashions of the court served as a reference and were involved in the spread of habits[2].

The consumption of chocolate spread in accordance with this latter model. For example, in France, under Louis XIV, consumption was essentially for aristocrats. In 1659 only one shop had the royal privilege of 'selling and supplying a certain composition called chocolate'. This privilege was renewed in 1666, and a second beneficiary appeared only in 1692 (10).

Chocolate was consumed because it was a model copied from the king, but also because it had numerous virtues. The abundant correspondence of Madame de Sévigné, virtual chronicler of the French Court, is proof of this: to begin with it was reputed to have medicinal and aphrodisiac qualities and above all that of being the product in fashion. These qualities were then decried, but again it

[1] The accessibility of sugar and its arrival in Europe in bulk also had economic, political and cultural consequences. I would like to note here, anecdotally, the development in Portugal, which was the main importer of sugar at that time, of a literary mode: the sonnets of sugar. These were collections of poems which were produced mainly by monks and nuns in monasteries and which lasted for about a century (end of the 16th to about the end of the 17th century). Love poetry, it has a specific feature: any allusion to love was replaced by allegories of sugar!

[2] This can still be found today, in another way. The inscription on the wrapping of certain products, certifying that these are brands recognized as supplied to such and such a famous person or reigning monarch, confer a definite marketing value on these products. It is as if the product has a higher quality just because of this. Thus, through the medium of a product, the consumer becomes a 'person of taste' and continues to perpetuate this model of society.

became the miracle beverage that one had to take every day. All these stages can be found in her correspondence for 1671 (12).

These beliefs regarding chocolate lasted for a long time and numerous references can be found in Brillat-Savarin (13). He was an amazing person: a magistrate, condemned to exile in 1794, he arrived in America where, after a short spell as a diplomat, he became the first violin at the New York Theatre. Curious about everything and above all about manners and gastronomy, he observed a great deal in the New World. When he returned to France in 1796 he became a magistrate again. But he was above all a chronicler of his time and one of the pioneers of today's gastronomic guides and guides to manners. Thus, in 1825, he reported a whole series of 'virtues of chocolate' which was said to have various properties: it would facilitate digestion, it would prevent minor everyday disorders, it would allow one to retain one's figure by limiting corpulence and of course, it was said, as with the Aztecs, to have restorative and aphrodisiac properties. In conclusion, Brillat-Savarin even asserted: 'What is health? It is chocolate!'

The appearance of chocolate as we know it today, most often in the shape of bars or delicacies such as soft-centred chocolates, pralines, etc., had to wait for the 19th century and the appearance of a whole series of industrial innovations. The industrial revolution which was beginning helped to facilitate and standardize manufacturing processes: hydraulic presses, hot tables, conching vats appeared. Within a very short time a whole series of chocolate factories bloomed, became trade names, some of which still exist. Thus for example, Menier set up a chocolate factory in France in 1824 at Noisiel, followed in 1848 by Poulain at Blois. In Switzerland the first factory was established at Vevey in 1819, followed by others: Suchard in 1824, Kohler in 1828, then Lindt, Tobler and Nestlé. The latter perfected the manufacture of milk chocolate in 1870. In the Netherlands, Van Houten, in 1828, invented processes for preparation of cocoas. Finally, much closer to our day, in 1923, the first filled chocolate-coated bar was created in the USA by the Mars chocolate factory. All these developments also resulted in a considerable drop in price, while quality was maintained and improved, so that chocolate became a widely available product.

In summary, almost five centuries have been needed for this miracle product to attain its present fame. This review is necessary for two reasons:

- On the one hand, because this has not happened to chocolate alone. Equally long time lapses can be found for other products which have 'been a success': maize, introduced early on into Europe, in 1493, took an equally long time to be accepted but has not become as common. It suffered from the fact that it was introduced first of all through the peasantry, who did not have the same prestige of 'a model to be imitated' as court society. The tomato, now so firmly incorporated as an item of southern and Italian cuisine in particular, only really spread in Europe at the end of the 18th and beginning of the 19th century. Historical time is often forgotten and measured in subjective terms rather than in real time.

- On the other hand, because even if it results in general acceptability, as in the case of chocolate, acceptance of dietary innovation experiences local adaptations. As a general assertion, chocolate is loved everywhere; but it is not exactly the same chocolate, from an organoleptic point of view, that people in different countries like. And this warrants consideration.

Society: individuals and their diet

One cannot live without eating. The act of eating is a result of biological and physiological determinants prescribed in the body. Can one then consider that eating is an act which occurs of its own, in view of the determinants which have just been mentioned? The answer is no. Eating in fact is learnt and this learning roughly depends on two inherited traits, the biological and the cultural.

The biological inheritance consists in particular of the genetic information, inscribed in each of our cells, from conception. This inheritance determines each of us in two ways: as a member of the species on the one hand, and as a unique, specific individual on the other.

The genetic inheritance of the species thus determines certain traits shared by all human beings: the omnivorous nature, the digestive and metabolic systems, and the very structure of the body. But it is also as a function of genetic information that differences between individuals appear: certain characteristics such as height, weight, bodily morphology, the colour of the eyes, the specific individual metabolism and also the major differences from one person to another in the sphere of sensory discrimination.

The cultural inheritance includes beliefs, behaviours, attitudes and practices established by human groups, a whole that can only be described by the term culture. Culture, which in its most advanced form is a specifically human concept, has emerged from a prolonged progression of accumulation and transmission of knowledge, opinions and attitudes. Culture, which is also evolving over time, is transmitted solely by learning, unlike biological information, transmitted by the genes and inscribed in the body.

The nature of these two inheritances is that they are given to us at the start, outside of any possible individual choice. They interact in programming feeding behaviour which is original and it is not possible later to decide which specific share is due to each (14, 15). The construction in this case is identical to what happens in spoken language: every human being is in fact genetically programmed to learn to speak. But this initial programming does not in any way predict which language he or she will speak. Paradoxically then, one can assert that speech is totally determined by genetics which lay the neurological and morphological foundations of the mechanisms of speech. But it is equally totally a function of the cultural framework, which alone determines which language is spoken; that is to say, the code employed for communicating. This is the same course that one finds in the establishment of feeding behaviour (16, 17).

It has been asserted above that eating is learnt, but what is there to learn? In the first place, what food is. As omnivores, our species can eat a very wide range of products. But while accepting a large variety, this range is much smaller at individual level. This can easily be illustrated by recalling different food prohibitions in force for large human groups. The prohibition on the consumption of pork for devout Jews and Muslims, of beef for the Hindus, of dogs for most Europeans, of horses which, eaten in France, cannot be eaten in the UK, no more than rabbit can elsewhere. Not to mention insects or rodents, which are widely consumed in one region of the world yet prohibited elsewhere.

In other words, for a given culture, all the products included in a given list and defined as such by this culture are food. It is through educational conduct, itself inscribed both in the culture and in an affective and relational context from birth, that the child establishes this list. In other words one thus learns what is 'food for me' or 'food for us'.

Among the different kinds of learning involved, it is above all by learning through imitation that this list is constructed. This is primarily learning which is not imposed but sought by the subject; doing as others do means being able to become a member of the social group, the group of peers. Imitating one's elders, which is another form of learning, particularly in children, which comes this time from a desire to acquire new skills and status, means that one has grown up. This learning is directly and strongly involved in the building of the identity of personality and social identity. It is the latter which makes it possible for one to recognize oneself as belonging to a given human group and at the same time to differentiate oneself from other groups (18, 19).

Learning what food is also involves the constructing of the *taste for food*. The taste for food is in fact a complex whole which involves nearly every sensory modality: taste, olfaction, vision, hearing, thermal perception, oral stereognosis (perception of forms and volumes) and touch. Each of these sensory modalities provides information about food. But the whole is more than the simple sum of the components. The result, called *synaesthesia*, incorporates all this sensory information as well as subjective components arising from previous emotional experience.

At the end of the day the taste for food includes two different and complementary approaches to information processing:

- One is cognitive – this is the one which makes it possible to give a meaning to the items of information, i.e. to name them and give them meaning.
- The other is hedonic – it processes the information from an affective point of view, giving a measure of pleasure or displeasure.

Here one finds specificity linked to gustatory sensation. This sensory modality is very early and functions even before birth: from the fourth month of pregnancy the fetus can distinguish between different tastes. But it is important to realize that there is a genetically determined differential processing, of pleasure or of displeasure regarding the different families of flavours. Thus, universally, and this can be found in the large majority of animal species, there is an attraction for sweet tastes and rejection of bitterness. To this hedonic rating, inscribed in the body, are added later emotional experiences linked to earlier experiences of food, which are remembered (20).

An extra set is added to this already complex picture: the programming which makes us prefer a sweet flavour predominates in early infancy. It may then be modified by later learning, particularly by cultural rules. But in the absence of learnt rules, initially a sweet flavour is attractive and facilitates the acceptance of new foods.

There is another form of learning then: that of the organoleptic properties of foods as valued by a given culture. These properties, developed over time, form a characteristic series of 'our food' or of 'our cuisine' and make it possible to identify foods as belonging to it or not by differentiating between them. Thus a sweet flavour is different, more or less intense, depending on whether one is dealing with what are termed 'Eastern' or 'European' pastries, for example. The use of pimentos, of characteristic spices, such as cumin, coriander, cinnamon, cardamom and dill, for example, characterizes certain cuisines. Similarly the nature of the fat, its quality and the quantity used play the same role.

Generally speaking these indicators are termed culinary markers or flavour principles (21). It is therefore the learning of these common rules/standards which also fashions personal preferences and which at the same time brings acceptance of new flavours or those initially rejected. Thus one learns to appreciate what is bitter, of which coffee (especially without sugar) is an example. In the same way certain cultures value the consumption of pimentos, although initially the burning sensation they produce is aversive.

This last example illustrates another fundamental concept in the construction of feeding behaviour, that of perception. Perception is a general psychological process. Simplifying, it consists in giving a meaning to the information contributed by the senses. To begin with, in fact, sensory messages are only a simple translation into electrical impulses carried towards the brain of external or internal stimuli of various origins (physical, chemical, etc.). It is their repetition, their association with other signals, the relationship with external conditions and the consequences of these signals, making it possible to give them a meaning, that can be articulated. The best example is that of learning a language: on the level of verbal communication, one is always dealing with physical, sound stimuli. Different combinations of them and the learning of the underlying code enable the meaning to be decoded, and to be understood – in short, communication.

Thus the burning sensation added by the pimento does not, from a physiological point of view, change with time in those who eat or do not eat different cuisines. What changes is the perception, that is to say the interpretation of the signal: for those who do not eat pimento, its detection always means pain and avoidance, while for those who have learnt to like it, it is the satisfaction of recognizing what characterizes 'my cuisine'.

In approaching the cultural aspects, another fact directly related to our condition as omnivores should be pointed out: food is not only consumed, it is also thought about. It should in fact be noted that food is not an object like others. It is unique along with drugs (in the broad meaning of the term) in having to be incorporated in order to fulfil its function. Eating means consuming things from the outside, which always involves a risk. However, this risk cannot be avoided because one cannot live without consuming food. This situation creates conflicts and anxiety and occurs, in fact, in all omnivorous species. However, it reaches a higher degree of sophistication in humans (2, 21–23).

According to Fischler (23), four socio-anthropological aspects characterize more particularly the human approach to food:

(1) Classificatory thought. One cannot grasp the world of food in particular without constructing classifications and categories. This then makes it possible to establish rules concerning the relationships between the categories and, beyond that, behaviour regarding consumption.

(2) The principle of incorporation. Emerging from the work of anthropologists, Tylore and Frazer in particular, it is based on the demonstration of the existence of a 'magic thought'. This thought is not the privilege of some populations only but coexists in all of us with logical thought (24, 25). The principle of incorporation, which comes from what Frazer has characterized as 'sympathetic magic', brings together beliefs and representations which can be summarized by the formula, 'one is what one eats'. In other words, by analogy the eater absorbs with their food not only its substance but also its qualities, virtues and defects, whether physical, moral or symbolic.

(3) The paradox of the omnivore. This concept, presented for the first time by P. Rozin (22), associates the constraint of a varied diet, biologically inscribed in the body, and freedom of choice, which permits the variety of cultural rules. It indicates the indissoluble link in our species between biological and cultural in our relationships with food.

(4) Dietary moralism. This concerns normative and moral judgements concerning food, its consumption, and the qualities of the individual who conforms or does not conform with social and cultural rules (26–28). This moralism plays a major although not consciously recognized role in the majority of the recommendations and advice circulated both by the media and by official bodies.

The succinct approach to these concepts and general mechanisms in perspectives that come more specifically from the human sciences is essential here. When stress is increasingly placed on information emerging from the exact sciences or from strictly normal or pathological medical approaches, it makes it possible to draw attention to other equally important facets. Today one can no longer ignore the importance of opinions, attitudes and perceptions in determining dietary behaviour.

Neither can one ignore the place of emotional and hedonic aspects, both in making decisions and in the motivation of our behaviour in general, and dietary behaviour in particular (29).

Now, to come back to chocolate, a product which has 'succeeded' in the West particularly because of its encounter with sugar. It is this association which has made it acceptable in a large number of cultures in as much as it sweetens the initial bitterness of the cocoa, by adding to it a sugary note. From an organoleptic point of view, chocolate also benefits from another intrinsic quality: the presence of fat. In fact, the association of fat and sugar easily confers on it qualities of texture which flatter the palate and make it a desirable product. This has been demonstrated by studies that have yielded a better understanding of the nature of the lubrication, which is so desirable in certain products (30). It is also true that manufacturing procedures, conching in particular, have considerably refined the grain of the product and its texture in the mouth.

Chocolate is also, and has long been, a product that is 'good to think about'. Good, first of all, because of its real or supposed properties, its virtually medicinal beneficial effects, its hidden virtues. The association with other products, also 'good to think about', such as milk, further strengthens these attitudes; because of this it has become a product which can and is widely recommended for children. But chocolate is a good product also simply because of the hedonic aspects, the sensory pleasure produced by eating it. Gifts of and eating of chocolate rapidly came to be associated with pleasant situations or events: birthdays, weddings, customs related to religious feasts, gifts made to children and adults, even more ambiguously (because of its aphrodisiac properties) gifts between lovers.

However, the virtually universal acceptance of chocolate should not allow one to forget that there are differences between one country and another. They are more particularly the results of culinary markers or local cultural features. The association of mint with chocolate is typically British. In Belgium, chocolate in the form of soft-centre sweets (pralines) is consumed more than elsewhere, with more fat and more sugar than in France, for example. In the latter country, although in market terms a lot of milk chocolate is consumed quantitatively, its taste is different from that of Swiss milk chocolate. Furthermore, in France for the past 25 years an increasingly marked liking for dark chocolate or bitter chocolate, which contains a higher proportion of cocoa, has been observed. This trend is interesting because it illustrates directly how a norm socially valued by an élite is spread and accepted in society. This list is not exhaustive and the producers are well aware of the phenomenon, and take it into account by diversifying output according to the populations targeted.

The image of chocolate: an experimental approach

Despite the universality of dietary behaviour, it is clear to all those concerned with dietary behaviour that there are many consumption strategies as well as commercial targets. It is clear today from many studies that three series of factors play a role very directly in the motivation and determination of the consumers' behaviour and must be taken into account systematically (14):

(1) The organoleptic aspects, i.e. the place of the sensory modalities which determine the taste of the food.

(2) The hedonic aspects, i.e. the dimension of pleasure, enjoyment or lack of enjoyment which a food can give us and also the social acceptability of this pleasure.

(3) The ideational aspects, i.e. what one thinks about the food, its virtues, qualities, defects or dangers. It is important to emphasize that these aspects may be based on objective scientific data or on beliefs. It is nonetheless true that in both cases they have a role to play in dietary behaviour.

That being the case, it is not enough to know that these three series of aspects are involved in the determinants and motivations of dietary behaviour. The next stage, which is of interest both to research workers and to industrialists, is the development of tools which can demonstrate and measure these different aspects.

There are many studies concerning the analysis of the organoleptic properties of food, particularly since the pioneering work of Pangborn (31, 32). They have made possible more particularly the development of numerous methods of quantitative and, partially, qualitative sensory analysis.

On the other hand, relatively few studies have approached the consumers' hedonic perception of the product. Critical reservations, which moreover are perfectly valid, have been expressed as to the difficulty of this approach as a general one (33). The most pertinent remarks concern two major aspects: on the one hand, the consumers' difficulty in explaining the reasons for the pleasure produced by a food; and on the other hand, the need to set these analyses strictly within the original cultural and social context.

As to the study of the ideational aspects, this is much more recent, and few people are as yet researching them at the moment. The number of experimental studies in this sphere is still quite small, but very promising.

Finally, until recently there were no studies which systematically approached the three aspects concerning one and the same food and one and the same population of subjects. Only with an approach of this kind can one examine not only each of the three aspects taken separately, but also the interactions between the three spheres and their reciprocal influences.

It is on the basis of this observation that the author has developed in the laboratory an integrated approach methodology, which aims to obtain a clearer understanding of the interaction of these three series of factors and, in addition, the consumers' perception of the products (34, 35). Without going into too many technical details here, it should be made clear that this is a series of three measurement scales, each concerning one of the three aspects mentioned above. Several populations have been investigated, each being questioned about several foods. Here only the findings concerning chocolate are presented.

The product investigated in this case is Lindt Dark Chocolate (70% cocoa). The study concerned a preliminary population of 41 adults (average age 34 years); a second study was duplicated on a population of 27 adults (same average age). The results presented here concern the second population only, knowing that the series of results from both studies agree.

In the case of the organoleptic data, the descriptors characterizing the different products came from a preliminary study on an ordinary population of adults who were asked to describe the different products. The statistically most consensual categories were adopted and thus formed the scale applied to the other populations. Thus in the case of chocolate, 35 classification categories from among the 269 spontaneously supplied by the consumers were chosen.

Perceived pleasure was estimated with the help of unstructured analogue scales in two ways: on the one hand a global assessment, and on the other hand for each organoleptic classification category perceived by the subject.

Finally, the ideational approach was investigated using a multiple-choice questionnaire, developed for the purposes of the study (Fig. 22.1).

The study of the relationship between perceived sensory intensity and hedonic rating is interesting. There is a strong and significant correlation between the two series of findings ($r^2 = 0.76$). This co-variation illustrates an important point in the construction of perception of the product. In fact, dark chocolate is now a product that is not only known but also appreciated. The clearer and more perceptible the apparent signs enabling him or her to recognize and identify it – colour, odour and texture first of all, aroma, taste and mouth-feel after that – the more the consumer appreciates it. The more easily perceived they are, the more intense the feeling of pleasure.

It should be emphasized that this is not an experimental artefact. On other products the same subjects did not show the same correlation. It is the agreement between expectation, sensory perception and the hedonic aspect which indicates in this case not only a certain consistency in the consumers, but also the fact that the chocolate in question benefits from a very specific and expected presentation.

Moreover, the average hedonic score attributed to the product is high: 8.07 before tasting, 7.17 after tasting on a 10-point scale, with a significant correlation

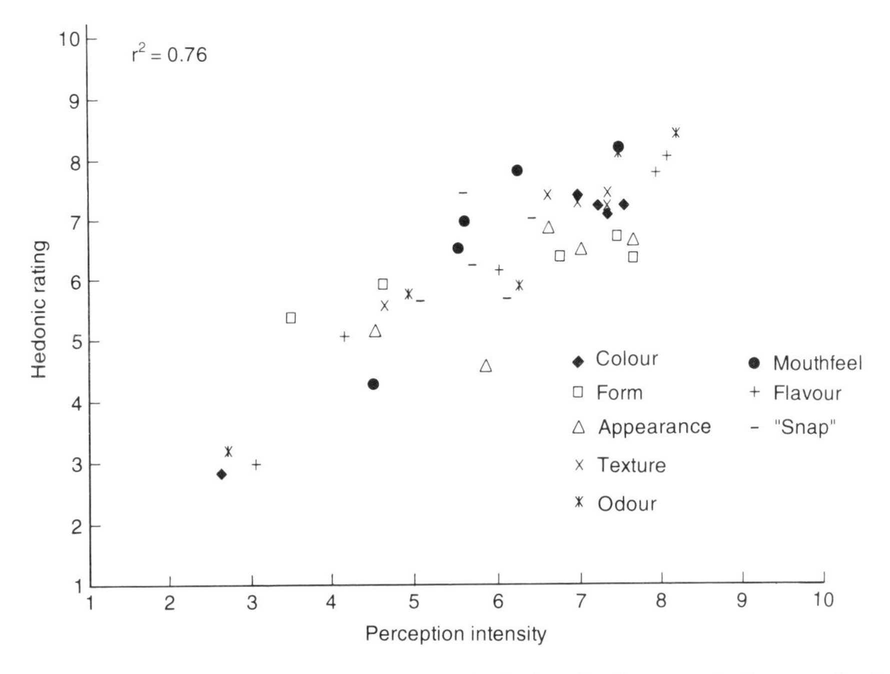

Fig. 22.1 Relationship between perception and hedonic rating for organoleptic properties for dark chocolate. *Source:* Manoury-Tomas (34).

between the two series of scores. We know from previous studies (19) that acceptance of a product requires its 'hedonization': the higher this score, the more the product is accepted and consumed. This is the case here.

The analysis of the ideational questionnaire was carried out making a distinction between three main topics, consumption, health and product image.

The study of ideal consumption preferences for dark chocolate, which is more bitter than other varieties, reveals in particular the very great range of situations and persons who may consume this chocolate (Fig. 22.2). Practically everybody, except babies and to some extent, the sick, showed a marked preference against eating it warm. This is logical since it is supplied as chocolate bars. Finally, it is not considered as a food for feasts, probably for the same reason.

The topic of health reveals the existence of opinions which may appear to be contradictory and which are expressed with equal force; but they do not prevent the consumer from greatly appreciating the product (Fig. 22.3). This dark chocolate is therefore considered to be full of calories, rich in fat, inducing weight gain and not beautifying the person who eats it. But these are the same people who consider it as useful to the body, good for health, fortifying and very energy-giving. The only reservation concerns the vitamin intake, on which opinions are divided[3]. This shows, if such a demonstration were needed, that contradictory

[3] This product contains no vitamin or other nutrient supplements, in accordance with French law.

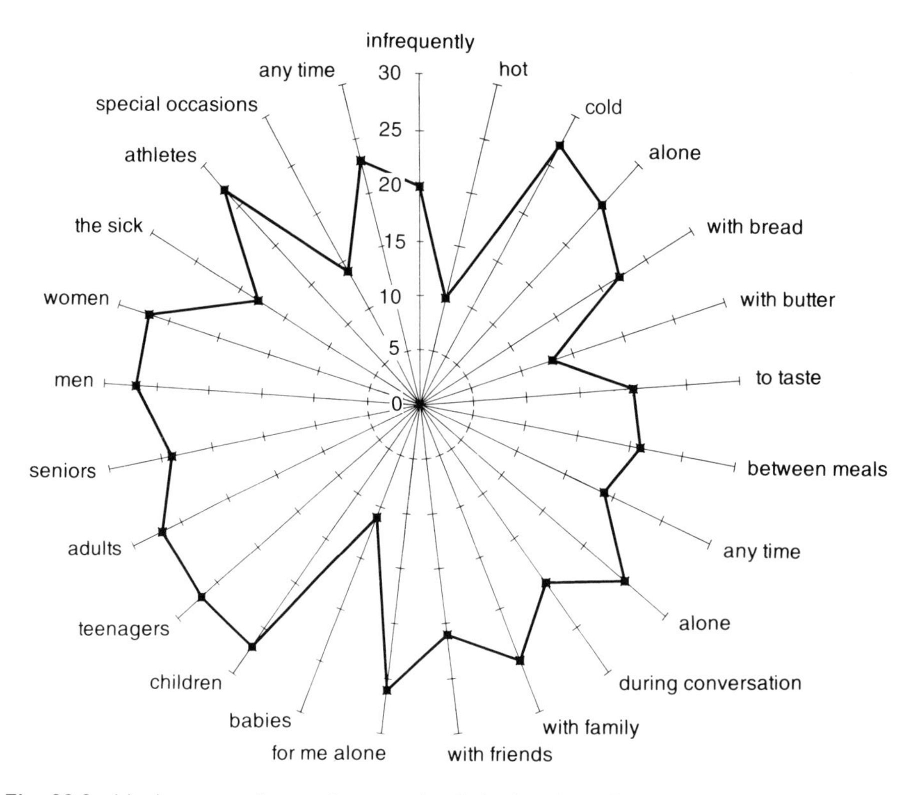

Fig. 22.2 Ideal consumption preferences for dark chocolate. *Source:* Manoury-Tomas (34).

perceptions can coexist in this sphere without necessarily provoking cognitive disagreements.

Finally, where the images of this product are concerned, there is first of all confirmation of the fact that this is a product for everyone: those who work, those who are lazy, those who are disciplined and those who are not (Fig. 22.4). It is a pure, clean product which is not discredited but which does not make one better for all that. Differences of opinion or rather complementary opinions appear here as well: there are practically as many people who consider that it has a moral strength as there are those who do not. The same thing can be found regarding the existence or not of 'powers', without being more specific, which one would attribute to this chocolate. The degree of prohibition of the product is fairly low.

The overall picture reveals a specific representation of the product, where apparently contradictory opinions coexist, but these do not in any way embarrass our subjects. Compared with the other products used in this study (apples, pale ham, for example) it was clear that the perception and representation of chocolate were different from other products. Furthermore, it was the only product for which no difference between hedonic and ideational aspects could be found. A statistical analysis (general linear model) revealed no disagreements between the two series of data.

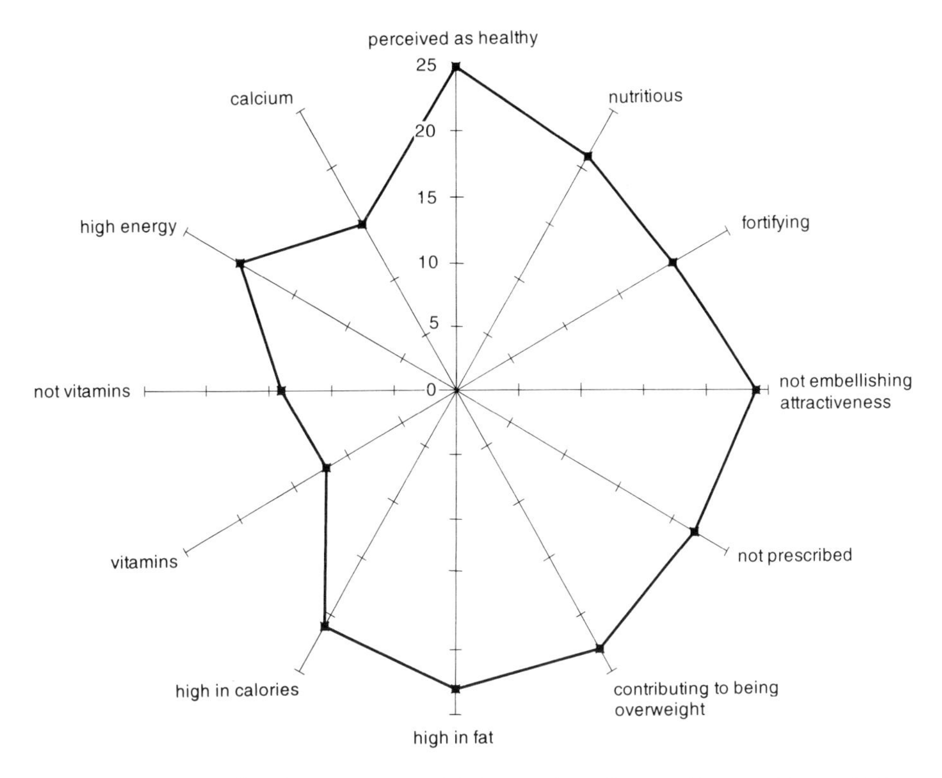

Fig. 22.3 Health profiles for consumers of dark chocolate. *Source:* Manoury-Tomas (34).

The purpose of the rapid presentation of this study is to show the possibility of grasping in both a detailed and a global manner the perception of the product by consumers. It is clear that it would be fully explained if one already had data which would enable one to compare different varieties of chocolate, for example, or again the same product as it may be perceived by subjects from different cultures. This is a methodology for both research and for possible immediate application in the spheres of industrial products.

Finally, it has the merit of providing an empirical confirmation of the extent to which chocolate is a 'pleasure product' of which the consumers' perception is very good.

By way of conclusion

Begun at the end of the 15th century, our voyage has brought us to the world of today. Chocolate, a divine product, a royal product, once a medicine and a piece of witchcraft, has become democratized. It is now accessible to all, within everyone's budget, and, despite the fact that it is very widely available, it continues to be perceived as a special product.

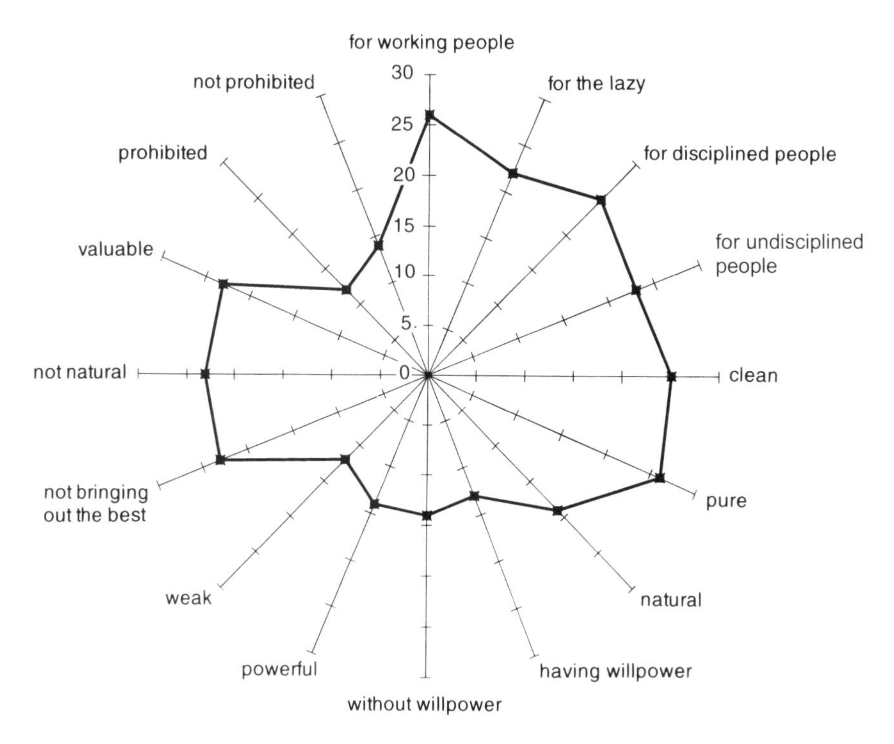

Fig. 22.4 Perceived imagery surrounding consumption of dark chocolate. *Source:* Manoury-Tomas (34).

We now have a better understanding not only of its composition but also of its real qualities when it first appeared. But one should not forget to take account as well of the imaginary properties, which have emerged from people's belief and which are also a reason for seeking it out. Because, despite all the advice, we are not yet eating on a doctor's prescription only.

Finally, despite the fact that it has become commonplace, it still remains a hedonic product, guarantee of a pleasure that anyone can easily obtain for themselves. More than ordinary products, which are available in great variety, chocolate is still the product which, when used by pastry cooks, chocolate makers and great chefs, lends itself to extreme refinements and becomes a work of art, in the real sense and figuratively.

But if one remains at plate level, culinary works of art are the only works of art which one has to destroy to appreciate. What remains then except the memory? And it is thanks to this memory that we will want to look for them again.

References

1. Barrau, J. (1983) *Les Hommes et leurs aliments*. Temps Actuels, Paris.
2. Fischler, C. (1990) *L'Homnivore. Le goût, la cuisine et le corps*. Odila Jacob, Paris.

3. Chiva, M. (1991) (Psychological aspects of dietary behaviour.) In *Alimentation et Nutrition Humaines* (Ed. by Dupin, H., Cuq, J.L., Maewiak, M.I, Leynaud-Rouaud, C. and Berthier, A.M.), pp. 417–444. ESF, Paris.
4. Flandrin, J.L. (1996) (Humanization of dietary behaviour.) In *Histoire de l'alimentation* (Ed. by Flandrin, J.L. and Montanari, M.), pp. 19–27. Fayard, Paris.
5. Perles, C. (1996) (Food strategies in prehistoric times.) In *Histoire de l'alimentation* (Ed. by Flandrin, J.L. and Montanari, M.), pp. 29–46, Fayard, Paris.
6. Chiva, M. (1997) Cultural aspects of meals and meals frequency. *Br. J. Nutr.* **777** (Suppl. 1), S21–S28.
7. Flandrin, J.L. and Montanari. M. (Eds) (1966) *Histoire de l'alimentation.* Fayard, Paris.
8. Mintz, S.W. (1985) *Sweetness and Power.* Viking Penguin, New York.
9. Moulin, L. (1988) *Les liturgies de la Table. Une histoire culturelle du manger et du boire.* Fonds Mercator and Albin Michel, Antwerp.
10. Huetz de Lemps, A. (1996) (Colonial beverages and sugar boom.) In *Histoire de l'alimentation* (Ed. by Flandrin, J.L. and Montanari, M.), pp. 629–641. Fayard, Paris.
11. Elias, N. (1974) *La société de Cour.* Calman-Lévy, Paris.
12. Girard, S. (1984) *Guide du chocolat et de ses à-côtés.* Messidor, Paris.
13. Brillat-Savarin, A. (1825) *Physiologie du goût.* Re-published Julliard, Paris, 1965.
14. Chiva, M. (1996) (The eater and the eaten: the complexity of a fundamental relationship.) In *Identités des mangeurs, images des aliments* (Ed. by Giachetti, I.), pp. 11–30. Polytechnica, Paris.
15. Chiva, M. (1987) Implication of sweetness in upbringing and education. In *Sweetness* (Ed. by Dobbing, J.), pp. 227–238. Springer-Verlag, London.
16. Chiva, M. and Fischler, C. (1986) (How does one learn to eat?) *Lieux de l'enfance* **6/7**, 89–104.
17. Loque, A.W. (1986) *The Psychology of Eating and Drinking.* H.W. Freeman, New York.
18. Chiva, M. and Fischler, C. (1986) Food likes, dislikes and some of their correlates in a sample of French children and young adults. In *Measurement and Determinants of Food Habits and Food Preferences.* EURONUT Report 7 (Ed. by Diehl, J.M. and Leitzmann, C.), pp. 137–156. The Netherlands Nutrition Foundation, Wageningen.
19. Rigal, W. and Chiva, M. (1995) (Cultural modelling of food preferences in children.) *Revue de Nutrition Pratique* **8**, 7–11.
20. Chiva, M. (1985) *Le doux et l'amer.* Presses Universitaires de France, Paris.
21. Rozin, E. (1973) *The Flavor-Principle Cookbook.* Hawthorn Books, New York.
22. Rozin, P. (1976) The selection of food by rats, humans and other animals. In *Advances in the Study of Behaviour* (Ed. by Rosenblatt. J.S., Hinde, R.A., Shaw, E. and Beer, C.), pp. 21–76. Academic Press, New York.
23. Fischler, C. (1996) (Food, morals and society.) In *Identités des mangeurs, images des aliments* (Ed. by Giachetti, I.), pp. 31–54. Polytechnica, Paris.
24. Fischler, C. (Ed.) (1994) *Manger magique.* Autrement, Paris.
25. Chiva, M. (1994) (Thought in construction.) In *Manger magique* (Ed. by Fischler, C.), pp. 51–61. Autrement, Paris.
26. Fischler, C. (1987) Attitudes towards sugar and sweetness in historical and social perspective. In *Sweetness* (Ed. by Dobbing, J.), pp. 83–98. Springer-Verlag, London.

27. Rozin, P. (1987) Sweetness, sensuality, sin, safety and socialization: some speculations. In *Sweetness* (Ed. by Dobbing, J.), pp. 99–110. Springer-Verlag, London.
28. Nemeroff, C. and Rozin, P. (1989) 'You are what you eat': applying the demand-free 'impressions' technique to an unacknowledged belief. *Ethos* **17**, 50–69.
29. Damasio, A. and Damasio, M.D. (1994) *Emotion, Reason and the Human Brain.* Putnam Books, New York.
30. Drewnowski, A., (1987) Sweetness and obesity. In *Sweetness* (Ed. by Dobbing, J.), pp. 117–192. Springer-Verlag, London.
31. Pangborn, R.M. (1960) Taste inter-relationships. *Food Res.* **25**, 245–258.
32. Pangborn, R.M. (1980) Sensory science today. *Cereals Foods World* **25**, 637–639.
33. Issanchou, S. and Hossenlopp, J. (1991) (Hedonic measurements: methods, scope and limits.) In *Plaisir et préférences alimentaires* (Ed. by Giachetti, I.), pp. 49–76. Polytechnica, Paris.
34. Manoury-Tomas, S. (1998) (Evaluation of hedonic perception of food: an experimental approach.) Thesis for Doctorate in Psychology, University of Paris X, Nanterre.
35. Manoury-Tomas, S., Chiva, M. and Touraille, C. (1996) (Hedonic evaluation of foods.) *Revue de Nutrition Pratique* **9**, 7–18.

Index